CASS SERIES ON POLITICS AND MILITARY AFFAIRS

CHURCHILL AND HITLER

CASS SERIES ON POLITICS AND MILITARY AFFAIRS

Series Editor
MICHAEL I. HANDEL
*U.S. Naval War College,
Newport, RI*

Leon Trotsky and the Art of Insurrection 1905–1917
H.W. Nelson

The Nazi Party in Dissolution: Hitler and the Verbotzeit 1923–1925
David Jablonsky

War, Strategy and Intelligence
Michael I. Handel

Cossacks in the German Army 1941–1945
Sam Newland

A Don at War
second edition
Sir David Hunt

Churchill, The Great Game and Total War
David Jablonsky

*Churchill and Hitler: Essays on the
Political–Military Direction of Total War*
David Jablonsky

Singapore, 1941–42
new edition
Louis Allen

CHURCHILL

AND

HITLER

ESSAYS ON THE POLITICAL–MILITARY DIRECTION OF TOTAL WAR

DAVID JABLONSKY

US Army War College

Routledge
Taylor & Francis Group
New York   London

First published in 1994 in Great Britain by
FRANK CASS & CO. LTD.
Newbury House, 900 Eastern Avenue, Newbury Park,
Ilford, Essex, IG2 7HH, England

and in the United States of America by
FRANK CASS
c/o International Specialised Book Services, Inc.
5804 N.E. Hassalo Street, Portland, Oregon 97213-3644

British Library Cataloguing in Publication Data

Jablonsky, David. Churchill and Hitler:Essays on the
  Political–military Direction of Total
  War. – (Cass Series on Politics &
  Military Affairs)
  I. Title   II. Series
  940.53

  ISBN 0-7146-4563-X
  ISBN 0-7146-4119-7 (paperback)

Library of Congress Cataloging in Publication Data

Jablonsky, David.
    Churchill and Hitler : essays on the political–military direction
  of total war / David Jablonsky.
        p.   cm. — (Cass series on politics and military affairs)
    ISBN 0-7146-4563-X — ISBN 0-7146-4119-7 (pbk.)
    1.   Militarism.   2. Military art and science.   3. Churchill,
  Winston, Sir, 1874–1965.   4. Hitler, Adolf, 1889–1945.   I. Title.
  II. Series.
  U21.2.J32   1994
  355.02'3—dc20   94-5767
                                                            CIP

Typeset by Regent Typesetting, London

Contents

For

Kyra Wiebke Jablonsky
Ithaca College, 1994

With great love and pride

For I dipt into the future,
  far as human eye could see,
Saw the Vision of the world,
  and all the wonder that would be;

Alfred, Lord Tennyson,
"Locksley Hall"

Preface

The essays collected in this volume deal with two of the greatest warlords in world history. Some may object to the connection of Winston Churchill with Adolf Hitler in this context. Certainly, despite some recent revisionist approaches to Churchill, there is still general agreement that the British statesman was truly the "Man of the Century" – a judgment that will continue to stand the test of time not only against such revisionist efforts, but also against the more extreme and ultimately counterproductive hagiographic approaches to Churchill. In a similar manner, there has been nothing in revisionist scholarship during the half century since Hitler's death that alters fundamentally the negative judgments of the Nazi leader and his criminal regime.

Nevertheless these two men were bound together in their roles as leaders of their nations during the most total war in history. National strategy in such a conflict calls for the use of all the elements of national power – not only military, but political, psychological and economic as well. Ultimately, success rests on the calculated relationship of these national means to the national ends, the essence of strategy. And ultimately, as these essays are designed to demonstrate, it was Churchill, not Hitler, who best understood the nature of that calculation.

I am very much indebted to the following for permission to reproduce previously published articles: US Army Strategic Studies Institute for Chapters 2 and 5; Frank Cass for Chapter 3; US Army War College for Chapter 4; Brassey's Defence Publishers for Chapter 6; and Sage Publications for Chapter 7. My thanks also to Mr Richard Langworth, President, International Churchill Society, United States; Colonel Karl Robinson, former Director, US Army Strategic Studies Institute; Dr Gary Guertner, former Director of Research, Strategic Studies Institute; Mrs Marianne Cowling, Editor, Strategic Studies Institute; Professor Arthur F. Lykke, Jr., US Army War College; Mr Lawrence Miller and his colleagues, Graphics Art Branch, US Army War College; and for typing support, Mrs Susan Blubaugh, Mrs Sandy Foote and, in particular, Mrs Rosemary Moore. Finally, I am especially grateful to Michael Handel, an office mate for much too short a time; and to Wiebke, as always, the keeper of the gate.

David Jablonsky
Carlisle, Pennsylvania

Cartoon published in *World's Press News*, 26 April 1945

1

Introduction:
The Clausewitzian Trinity

Winston Churchill and Adolf Hitler almost met in 1932. In August of that year, the future British leader mounted a family expedition to the Low Countries and Germany as part of his research on the *Life of Marlborough*. The trip traced the 1705 march by Churchill's celebrated ancestor from the Netherlands to the Danube and concluded with a week-long stay at the Regina Hotel in Munich. At that point, Churchill was approached by Ernst "Putzi" Hanfstaengel, a long-time friend and benefactor of Hitler, with an offer to arrange a meeting with the Nazi leader. Churchill agreed to the meeting, but nothing came of the offer.[1]

Both men were out of power at the time. Churchill's "wilderness years" had begun in 1929 with the fall of the Conservative government and would continue until May 1940 when he kissed hands as Prime Minister of Great Britain. Hitler, on the other hand, had never known anything but the "wilderness." By August 1932, however, events were moving his way. In the general election of 31 July, the Nazis had become the largest party in the state and, shortly after the aborted Munich meeting with Churchill, Hitler turned down the offer of becoming Vice-Chancellor in the von Papen government. His was a higher goal, which he attained at the beginning of the next year in a process as legal as that which ushered Churchill into office seven years later.

By this time, the last total war of the twentieth century was underway, and Churchill and Hitler were the dominant political–military leaders of their nations, truly the embodiment of Carl von Clausewitz's dictum that "policy is the guiding intelligence and war only the instrument."[2] The essence of such leadership was an understanding of the massive changes in the nature of war and strategy which began in the Napoleonic era, continued throughout the nineteenth century and the formative years of both leaders at the end of that century, and culminated in the First World War, the first modern total war. How well both of these warlords responded to these changes and what experiences conditioned their responses are the central themes of the essays in this volume.

The Changing Nature of War and Strategy

Strategy is the calculated relationship of ends, ways and means. In a generic sense this definition is practiced every day, whether it be the housewife determining what to buy at the local store or the small unit military leader responding to the inevitable cry of "what would you do in this situation?" At the highest level of decision-making concerned with a nation's security, however, the process so briefly mentioned in the definition of strategy has become infinitely more complex in the last 200 years.

In the nineteenth century, this complexity began with two upheavals, the French Revolution and the industrial/technological revolution. Before the French Revolution, eighteenth-century rulers had acquired such effective political and economic control over their people that they were able to create war machines which were separate and distinct from the rest of society. The Revolution changed all that with the appearance of a force, as Clausewitz described it,

> that beggared all imagination. Suddenly, war again became the business of the people – a people of thirty millions, all of whom considered themselves to be citizens .... There seemed no end to the resources mobilized; all limits disappeared in the vigor and enthusiasm shown by governments and their subjects .... War, untrammelled by any conventional restraints, had broken loose in all its elemental fury. This was due to the peoples' new share in these great affairs of state; and their participation, in its turn, resulted partly from the impact that the Revolution had on the internal conditions of every state and partly from the danger that France posed to everyone.[3]

For the Prussian philosopher, the people greatly complicated the formulation and implementation of strategy by adding "primordial violence, hatred and enmity, which are to be regarded as a blind natural force" to form with the army and the government what he termed the remarkable trinity (see Figure 1). The army he saw as a "creative spirit" roaming freely within "the play of chance and probability," but always bound to the government, the third element, in "subordination, as an instrument of policy, which makes it subject to reason alone."[4]

It was the complex totality of this trinity which, Clausewitz realized, had altered and complicated strategy so completely. "Clearly the tremendous effects of the French Revolution," he wrote "... were caused not so much by new military methods and concepts as by radical changes in policies and administration, by the new character of

government, altered conditions of the French people, and the like ....
It follows that the transformation of the art of war resulted from the
transformation of politics."[5] But while that transformation had made
it absolutely essential to consider the elements of the Clausewitzian
trinity within the strategic framework, the variations possible in the
interplay of those elements moved strategy even further from the

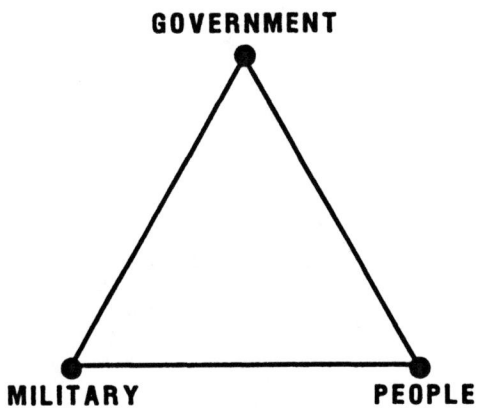

GOVERNMENT

MILITARY   **PEOPLE**

Figure 1
The Remarkable Trinity

realm of scientific certitude. "A theory that ignores any one of them or
seeks to fix an arbitrary relationship between them," Clausewitz
warned in this regard, "would conflict with reality to such an extent
that for this reason alone it would be totally useless."[6]

Like most of his contemporaries, Clausewitz had no idea that he
was living on the eve of a technological transformation born of the
Industrial Revolution. But that transformation, as it gathered
momentum throughout the remainder of the nineteenth century,
fundamentally altered the interplay of elements within the Clausewitzian trinity, further complicating the formulation and application
process within the strategic framework (see Figure 2).

In terms of the military element, technology would change the basic
nature of weapons and modes of transportation, the former having
been stable for 100 years, the latter for 1,000. Within a decade of
Clausewitz's death in 1831, that process would begin in armaments
with the introduction of breech-loading firearms, and in transportation with the development of the railroads.[7]

Technology had a more gradual effect on the role of the people.
There were, for example, the great European population increases of
the nineteenth century as the Industrial Revolution moved on to the

continent from Great Britain. This trend led, in turn, to urbanization: the mass movement of people from the extended families of rural life to the "atomized," impersonal life of the city. There, the urge to belong, to find a familial substitute, led to a more focused allegiance to the nation-state, manifested in a new, more blatant and aggressive nationalism.

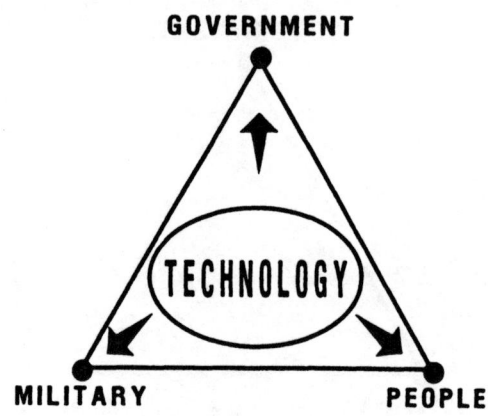

Figure 2
The Impact of Technology

This nationalism was fueled by the progressive side-effects of the Industrial Revolution, particularly in the area of public education, which meant, in turn, mass literacy throughout Europe by the end of the nineteenth century. One result was that an increasingly literate public could be manipulated by governments as technology spawned more sophisticated methods of mass communications. On the other hand, those same developments also helped democratize societies which then demanded a greater share in government, particularly over strategic questions involving war and peace. In Clausewitz's time, strategic decisions dealing with such matters were rationally based on *Realpolitik* considerations to further state interests, not on domestic issues. By the end of the nineteenth century, the Rankeian *Primat der Aussenpolitik* was increasingly challenged throughout Europe by the need of governments for domestic consensus – a development with far-reaching implications for the conduct of strategy at the national level within the basic ends-ways-means frame-work.[8]

During much of that century, as the social and ideological up-heavals unleashed by the French Revolution developed, military leaders in Europe generally attempted to distance their armed forces from their people. Nowhere was this more evident than in the Prussian cum German military, where the leaders worked hard over the years to prevent the adulteration of their forces by liberal ideas. "The army is now our fatherland," General von Roon wrote to his wife during the 1848 revolutions, "for there alone have the unclean and violent elements who put everything into turmoil failed to penetrate."[9] The revolutions in industry and technology, however, rendered this ideal unattainable. To begin with, the so-called *Technisierung* of warfare meant the mass production of more complex weapons for ever-larger standing military forces. The key ingredients for these forces were the great population increases and the rise of nationalism as well as improved communications and governmental efficiency – the latter directed at general conscription of national manhood which, thanks to progress in railroad development, could be brought to the battlefield in unlimited numbers.

At the same time, this increased interaction between the government/military and the people was also tied to other aspects of the impact of technology on the Clausewitzian trinity. Technological innovations in weaponry during this period, for example, were not always followed by an understanding of their implications, societal as well as military. Certainly, there was the inability on the part of all European powers to perceive the growing advantage of defensive over offensive weapons demonstrated in the Boer and Russo-Japanese Wars. That inability was tied in with a trend in Europe at the time to combine *élan* with a military focus on moral force, bloodshed and decisive battles. The result was that the military leaders of France, Germany and Russia all adopted offensive military doctrines in some form.[10]

The fact that these doctrines led to the self-defeating offensive strategies of the First World War ultimately had to do with the transformation of civil–military relations within the Clausewitzian trinity in their countries. In France, as an example, the officer corps distrusted the trend by the leaders of the Third Republic towards shorter terms of military service, which it believed threatened the army's professional character and tradition. Adopting an offensive doctrine and elevating it to the highest level was a means to combat this trend, since there was general agreement that an army consisting primarily of reservists and short-term conscripts could only be used in the defense. "Reserves are so much eyewash," one French general wrote at the time, "and take in only short-sighted mathematicians who equate the value of armies with the size of their effectives, without considering their moral value."[11] Although these were setbacks for those who shared this sentiment in the wake of the Dreyfus Affair and the consequent military reforms, it only required the harsher international climate after the Agadir crisis of 1911 for General Joffre and his young Turks to gain the ascendancy. Their philosophy was summed up by their leader who explained that in planning for the next war he had "no preconceived idea other than a full determination to take the offensive with all my forces assembled."[12]

Under these circumstances, French offensive doctrine became increasingly unhinged from strategic reality as it responded to the more immediate demands of domestic and intragovernmental politics. The result was France's ill-conceived strategic lunge in 1914 toward its former possessions in the East, a lunge which almost provided sufficient margin of assistance for Germany's Schlieffen Plan, another result of military operational doctrine driving policy. In the end, only the miracle of the Marne prevented a victory for the Germans as rapid and complete as that of 1870.[13]

There were other equally significant results as the full brunt of technological change continued to alter the relationship among the elements of the Clausewitzian trinity for all the European powers. The larger, more complex armies resulted in a growing specialization and compartmentalization of the military – a trend which culminated in the emulation of the German General Staff system by most of the European powers. Significantly, Clausewitz had ignored Carnot, the "organizer of victory" for Napoleon, when considering military genius. Now, with the increase in military branches as well as combat service and combat service support organizations, the age of the "military-organizational" genius had arrived. All this, in turn, affected the relationships in all countries between the military and the government since the very increase in professional knowledge and skill caused by technology's advance in military affairs undermined the ability of political leaders to understand and control the military, just as technology was making that control more important than ever by extending strategy from the battlefield to the civilian rear, thus blurring the difference between combatant and non-combatant.[14]

At the same time, the military expansion in the peacetime preparation for war began to enlarge the economic dimensions of conflict beyond the simple financial support of Clausewitz's era. As Europe entered the twentieth century, new areas of concern began to emerge, ranging from industrial capacity and the availability and distribution of raw materials to research and development of weapons and equipment. All this, in turn, increased the size and role of European governments before the First World War, with the result, as William

James perceptively noted, that "the intensely sharp competitive preparation for war by the nation is the real war, permanently increasing, so that the battles are only a sort of public verification of mastery gained during the 'peace' intervals."[15]

Nevertheless, the full impact of the government's strategic role in terms of national instruments of power beyond that of the military was generally not perceived in Europe, despite some of the more salient lessons of the American Civil War. In that conflict, the South lost because its strategic means did not match its strategic ends and ways. Consequently, no amount of operational finesse on the part of the South's great captains could compensate for the superior industrial strength and manpower that the North could deploy. Ultimately, this meant for the North, as Michael Howard has pointed out, "that the operational skills of their adversaries were rendered almost irrelevant."[16] The Civil War also illustrated another aspect of the changes within the strategic framework: the growing importance of the national will of the people in achieving political as well as military strategic objectives. That social dimension of strategy on the part of the Union was what had prevented the early Southern operational victories from being strategically decisive and what ultimately allowed the enormous industrial-logistical potential north of the Potomac to be realized.

The Great War

The First World War was a common experience for both Churchill and Hitler. And it was in that conflict that strategy and the nature of war changed irrevocably with the full confluence of the trends set in train by the Industrial and French Revolutions. In particular, the technology in the war provided, as Hanson Baldwin has pointed out, "a preview of the Pandora's box of evils that the linkage of science with industry in the service of war was to mean."[17] How unexpected the results of that linkage could be was illustrated by a young British subaltern's report to his commanding general after one of the first British attacks in Flanders. "Sorry sir," he concluded. "We didn't know it would be like that. We'll do better next time."[18] But, of course, there was no doing better next time, not by British and French commanders in Flanders, not by Austrian troops on the Drina and Galician fronts in 1914, not by the Russian officers on the Gorlice–Tarnow line in 1915. The frustration at this turn of events was captured by Alexander Solzhenitsyn in his novel, *August 1914*. "How disastrously the conditions of warfare had changed," he wrote, "making a commander as impotent as a rag doll! Where now was the battlefield ... across which he could gallop over to a faltering commander and summon him to his side."[19]

At the same time, the full impact of the new technology on the Clausewitzian trinity in each of the combatant states during the war created an infinitely more complex concept of national strategy. To begin with, the growing sophistication and quantity of arms and munitions, as well as the vast demands of equipment and supply made by the armies, involved the national resources of industry, science and agriculture – variables with which the military leaders were not prepared to deal. To cope with these variables, governments were soon forced to transform the national lives of their states in order to provide the sinews of total war.

All this vastly complicated the interaction between the elements of the remarkable trinity. No nation at the beginning of the First World War, for example, contemplated a shift from Clausewitz's strategy of annihilation to one of attrition. The primary reason, as Michael Howard has pointed out, was that "rapid victory was seen as a social imperative as well as a military one."[20] The cause and effect lessons of the Russo-Japanese War and the 1905 Russian Revolution were obvious in this regard to all the European and governmental elites. And if that were not enough, there was the burgeoning socialist movement cutting across national lines with an increasingly restive and threatening working class – a constant reminder to these elites of the fragility of the social, and thus strategic, consensus in all European societies.

Nowhere did this fragility have a greater effect than on the relationships between the other two-thirds of the Clausewitzian trinity: the European governments and their military establishments. In the Second German Empire, the Schlieffen Plan which launched the war was designed for a quick victory that would keep the conflict autonomous within limited military means. The preservation of war as a professional domain thus became the overriding rationale for the plan. And although before the First World War German government leaders generally knew the features of such war plans, they were not aware of many of the crucial details. Unlike Bismarck, however, these politicians made no attempt to gain the details; and, because there existed no formal mechanism to co-ordinate military strategic planning with the emerging complexities of national strategy, German strategic thought preceding the First World War focused primarily on the operational and tactical levels of war and only rarely on national policy. "The German military establishment," Dennis Showalter has pointed out in this regard, "developed plans to win campaigns, not wars."[21]

That these plans were elevated to the level of national policy is an indication of how far civilian deference to increased military specialization could reverse Clausewitz's most fundamental dictum of policy dominating strategy. The result was an operational mix of ends, ways and means that failed to achieve its theater and national military objectives because of the fundamental disconnects with all the instruments of natural power that comprised the sinews of national strategy. In terms of the political dimension, for instance, the violation of neutral Belgium as the Schlieffen Plan unfolded brought world-wide condemnation and, most significantly, the entry of Great Britain into the war on the side of the Entente. It was this entry which hindered the great turning movement of the German forces, a considerable cumulative contribution to the miracle of the Marne. Even more important, however, was the psychological aspect of national strategy involving the people and national will which made it very unlikely that even if the initial German offensive had succeeded, the members of the Entente would have given up the fight. In this way, a plan for a quick, decisive operational victory to achieve theater military strategic objectives made a two-front war more likely at the national strategic level – the conflict the German military most wished to avoid.[22]

The German experience with the relationship between the government and the military was not unique as at the beginning of the conflict political leaders throughout Europe attempted to come to grips with the new demands of total war. In fact, all the European civil keepers of national strategy were universally deferent to the expertise of the military strategists, generally accepting the basic view of the elder Moltke that "the politician should fall silent the moment mobilization begins, and not resume his precedence until the strategist has informed the King, after the total defeat of the enemy, that he has completed his task."[23] As the war proceeded, however, there was a general disillusionment on the part of governmental elites about the capacity of the military to deal with all the complexities of the ends-ways-means framework at the national strategic level. "The General no doubt was an expert on how to move his troops," Churchill observed in this regard after the war, "and the Admiral upon how to fight his ships .... But outside this technical aspect they were helpless and misleading arbiters in problems in whose solution the aid of the

Statesman, the financier, the manufacturer, the inventor, the psychologist, was equally required."[24]

The French reaction to all this ultimately resulted in a dictatorship of the policy-makers over the strategists in keeping with a civilian radical tradition stretching back through Gambetta to the ruthless national strategic rigors of Danton's efforts in the revolutionary wars. In Germany, on the other hand, strategy achieved an even stronger ascendancy over policy by 1916. The Hindenburg–Ludendorff military dictatorship of that year mediated between the people in terms of social mobilization, industry and the military, while providing direction and purpose utterly devoid of normal political expression. The result was that the national strategy for Germany in the last years of the conflict was limited to the social and technical mobilization of the nation for unlimited war – a logical development in a leadership that had reversed Clausewitz's basic extension of the strategic framework into the realm of policy. "All the theories of Clausewitz should be thrown overboard," Ludendorff wrote. "Both warfare and politics are meant to serve the preservation of the people, but warfare is the highest expression of the national 'will to live,' and politics must, therefore, be subservient to the conduct of war."[25] Means, in such absence of policy direction at the national strategic level, were now dominant in that framework. The greater the social and technological mobilization, the greater and more encompassing became the goals – a war without limits that could only end with the destruction of the enemy nation.[26]

The developments in Germany also demonstrated how important the third element of the Clausewitzian trinity had become as the full scope and complexity of national strategy were revealed in the war. For the first time, strategy absorbed the full energies of the societies engaged in conflict – creating total war in the sense that very few people in the combatant nations were allowed to remain unaffected. In the past, even after the French Revolution, this had not been the case. Jane Austen, for example, could write her novels about the English rural gentry without once mentioning the French Revolutionary and Napoleonic Wars that swirled around her for 25 years. This type of detachment, which remained generally in effect throughout the European conflicts of the nineteenth century, ended with the First World War.[27]

To begin with, there was the breakdown of the distinction between soldiers and civilians – a development, as we have seen, partially due to technological innovations that expanded warfare beyond the combat front to the so-called Home Front. With the advent of dirigible bombings of cities and submarine attacks on merchant ships, the battle line was everywhere. Most important, however, was that government leaders in all the belligerent countries began to realize by the winter of 1914–15 that the war could not be won by military means alone, that it would require the effective national strategic mobilization of all the civilians on the Home Front feeding into the political and economic instruments of power as well as that of the military. This meant, in turn, that the psychological element of power became an even more critical factor in strategic calculations. Now, the morale and national will of the people on the Home Front became as important in national strategic terms as the spirit and determination of the soldier in the trenches.

The awareness of this phenomenon gradually affected the war aims of all the belligerents and thus the process of conflict termination. Those aims were generally limited at the beginning of the war, with the parameters for the making of peace, like those for the conduct of war, governed by the nineteenth century concept of strategy. For example, defeat of the German field armies in 1914 would probably have meant the restoration of key territorial acquisitions like Alsace–Lorraine. No member of the Entente at the beginning of the First World War even remotely associated victory with the domestic political and social transformation of German society that was to be a mainstay of Allied efforts in the next war.

In the last years of the war, however, the full evolution of national strategy meant not only the attrition of enemy armed forces, but the moral and social attrition of the populations that provided those forces with men and material. As a consequence, all the belligerents to some extent targeted enemy governments and the very social structure on which their stability was based. Thus, for example, the German government in April 1917 facilitated the passage of Lenin in a sealed railroad car to the Finland Station in order to overthrow the provisional Kerensky government. And the British government that same year began to conduct a political warfare campaign in support of dissident Slav minorities in the multinational Austro-Hungarian Empire. Finally, in November 1918, President Wilson decreed that the most basic conditions for peace with the Germans must include the overthrow of the Hohenzollern monarchy, thus creating a void that in the coming years was to leave the competition for the nation's loyalty open to more sinister forces.[28]

There was another equally fateful result of the new combination of the Clausewitzian trinity when, in order to mobilize the political, economic and psychological wherewithal of their Home Fronts, the

European wartime governments subjected their citizens to wide-ranging disciplines and deprivations not unlike those undergone by the soldiers in the trenches. The hardships were accompanied in all countries by the ceaseless drumbeat of propaganda. On the combat front, these one-dimensional depictions of the foe were leavened by the soldier's normal empathy for his enemy counterpart living under the same horrific conditions. Thus, there were the soccer games between belligerents in the no-man's land of Flanders, the singing of Christmas carols and the exchange of presents between the trenches.

There was no such empathy on the Home Front. There, the economic and social hardships as well as the propaganda combined with the prolongation and totality of the war to bring the citizenry of all countries to an ideological, pseudo-religious peak that had not been seen in Europe since the Thirty Years War. The result was that for the people of Europe, as the war continued, the enemy was increasingly considered to be capable of any enormity and thus not subject to compromise. Consequently, the foe must be beaten to his knees, no matter the cost, and, once defeated, kept in permanent subjugation by means of the peace terms. All this had enormous national strategic consequences in terms of conflict termination, a rational strategic product in Clausewitz's time solely of efforts by the other two-thirds of the Clausewitzian trinity. The Prussian philosopher had written,

> As war is no act of blind passion, but is dominated by the political objective, therefore the value of that objective determines the measure of sacrifices by which it is to be purchased ... not only as regards extent, but also as regards duration. As soon, therefore, as the required outlay becomes so great that the political object is no longer equal in value, the object must be given up, and peace will be the result.[29]

The reality, of course, was much different during the First World War since governmental and military elites in all nations had to take the popular passions they had aroused into consideration if they expected to remain in power. Thus, a German chancellor in the third year of the war could be forced from office because he advocated a peace short of total victory. And when a former foreign secretary in Great Britain urged a negotiated peace in a November 1917 letter to the *Daily Telegraph*, he was subjected to a flood of invective by the British press and public. All this fueled a tendency by the political and military leaders as the conflict continued to escalate war aims in order to justify to the people all the carnage and hardships that had gone before. For the Central Powers, this escalation took the form of annexations, culminating in the 1918 treaty of Brest–Litovsk by which Russia ceded its western territory to Germany. For the Entente, it took a more ideological form, with the struggle becoming one to save democracy, a war to end all wars.

It was this ideological foundation that the Allied leaders took to the peace conference at Versailles. There, pursued by the public passions they had unleashed, the Western political and military elites ensured by their vindictive treaty that the recently ended conflict would be linked to the next in what Churchill later described as the Thirty Years War of the twentieth century. "The peoples, transported by their sufferings and by the mass teachings with which they had been inspired," he wrote, "stood around in scores of millions to demand that retribution should be exacted to the full. Woe betide the leaders now perched on their dizzy pinnacles of triumph if they cast away at the conference table what the soldiers had won on a hundred blood-soaked battlefields."[30]

The Warlords: The Operational Code

The cognitive map of a political leader, what Alexander George has called the "Operational Code," is a key ingredient in examining how Churchill and Hitler approached the elements of the Clausewitzian trinity in leading their nations through total war. The code would include the actor's personal character and traits, his decision-making style and his philosophical and instrumental beliefs, which might be better labeled, in George's estimation, "approaches to political calculation."[31] These approaches can range from orientation on a leader's formative experiences and notions about work and loyalty to strategic vision and the ability and commitment to implement that vision. The operational code is not a simple key to explanation and prediction, but as George has pointed out, "it can help ... to 'bound' the alternative ways in which the subject may perceive different types of situations and approach the task of making a rational assessment of alternative courses of action."[32]

The Egocentric: Destiny, Intuition and Fate

Traits of personality and character were formed for both men in the crucibles of unhappy childhoods. Hitler was constantly at odds with his domineering father, although he remained devoted to his mother, the long-suffering Klara. And Churchill was neglected by both his parents to a degree that was remarkable even for the patricians of the late Victorian era. Later in life, the future British leader told his son Randolph, during a vacation from Eton: "I have talked to you more in this holiday than my father talked to me in his whole life."[33] Each man reacted differently to the early adversity. Hitler typically blamed his childhood experiences for many of his problems and lack of achievement during early manhood in Vienna, where, as described by his companion during those years, he "brought starvation to a fine art."[34] Churchill, on the other hand, saw his childhood as a positive force in his development. "Solitary trees, if they grow at all, grow strong," he wrote as early as 1898, "and a boy deprived of a father's care often develops ... an independence and vigour of thought which may restore in after life the heavy losses of early days."[35]

Both men emerged from childhood as extreme egocentrics with essentially rhetorical approaches to communication. This, combined with the normal fixity with which they adhered to points once they had made a decision, did not bode well for dialogue. Thus there was Hitler's rambling *Tischgespräche* in the Second World War, occupying listless hours of trivial emptiness at the Obersalzberg in what Albert Speer referred to as "a curious vacuity."[36] In sharp contrast, there was the witty, knowledgeable, eclectic conversation that characterized Churchill's meetings. Nevertheless, there was still the unmistakable centrality of a man who once said that his idea of an entertaining dinner was "to discuss a good topic with myself as chief conversationalist."[37] And with that centrality came dominance, as Eisenhower recalled of Churchill's only half-humorous edict at a wartime meeting: "All I want is compliance with my wishes, after reasonable discussion."[38]

In both men, this egocentric outlook combined with a strong sense of destiny to allow them to endure the exceptional vicissitudes of their careers. Between 1897 and 1900 on three different continents, for example, Churchill asserted to three very different people his passionate conviction that he would one day be England's Prime Minister.[39] And in 1906, he told Violet Asquith: "We are all worms. But I do believe that I am a glow-worm."[40] It would be a lifelong conviction. "This cannot be accident," he told his physician during the Second World War, "it must be design. I was kept for this job."[41] In a similar manner, Hitler's sense of destiny provided a foundation for a persistence that was necessary, as he pointed out in his *Secret Book* of 1928, "in order finally to be able to achieve the vitally necessary aim on a large scale."[42] It was a sustaining sense that never left the Nazi leader, as Walter Langer noted in his perceptive wartime psychoanalytical study: "Hitler believes himself destined to become an Immortal Hitler, founder of a new social order for the world. He firmly believes this and is certain that in spite of all the trials and tribulations through which he must pass he will finally attain that goal."[43]

Within a broad sense of destiny, both leaders operated intuitively, able to seize opportunities, as C. J. Jung noted in his *Psychological Types*, not "among the genuinely recognized reality values, but ... where possibilities exist."[44] One colleague of Churchill described it as his "zigzag streak of lightning in his brain." And General Alanbrooke, the Chief of the Imperial General Staff (CIGS) for much of the Second World War, was constantly astonished at Churchill's "method of suddenly arriving at some decision as it were by intuition."[45] The British leader, in fact, could have been describing his own character and decision-making traits in his analysis of Lloyd George during the First World War, when he noted: "His intuition fitted the crisis better than the logical reasoning of more rigid minds."[46] Such intuitiveness was even more pronounced in Hitler. "Trust your instincts, your feelings, or whatever you like to call them," he admonished in 1934. "Never trust your knowledge."[47] This approach provided the Nazi leader with a sixth sense that often led him to find and exploit the primary weaknesses and vulnerabilities of his opponents with complete assurance. "I follow my course," he stated during the Rhineland crisis, "with the precision and security of a sleepwalker."[48]

Centuries earlier, Pascal in his *Pensées* had described the complexity of the elements on which the intuitive mind operates and concluded that the intuitive person "must see the matter ... at one glance and not by a process of reasoning."[49] In his essay on painting, which he likened to conducting a war, Churchill described a similar process. "There must be that all-embracing view," he wrote, "which presents the beginning and the end, the whole and each part, as one instantaneous impression relatively and untiringly held in the mind."[50] It was an approach with which Hitler could empathize. "When I go to the Obersalzberg," he reminisced during the Second World War,

> I'm not drawn there merely by the beauty of the landscape. I feel myself far from petty things, and my imagination is stimulated. When I study a problem elsewhere, I see it less clearly, I'm submerged by the details. By night, at the *Berghof*, I often remain for hours with my eyes open, contemplating from my bed the mountains lit up by the moon. It's at such moments that brightness enters my mind.[51]

Clearly allied to a feeling of destiny and an intuitive approach to decision-making was a sense of *Schicksal* or fate ordained at a higher level. Typically, as *Mein Kampf* endlessly illustrated, Hitler perceived a "divine providence" *(Vorsehung)* working in his favor. It was fate that caused him to be born so close to the German frontier; that sent him to Vienna to suffer with the masses; that spared him in the First World War, etc. – all for a larger purpose. "Divine Providence," he concluded, "has willed it that I carry through the fulfillment of the German task."[52] For Churchill, it was a process that ultimately was out of his hands and consequently could provide a calming sustainment. "He hoped for immortality," the protagonist in his only work of fiction states, "but he contemplated annihilation with composure."[53] Thus on 26 May 1940, after the British leader had made the decision to abandon the besieged British garrison at Calais, he asked one of his personal secretaries to provide him with a quotation from George Borrow's *The Bible in Spain* – a nineteenth-century prayer for England at Gibraltar: "Fear not the result, for either shall thy end be majestic and an enviable one, or God shall perpetuate thy reign upon the waters."[54]

Personal Factors

Hitler at five feet, nine inches was three inches taller than Churchill. Neither was physically prepossessing. Churchill loved food, drink and a good cigar. Nevertheless, despite at least one heart attack and a serious bout of pneumonia during the Second World War, he remained relatively healthy in that conflict, able to maintain work and travel schedules that were astonishing for a man entering his eighth decade. On the other hand, except for a fondness for sweet creamy cakes, Hitler's culinary tastes were spartan, fueled in part by his increasing hypochondria and medical quirks. He hated body odors, for instance, and was particularly bothered by his own tendency toward flatulence. For this, he received huge quantities from Dr Theodor Morell, his personal physician, of "Dr Koester's Anti-Gas Pills" which contained strychnine and atrophene. Equally serious, Morell also prescribed injections of pulverized bull testicles in grape sugar as well as massive doses of dexedrene, pervatin, caffeine, cocaine, prozymen and ultraseptyl. The result for the Nazi leader, who was 14 years younger than Churchill, was described by General Guderian in the last six months of the war: "It was no longer simply his left hand, but the whole left side of his body that trembled .... He walked awkwardly, stooped more than ever, and his gestures were both jerky and slow. He had to have a chair pushed beneath him when he wished to sit down."[55]

Both men also suffered from recurrent fits of depression. Typically, Hitler's black moods would cause lethargy and indolence, conditions painfully at odds with the systematic routine required of the Nazi leader once the war began. "When peace has returned," Hitler remarked longingly in 1942, "I'll begin by spending three months without doing anything."[56] Churchill, on the other hand, coped with the symptoms of what he nicknamed "Black Dog" by turning to work. Incessant activity was the key, he told Violet Asquith; without it he would relapse into "dark moments of impatience and frustration."[57] As a consequence, there was the bricklaying at Chartwell or the painting at Marrakesh. And when he was fired from office after the Dardanelles fiasco in 1915, the sudden cessation of work and authority was devastating. "Like a sea-beast fished up from the depths," he wrote, "or a diver too suddenly hoisted, my veins threatened to burst from the fall in pressure."[58]

These dissimilarities also extended to their dealings with families and friends. Hitler remained essentially a loner who, except for a liaison with his niece in the inter-war years, kept the members of his family at an embarrassed distance. Moreover, he only used the informal "Du" with four *Kampfzeit* comrades, and even then in later years formally ended the *Brüderschaft* with one, Hermann Esser, and ordered the murder of another, Ernest Röhm. "He could not let anyone approach his inner being," Speer concluded, "because that core was lifeless, empty."[59] Churchill, on the other hand, gloried in and was reinforced by his large family and a continuous flow of old and new friends who refreshed and stimulated him, often compelling him to adjust his own ideas. During the war, weekends at Chequers with these friends and family members as well as other guests had an air, as one Private Secretary noted, "of calm and happiness about it."[60] That could hardly describe Hitler's retreat at the Eagle's Nest on the Kehlstein near Berchtesgarden where, as Speer noted, the Nazi leader "was more affected by the awesomeness of the abysses than by the harmony of a landscape," and where there were no friends, much less advisers, to supply the uninhibited criticism so necessary for an intuitive to avoid mistakes.[61] The result was a myopic and suspicious self-assurance.

In their cultural tastes, the differences between the two statesmen also tended to dominate. Hitler was genuinely inspired by the works of Wagner and was generally knowledgeable about classical music – a knowledge for which he typically claimed an all encompassing expertise. Churchill's taste in music was more pedestrian, with a penchant for Harry Lauder, military marches, Gilbert and Sullivan, and the

Harrow School songs. And while both men were artists, after he entered politics Hitler never returned to painting, an interest that had dominated his early years. Thereafter, his artistic endeavors were confined to discussions with Speer concerning the mammoth architectural designs planned for the cities of the Third Reich and to his patronage of artists who recreated the banal genre painting of the earlier, simpler Biedameier period. Churchill reversed the pattern, beginning a painting pastime in his forties which remained undiminished in output and enthusiasm throughout his long life. "Happy are the painters, for they shall not be lonely," he wrote in the inter-war years. "Light and colour, peace and hope, will keep them company to the end, or almost to the end of the day."[62]

Only in their early reading habits was there a cultural confluence for Churchill and Hitler, since both liked the popular fiction and histories of their day. Each was fascinated by the Red Indians of North America as portrayed in James Fenimore Cooper's novels and by the technological gadgetry in the books of Jules Verne. To this were added the distinctly German westerns of Karl May for Hitler and the imperial romances, stories and poems of H. Rider Haggard and Rudyard Kipling for Churchill.[63] But Churchill also had an ear for poetry, reinforced by prodigious feats of memorization. This reinforced his feel for the English language which animated the matchless prose in his monumental collection of books and articles – all dictated either entirely or in substantial part, or, as he described it, living "from mouth to hand."[64] Hitler also dictated his two major works, but as the turgid, often unreadable prose of those books illustrated, he never mastered the written word. In *Mein Kampf*, for example, the sentences were often clogged with the useless particles endemic to the Lower Bavarian dialect such as *wohl, ja, denn, schon* and *eigentlich*. And in contrast to Churchill's rich prose, there was no color and movement in Hitler's style. Only rarely did the Nazi leader use images, and even then they tended to be verbal and thus impossible to visualize, such as forcing "the less strong and less healthy back into the womb of the eternal unknown."[65]

War and Action

"What has he done all his life?" Hitler once asked concerning Churchill.[66] The answer, of course, included participation as a soldier and a journalist in a series of late Victorian wars and insurrections and as a soldier and a statesman in the First World War. The latter conflict was a shock to the future British leader, reflecting the belief of his generation in progress and the benign effects of technology in a pre-1914 world in which the word "machine" was not yet invariably coupled with the word "gun." Despite this, Churchill remained to the end the type of courageous, combative man of action he had described in his only fictional work, with a "cast of mind" which was "[v]ehement, high and daring"[67] In August 1953, for example, the SIS in cooperation with the CIA helped overthrow the anti-Western Iranian prime minister, Mohammed Mossadeq. The CIA leader visited Downing Street to brief Churchill on the matter, only to find him in bed recovering from a stroke. Nevertheless, the Prime Minister insisted on hearing the story. "Young man," he said at the end of the briefing, "if I had been but a few years younger, I would have loved nothing better than to have served under your command in this great venture."[68]

Hitler also emerged from the First World War as a courageous, combative soldier. Just before a battle in 1915, for example, a field telephone operator told him that he did not care (*ihm sei es Wurst*) if Germany won or lost the war. Hitler became so infuriated that only the interference of others saved the soldier from a beating.[69] After the war, he remained devoted to action, to avoiding what he described in *Mein Kampf* as one of the worst symptoms of national decay: the "*habit of doing things by halves*."[70] It would be a life-long devotion. "We approach the realities of the world only in strong emotion and in action," he stated decades later.

> I have no love for Goethe. But I am ready to overlook much in him for the sake of one phrase – "In the beginning was action." Only the man who acts becomes conscious of the real world. Men misuse their intelligence. It is not the seat of a special dignity of mankind, but merely an instrument in the struggle for life. Man is here to act. Only as a being in action does he fulfill his natural vocation. Contemplative natures, retrospective like all intellectuals, are dead persons who miss the meaning of life.[71]

But there were also some key differences in the Great War experiences of the two future leaders. For in addition to combat duty as a battalion commander in that conflict, Churchill acquired first-hand knowledge of the political–military direction of total war in a series of high-level positions, ranging from First Lord of the Admiralty during Asquith's reign to Minister of Munitions in the Lloyd George government. That direction dealt with an effort, he explained to workers at the Ponders End munitions factory in 1917, which "was not a war only of armies, or even mainly of armies. It was a war of whole nations. . . ."[72] Hitler, on the other hand, had only a *Gefreiter*'s "worm's-eye view" of the first modern total war – a view that reinforced his penchant for action at the lowest level without adding to his understanding of political–military interaction. "I distrust officers who have exaggeratedly theoretical minds," he could still contend during the next war. "I'd like to know what becomes of their theories at the moment of action."[73] Equally important, at the end of the Great War Churchill returned as an insider to a society still governed by a small, select, homogeneous patriciate. At the same time, Hitler returned as the eternally disillusioned, disenfranchised outsider, a member of the petit bourgeois, lacking even German citizenship – one who could identify with the soldier in *All Quiet on the Western Front* who believed that once he and his comrades returned to their homeland "men will not understand us – for the generation that grew up before us, though it has passed these years with us here, already had a home and a calling; now it will return to its old occupation, and the war will be forgotten."[74]

Style: Work and Decision-Making

Churchill and Hitler had unorthodox work habits, with both leaders typically during the war not retiring to sleep until the early morning hours. Throughout the day, Churchill would devour reports and papers on every aspect of national strategy, from economics to politics, spewing forth a torrent of "Action This Day" memoranda and directives, which like his public and private speeches had self-conscious literary qualities of their own. Some of the directives, in fact, were so magisterial in tone that they became known in government circles by their opening words in the manner of papal encyclicals. One directive that began "Renown awaits the commander who first . . . ." was simply titled "Renown Awaits."[75] On the other hand, Hitler rarely put things in writing and even occasionally directed a secretary not to take notes on what he would say in a meeting. Moreover, the Nazi leader read very few documents in the course of a day, normally focusing exclusively on those dealing with the day-to-day battles and engagements in the theaters of operation. The result was a general lack of follow-through except in the case of military affairs at the operational and tactical level – all in sharp contrast to Churchill who, despite his unorthodox work habits, was efficient and adept at follow-up actions to ensure that all decisions concerning national strategy were implemented.

In these approaches to political–military leadership, the concept of loyalty was a key ingredient for both leaders. In Churchill's case, Violet Asquith touched in later years on the passionate, almost irrational intensity of his feelings concerning that concept. "There was an absolute quality in his loyalty, "she wrote, "known only to those safe within its walls. . . . In a friend he would defend the indefensible, explain away the inexplicable – even forgive the unforgivable."[76] Such an approach could have disastrous results, as the romantic gesture of his loyalty to Edward VIII during the abdication crisis demonstrated. "He blundered into a very difficult and delicate situation," Robert Rhodes James pointed out, "heedless of the warnings of his wife and his friends, too late to save the King but not too late to destroy himself."[77] Normally, however, once Churchill assumed office, the British system of government mitigated the Prime Minister's tendency, as one Private Secretary observed, "to disregard failings that less tolerant people found objectionable."[78] Thus, even an old confidant and Victoria Cross winner like Roger Keyes, one of those "energetic adventurers" to whom Churchill was always attracted, was relieved of his duties as Director of Combined Operations; but only after General Dill had advised the Prime Minister that "you can't win World War II with World War I heroes."[79]

Personal loyalty also helped ease the strain of leadership that sometimes colored Churchill's dealings with his subordinates during the war years. At one point, a Private Secretary noted in his diary that the Prime Minister's "inconsiderate treatment of the Service Departments would cause trouble were it not for the great personal loyalty of the Service Ministers to himself."[80] And after the British evacuation of the continent, Mrs Churchill felt compelled to write her husband admonishing him for a "deterioration" in his manner toward subordinates, pointing out that "you won't get the best results by irascibility and rudeness. They will breed either dislike or a slave mentality. . . ."[81] Part of the problem was a lack of empathy and sensitivity – a remoteness illustrated by Churchill's role in the choice of an American as Supreme Commander for Overlord, a position he had promised to Alanbrooke. If this was a rational acceptance of what was inevitable, given increasing American predominance in overall strategy, it also revealed a self-absorption on the part of the Prime Minister, who apparently assumed that his CIGS also realized that inevitability and did not require a modicum of tea and sympathy. "He bore the great disappointment with soldierly dignity," Churchill wrote of Alanbrooke in his memoirs.[82] The general's perception, however, was somewhat different. "Not for one moment did he realize what this meant to me," he wrote. "He offered no sympathy, no regrets at having had to change his mind, and dealt with the matter as if it were of minor importance."[83]

For Hitler, loyalty was absolutely essential to the *Führerstaadt*. This was primarily because of his "charismatic" form of leadership which in essence rejected the institutional and bureaucratic norms required for the "rational" governing of a modern state in favor of dependence on personal loyalty to the Führer as the base support block of authority. Throughout his time in power, as an example, Hitler's relations with his *Gauleiter* demonstrated this emphasis on personal loyalty which, in terms of those regional chieftains, eliminated any semblance in their provinces of ordered government. The Nazi leader, in Herman Rauschning's words, "never ran counter to the opinion of his Gauleiter," invariably taking their side in any dispute with central government agencies or institutions, thus ensuring that a large residue of personal loyalty and concomitant chaos existed in key positions throughout the Reich.[84] Behind all such dealings, there was the absolute centrality of Hitler who, in Rudolf Hess's description, was "simply pure reason incarnate."[85] A few days before the Blood Purge of 1934, the Deputy Führer emphasized this centrality to Ernest Röhm, who would pay with his life in the purge for his refusal to accept that basic fact.

> With pride we see that one man remains beyond all criticism, that is the Führer. This is because everyone feels and knows: he is always right, and he will always be right. The National Socialism of all of us is anchored in uncritical loyalty, in the surrender to the Führer that does not ask for the why in individual cases, in the silent execution of his orders. We believe that the Führer is obeying a higher call to fashion German history. There can be no criticism of this belief.[86]

In this manner, the transference of the Nazi Party ethos from the *Kampfzeit* to a modern government also caused a preoccupation with maintaining Hitler's prestige and image. Thus there was his constant concern with his appearance, insisting, for instance, that there be no photographs of him wearing glasses. Moreover, the need for infallibility only reinforced a growing hubris on the part of the Nazi leader, already pronounced in the early days of struggle. "I cannot be mistaken," Hitler remarked to a party confidant in those days. "What I do and say is historical."[87] The early victories in the Second World War only reinforced this feeling. One day during that conflict he was whistling a classical tune when a secretary summoned the nerve to suggest that he had made a mistake in the melody. "I don't have it wrong," Hitler raged. "It is the composer who made a mistake in this passage."[88]

Churchill had few such concerns with appearances except when it applied to continuity and tradition. Like his Victorian predecessor, for instance, as Prime Minister he never called at Buckingham Palace or attended Parliament wearing anything but a frock coat. Otherwise, the British leader often indulged his penchant for the most outrageous of hats and ad hoc uniforms from his vast collection. And unlike Hitler, who would never let his servants see him in his underwear, Churchill would often conduct business and conversations in various stages of undress, even receiving visitors while taking baths.[89] Most important, the British leader had no sense of infallibility. In his autobiography, he reflected on a period when as a young subaltern he was to be at a dinner attended by the Prince of Wales, the future Edward VIII: "I realized that I must be upon my best behaviour: punctual, subdued, reserved, in short – display all the qualities with which I am least endowed."[90] He was equally frank concerning military blunders. "I certainly bore an exceptional measure of responsibility for the brief and disastrous Norwegian campaign," he wrote to Ismay after the Second World War, "– if campaign it can be called."[91] And it is hard to imagine Hitler at any time making a self-deprecating, humorous speech such as Churchill presented in that same conflict: "I have stuck hard to my blood, toil, tears and sweat *to which I have added muddle and mismanagement*, and that, to some extent, is what you have got out of it."[92]

The results of the two styles were striking in their dissimilarities. In the Third Reich, Hitler was generally surrounded by the most egregious of toadies, those habitués of his *Stammtisch* who accompanied him to power, a "collector's corner," in D.C. Watt's description, "of failure, inadequacy, empty pomposity, and sheer intellectual loopiness," ever ready to agree to whatever the Nazi leader proposed.[93] "Only a few individuals among those around Hitler," Speer observed in this regard, "... withstood the temptation to sycophancy."[94] For Churchill, although he liked to dominate discussions, the give and take of argument was essential to decision-making in war. "The temptation to tell a chief in a great position the things he most likes to hear," he wrote after the First World War, "is the commonest explanation of mistaken action."[95] Early in the next war, General Ismay, who had memorized long passages from Churchill's works, made a vow that he would not be a "yes man" to the British leader on matters of importance. Later in that conflict, after Churchill had reproached him for sticking to his opinion in an argument against a course of action favored by the Prime Minister, Ismay told him of the vow. "You should forget these outmoded Staff College shibboleths," Churchill growled with simulated ferocity.[96]

To all this was added the major differences in the backgrounds of the two men, which helped determine their approaches to leading their nation-states in war. Hitler, the natural conspirator, the master of the *coup de main*, revealed very little of his inner thoughts despite the fact that he was talkative in his private conversations with the paladins of his inner circle and in his dealings with the military leadership and foreign visitors. "You will never learn what I am thinking," he told General Halder. "And those who boast most loudly that they know my thought, to such people I lie even more."[97] Even Speer who was close to the Nazi leader, was in agreement: "Never in my life have I met a person who so seldom revealed his feelings."[98] In contrast, Churchill, the Victorian, was open and frank with his subordinates, his interaction with them often suffused with emotionalism and, like his paintings, never simply a monotone. During a break in a wartime conference at Chequers, for instance, one of his generals began playing "The Blue Danube" on a piano. To the amazement of the assembled military leaders, the Prime Minister began to waltz alone, dreamily around the floor. The incident was symptomatic of an openness summed up by Sir Edward Bridges, the Secretary of the War Cabinet, who recalled how

> significant are the frankness and freedom with which he would discuss things with us, or in our presence. When his mind was occupied with some important issue, he would often discuss it off and on for two or three days with those who happened to be summoned to his work-room .... In this sort of discussion he would keep nothing back. He would express the most outspoken views about the public reactions or attitudes of the most important persons, or about the various ways in which the situation might be expected to develop. And these confidences were not prefixed by "You must not repeat this."[99]

The Enemy

Magnanimity was a concept totally alien to Hitler, a man consumed by hate, resentment and vindictiveness, who never forgave a slight. Thus there was his wider roll call of revenge during the 1934 "Night of the Long Knives" and his delight a decade later in the pictures from the Plötzensee Prison of the members of the assassination conspiracy against him, garroted and hanging by wire from meat hooks. Added to this when war came was his constant call to exterminate the "enemy," an all-embracing term which included Jews and foreigners as well as opposing armies: "Gentlemen, let us not play at being heroes, but let us destroy the enemy. Generals, in spite of the lessons of the war, want to behave like chivalrous knights. They think war should be waged like the tourneys of the Middle Ages. I have no use for knights."[100]

For Churchill, magnanimity was an essential element of his character. As one who had been bullied and misused as a child, and as a public school product, he had an instinctive sympathy for the underdog. "'Never despise your enemy' is an old lesson," he wrote after fighting in the 1897 Indian campaign, "but it has to be learnt afresh, year after year, by every nation that is warlike and brave."[101] Throughout his long life, Churchill continued to display a fairness and generosity to opponents whether in the fictional *Savrola* or in his historical writings dealing with disparate personalities ranging from the Khalifa to Michael Collins. "It is extraordinary," he wrote of his negotiations in the inter-war years with Collins on the Irish Treaty, "how rarely in history have victors been capable of turning in a flash to ... that utterly different mood, which alone can secure them forever by generosity what they have gained by force."[102] And, in a speech during the Second World War, the British statesman extended this philosophy to Erwin Rommel: "We have a very daring and skillful opponent against us, and, may I say across the havoc of war, a great general."[103]

At the conclusion of that war, Churchill could even find some positive points in Hitler, whose earlier misfortunes, the British statesman pointed out in his Second World War memoirs, at least "did not lead him into Communist ranks. By an honorable inversion he cherished all the more an abnormal sense of racial loyalty and a fervent and mystic admiration for Germany and the German people."[104] In the 1920s, with the many failures of parliamentary democracy, Churchill had responded to the first blush of Fascist movements, with their proclamations of national regeneration, racial pride, patriotism, codes of behavior, respect for order and tradition, and, above all, hostility to Communism, the great threat to the traditional order. "(I)n the conflict of Fascism and Bolshevism," he noted of that period, in which he had twice met with Mussolini, "there was no doubt where my sympathies and convictions lay."[105] This admiration for authoritarian regimes also colored Churchill's earlier

This type of historical philosophy was not categorical and systematic; nevertheless, it left the future British leader with firm principles within which his ideas might change as they often did. To these principles was added a Gothic, almost apocalyptic outlook, in which, as he did with Hitler, Churchill could visualize the deep, dark, destructive forces that lay just below the veneer of human civilization. There was, in short, more to villainy than the typical bounder and cad of the Victorian and Edwardian eras. It was an approach to history that early on reinforced Churchill's deeply rooted patriotism. "I shall devote my life," he wrote his mother in 1897, "to the preservation of this great Empire and to trying to maintain the progress of the English people."[117] Implicit in all that was a conservative, essentially reactionary and defensive outlook concerning history. "Fortune is rightly indignant to those who break with the customs of the past."[118]

As a revolutionary, Hitler had a considerable understanding of certain historical factors, formed into a systematic and categorical philosophy of history that caused him to think of ideas as principles. This approach, as his allusions to the Franco-Prussian War in both of his books demonstrated, led him to nationalism, a more shallow force than Churchill's patriotism and one that was an essentially aggressive, more offensive development, whether focused against perceived domestic or foreign enemies or both. It was also an historical outlook devoid of Churchill's sense of progress founded on freedom and the perfection of institutions and man. "I have not come into the world," Hitler was fond of saying, "to make men better, but to make use of their weaknesses."[119]

All this flowed into a *Weltanschauung* based on an ideological admixture of social Darwinism, Völkisch themes and Pan-Germanism – all buttressed by the romantic hostility to civilization and glorification of conflict that was symbolized in one of Hitler's favorite works of Nietzsche, *Also sprach Zarathustra*: "You say a good cause can even sanctify war? I say unto you: it is a good war that sanctifies any cause."[120] By the time Hitler had finished his *Secret Book* in 1928, his world view was complete, centered on three *idées fixes*: the need for additional living space, the destruction of the Jews and eventual global domination by Germany. Unlike Churchill, who continued to supplement his views with intellectual and personal growth throughout his long life, this primitive and nihilistic outlook remained the philosophical basis for the Nazi leader throughout his career. "The Fuehrer does not change," Goebbels wrote in 1934. "He is the same now as he was when he was a boy."[121]

There was another, equally important aspect of Hitler's character that also did not change: a pragmatic approach to issues. It was an approach that caused him to lose patience with the theological niceties of Alfred Rosenberg, the high priest of the Nazi party, who believed that "National Socialism stands or falls by its *Weltanschauung*." Rosenberg failed to realize that the ideology in which he believed so fervently carried no weight in the centers of power. "The ideas behind our programme," Hitler pointed out, "do not oblige us to act like fools."[122] In the interplay of ideology and power seeking, there was hardly any part of the National Socialist *Weltanschauung*, with the exception of the Führer's maxims, that could not be at least temporarily set aside if not abandoned for the sake of gaining or holding a position. This opportunism was demonstrated in the Nazi concordat with the Holy See and the Nazi–Soviet Non-Aggression Pact as well as in the hodgepodge of concepts that were quickly acquired and often just as quickly discarded during the Third Reich. "Any idea, even the best," Hitler concluded in *Mein Kampf*, "becomes a danger when it imagines it is an end in itself, whereas in reality it represents merely a means to such."[123]

This was the pragmatism of the born revolutionary, one who would continue the *coup d'état* by other means after he came to power legally, "without ... the breaking of a window pane."[124] Thereafter, the quasi-legality of Hitler's coordination (*Gleichschaltung*) of the state could not hide the essentially revolutionary nature of his regime: the refusal to play by the rules of the international state system symbolized by his often repeated remark that "for Germany there is no lie I would not utter."[125] For those who played by those rules like Churchill, he had only contempt: "It is characteristic of the narrowness of these outlived classes that they should be indignant with me, indignant at our contempt for past customs and assumptions in political life."[126] To the end, Hitler remained a living contradiction to the truism of revolutions devouring their children. He was as H. R. Trevor Roper has noted,

> the Rousseau, the Mirabeau, the Robespierre, the Napoleon of his revolution; its Marx, its Lenin, its Trotsky, its Stalin. In character and mentality he may have been far inferior to most of these men, but at least he did what none of them did: he controlled his revolution through all its stages even in defeat.[127]

Churchill was very much aware of the threat that this modern, revolutionary force posed to his traditional, conservative world view grounded as it was in the Whig approach to British history, the concept of the British empire, and, above all, the maintenance of

freedom. There had been in the past, of course, revolutionary
hegemonies such as Napoleon's France bent on expansion. But as
Churchill realized, the French leader had operated within the bounds
of international rules; Hitler had no such constraints. "A Nazi victory," the British leader wrote in an introduction to the speeches of
Pitt the Younger in October 1940, "would be an immeasurably worse
disaster for us and for all mankind than Napoleon's victory could ever
have been .... Napoleon could construct as well as destroy."[128] Under
such conditions, a breathing space such as Britain had concluded in
the 1802 Peace of Amiens with Napoleon was not possible with Hitler.
Without a basic respect on the part of the Nazi state for international
norms and the continuing independence of opponents, all implicit in a
world view based on a concept of an inferior species of mankind, there
could be no peace party in 1940 that approached the caliber of the
movements led by Charles James Fox or Lord and Lady Holland in
1802.[129]

   In the end, it was Churchill's deep belief in the continued viability of
progress under freedom and justice that triumphed. His tenacious
defense of the past, embodied in Britain's status as a great power and
in the notion of the British Empire, would not long survive the great
victory in the Second World War. But his devotion to the human
spirit, which so colored his outlook on events in the world, would live
on in the gesture of his two outstretched fingers. Unlike the Nazi
salute, hated and mocked in its own time and still so despised that it
appears only in movies, Churchill's V-sign has become an almost
universal symbol of either freedom leading to victory or victory
leading to freedom.[130]

The Warlords and the Trinity

Because of the First World War experience, the totality of the Second
World War was not unexpected. As a result, the concept of national or
grand strategy in wartime was a familiar fixture in all the combatant
nations. What was new, however, from the inception of the Second
World War, was the dominance in national strategy of political
leaders, whether it was Franklin Roosevelt and Adolf Hitler or
Joseph Stalin and Winston Churchill. The personal approaches of
these leaders to national strategy were vastly different, adding new
complexities at every level and in every area to the mix of elements
within the strategic framework.[131]

   Both Churchill and Hitler, as the essays in this volume demonstrate in detail, were very much concerned as modern warlords with the interaction of the elements of Clausewitz's remarkable trinity. Hitler's authority in wartime was an extension of the dictatorial powers he had wielded for over six years as the peacetime head of the Third Reich. And yet, despite the pervasive all-encompassing character of those powers, the Nazi leader repeatedly tied governmental wartime policy to the German people, the *Geführten* of his *Führerstaadt* whom he never really trusted, and to the German military, with whom he had an enduring *Hassliebe* relationship. For Churchill, the elected head of a parliamentary democracy, the ultimate judgment of how well he prosecuted the war and therefore how long he would remain in power rested with the people. The votes of confidence sustained by the British leader earlier in the war stand as vivid reminders of this power source. In the end, the elements of the trinity held for both leaders: Churchill in victory, Hitler in the *Gotterdämmerung* of defeat in total war. Ironically, however, the victory in Europe dissolved the links of the trinity for the British statesman, and he was driven from office. It would be a few years before the citizens of Britain looked once again to Churchill for leadership, just as he had described in his earlier work of fiction when the people of Laurania turned again to his protagonist, Savrola, "the illustrious exile who had won them freedom and whom they had deserted in the hour of victory."[132]

The Government

Political leaders of nations in total war normally tend to dominate government, no matter what the form of that government. In a dictatorship, as in Hitler's case, this type of war can reinforce an already pervasive totalitarian control. And in a parliamentary government like England fighting for its survival, there can be a very fine line between authoritarianism and dictatorship. That Churchill walked the line successfully throughout the war was due in part to what C. P. Snow described as "the small size, the tightness, the extreme homogeneity, of the English official world."[133] The British leader was at home with both the civil and military elements of that essentially patrician world and would not ultimately go beyond what those elements allowed in his decisions as warlord.

Restraint of power was also linked to the structure of governmental institutions and the attitude of the warlords to those institutions. For Hitler, traditional institutions were merely a means to an end in a revolutionary world, to be discarded once that end had been achieved. Thus, early in his reign, there was the burning of the Reichstag, the hated symbol of democratic growth in the Second

Reich and the parliamentary repository for the despised post First World War republic, once the capital was moved from Weimar to Berlin. And although the Nazi leader apparently had nothing to do with the destruction of that building, he also never rebuilt it. Instead, he used the event as the excuse for his infamous February 1933 "Emergency Laws," the beginning of the *Gleichschaltung* process that would bring him to ultimate power in Germany long before the Second World War began.

Churchill, on the other hand, never deviated from a romantic Whig concept of British government institutions that translated into maintenance of their traditional forms no matter what the exigencies of total war might be. In this context, Parliament remained a major source of strength for the British leader throughout the war. Despite the advent of radio, Parliament was his conduit to the people; and it was no accident that many of his radio addresses to the nation were barely reworked parliamentary speeches. And when the empty buildings of that institution were destroyed in a night bombing attack during the spring of 1941, Churchill's reaction typified the vast gulf between his approach to government and that of his Nazi counterpart. The morning after, he wandered in the still smoking ruins of the House of Commons, savagely poking the end of his walking stick into the cinders that had once been the doors leading from behind the Speaker's Chair. Tears ran unchecked down his cheeks. Brushing the sleeve of his coat across his eyes in the manner of a schoolboy ashamed of his tears, the Prime Minister turned abruptly on an official and, in a voice controlled only with effort, said with quiet ferociousness: "This Chamber must be rebuilt – just as it was! Meanwhile, we shall not miss a single day's debate through this!"[134]

In terms of running the government, Churchill as both Prime Minister and Minister of Defence took responsibility for everything concerned with grand strategy, insisting on written evidence for any decision. This organization became a more rational and efficient instrument for coordinating all the elements of national power as the conflict proceeded, creating, as he rightly claimed, "a stream of coherent thought capable of being translated with great rapidity into coherent action."[135] In the end, the organizational machinery that evolved for prosecuting the war achieved an important balance. On the one hand, it allowed Churchill to bring his strategic vision bolstered by the full brunt of his personality and eloquence to bear in decision-making sessions with his advisers. On the other hand, there were the restraints that could be imposed on the Prime Minister from a variety of military and civilian sources – "one of the reasons," a member of the Defence Secretariat later told Martin Gilbert, "why we won the war."[136] The fact that those restraints worked, however, was also due in large part to Churchill's respect for the system when his own forces of persuasion or intimidation failed to win the day. "I cannot recollect a single minister, serving officer or civil servant," Lord Bridges recalled in this regard, "who was ever removed from office because he stood up to Churchill."[137]

There was no such restraint in the Third Reich under the *Führerprinzip*, the essence of which was captured in a caption under Hitler's portrait in the Munich headquarters of the Nazi Party: "Nothing happens in this movement, except what I wish."[138] This charismatic principle continued to apply after 1939, extending into the conduct of total war the peacetime chaos of the unbureaucratic, almost neo-feudalistic system of government that marked Hitler's Germany. In this context, the Nazi leader remained the final arbiter in the countless civil and military rivalries produced by this form of government, the difference from his role over the various party factions in the *Kampfzeit* being essentially only one of scale. Added to this chaotic state was Hitler's idiosyncratic style of governing, basically reflecting the bohemian and artistic temperament of his Vienna days in his frequent hesitations to make decisions, his complete lack of punctuality, his detestation and avoidance of routine work, and his chronic procrastination. This wartime vacuum was filled by Martin Bormann, the "Brown Eminence," who by 1942 became the sole channel to the Nazi leader. In his almost total concern with possible rivals, Bormann only added to the chaos and inefficiency of the wartime German government – a marked contrast to the outstanding efforts of Lord Bridges and General Ismay in coordinating the offices of the British Prime Minister and Minister of Defence for the conduct of grand strategy.[139]

The paradox was complete. Hitler's government, over which he had such complete authority that it included a military bound by a personal oath to the Führer, strangled in its own compartmentalized, totalitarian inefficiencies. In such a milieu, a grand strategic calculation of ends, ways and means was impossible. Thus there was the wartime production of German armaments hampered by divisions among jealous rivals and the restriction on thought that inhibited the evolution of ideas and the interchange of speculations. Above all, there was the increasingly isolated, hubristic and indolent supreme authority, in many cases passively allowing the centrifugal forces of his government to play out day by day as the Second World War moved toward its final denouement.

By contrast, the British system of government was an evolutionary product based on authority tempered by checks and balances. Within that system, Churchill actively and incessantly exercised his superb leadership and organizational traits to optimize the British government for total war. As an example, there were his efforts against the growing U-boat menace in 1941 beginning with his inspirational proclamation of "the Battle of the Atlantic." As back up for this abstract battle cry, however, there was also the Prime Minister's creation of a hard-headed, top down organization coupled with coordinated decentralization and supervision. In March 1941, Churchill published his campaign directive which outlined civil and military responsibilities for the many agencies concerned. The efforts of those agencies were to be directed at the highest level by means of the newly created Battle of Britain Committee, chaired by the British leader and composed of the War Cabinet, naval and air chiefs, scientists, and in Churchill's own words, any other "high functionaries concerned." By 22 October 1941, the date of its last meeting, the committee had succeeded in the major political, economic and military tasks outlined in Churchill's grand strategic campaign plan.[140]

The People

Both Churchill and Hitler personally experienced the negative popular reactions to the Great War, as that first encounter with modern total war ground out over a bloody four-year period. The British people were not used to rationing; and the myriad regulations that came to govern their daily lives were intensely unpopular, the all encompassing DORA, the Defence of the Realm Act, being the most egregious example. The Russian Revolution only fueled this discontent, and there was increased industrial unrest in Britain toward the end of the conflict that sometimes took on an anti-war character, as Churchill knew first-hand from his vantage point in the Lloyd George government. The turning point for Hitler in that conflict was the 1918 November revolution in Germany. And although in his version of the *Dolchstoss*, the so-called "stab in the back" that caused the revolution, there was no mention of the social and economic causes for that upheaval, he was acutely aware of them. "Clearly there was dire misery everywhere," he wrote in *Mein Kampf* of a leave spent in Berlin in the last year of the war. "Discontent was great." And when a general strike broke out just as the Ludendorff offensive began, Hitler noted that "at home the revolution was before the door, and not the victorious army." "What was the army fighting for," he concluded," if the homeland itself no longer wanted victory."[141]

The responses of the two leaders to these earlier experiences were markedly different. For Churchill, the First World War clearly demonstrated the centrality of the people, whether serving in the military or on the Home Front. Total war, however impersonal and technical in nature, was still dependent on the anonymous individual citizen who would, the British leader believed, like him, undergo any hardship to preserve Britain. "This is no war of chieftains or of princes, of dynasties or national ambition," he told the nation at the beginning of the next conflict; "it is a war of peoples .... This is a war of the unknown warrior." The effect of such an appeal was immediate. "But really he has guts that man," Harold Nicolson wrote to his wife after the speech, reflecting the national impression. "... I felt a great army of men and women of resolution watching for the fight. And I felt that all the silly people were but black-beetles scurrying into holes."[142]

Nicolson's reaction demonstrated how important a sense of participation in total war was to members of the Home Front and how perceptively Churchill had gauged that need. The Blitz, of course, more than satisfied this need for the people of London and the other major cities that bore the brunt of German air attacks. Nevertheless, the British leader pointed out in October 1940, the air attacks were a manifestation of total war "which would suit the English people once they got used to it. They would prefer all to be ... taking part in the battle rather than to look on hopelessly at mass slaughters like Passchendaele."[143]

Hitler had no such faith in the German people. *Mein Kampf* was studded with contempt for the "great masses, who ... have more understanding for beer and milk regulations than for farsighted plans ... " and who "cannot see the whole road ahead of them without growing weary and despairing of the task."[144] It was the contempt of the manipulator, the great simplifier, who combined oratorical slogans of indignation and the consciously primitive apportioning of blame with, as he put it, "gigantic mass demonstrations, those parades of hundreds of thousands of men, which burned into the small, wretched individual the proud conviction that, paltry worm as he was, he was nevertheless a part of a great dragon."[145] Most important, that contempt was tinged with fear and insecurity. In his 1928 *Secret Book*, the Nazi leader stressed the need not to allow the living standards of "people of a definite cultural capacity" to decline:

The broad masses especially will show no understanding of this. They feel the hardship; either they grumble against those who in

their opinion are responsible – something which is dangerous ...
since thereby they provide the reservoir for all attempts at
revolutionary upheavals – or through their own measures they
try to bring about a rectification.[146]

After the *Machtübernahme*, as a consequence, the domestic
policies of the Nazi regime swung inconsistently back and forth
between vicious attacks on personal and political rights and fearful
circumspection inspired by the need to gain and maintain popularity
with the masses. When the war came and the conflict turned against
him, Hitler's contempt and indifference for the German people
became more pronounced. He ceased to address them by radio and
never once visited the ruins of towns and cities after the Allied
bombing offensive began. And when the war came to an end, there
was only his vision of national resistance until the final *Untergang* of
an undeserving nation, a belief to which Hitler held with remarkable
consistency. "Unfortunately," he wrote in 1925, "the military defeat
of the German people is not an undeserved catastrophe, but the
deserved chastisement of eternal retribution. We more than deserved
this defeat."[147] And 20 years later in the closing days of the war, the
Nazi leader again returned to this theme. "If the war is to be lost, the
nation will also perish," he stated, adding that "... it is better to
destroy it and to destroy it ourselves."[148]

Churchill's concept for the termination of total war was no less
apocalyptic, founded as it was on the Victorian concept of the Last
Stand. "What is the position about London?" he queried Ismay on
2 July 1940 in this regard. "I have a very clear view that we should fight
every inch of it, and that it would *devour* quite a large invading
army."[149] The difference was that his outlook was based on his high
expectations for the British people who, in response to his intense
eloquence, soon saw themselves acting, as the British Statesmen saw
them, in a larger, heroic litany of great deeds ranging from Thermopylae to the Spanish Main. "I don't know about oratory," Churchill
commented at the time, "but I do know what is in people's minds."[150]
This was the essence of Britain's defiance in 1940 which appeared so
irrational to Hitler; and this is why when asked later which year of his
life he would like to relive, Churchill replied: 1940, "every time, every
time."[151] It was a linkage that understandably baffled the Nazi leader,
a linkage epitomized in a visit that Churchill made in 1940 to an air-
raid shelter on the London docks after an attack in which 40 people
had been killed. The survivors, a mix of young and old, male and
female, but all poor, were still milling around. "One might have

expected them to be resentful against the authorities responsible for their protection," Ismay later recounted;

> but, as Churchill got out of his car, they literally mobbed him. "Good old Winnie," they cried. "We thought you'd come and see us. We can take it. Give it 'em back." Churchill broke down, and as I was struggling to get him through the crowd, I heard an old woman say, "You see, he really cares; he's crying."[152]

For Churchill, the Blitz was just one example of how the British people would respond positively, no matter how grim the event, if they were properly prepared. In June 1940, he emphasized that air raids should be treated "in a cool way," adding that the "people must learn to take air raids as if they were no more than thunderstorms."[153] The maintenance of public morale, he realized, was to a large extent dependent on his frankness and openness with the people concerning the tasks and hardships that lay ahead. This was particularly true for his radio broadcasts. Censorship aided him in those broadcasts, because the public was starved for news and information. Moreover, censorship always left the lingering suspicion that what news remained might be altered for policy purposes. But with Churchill, the public felt that he would give them the facts straight, that he would not try to disguise bad news. Furthermore, there was always the hope that he might give more information than planned, since he was one of the few men in England who could not be censored, and particularly since he revised his texts right up to broadcast time. As a consequence, Churchill had a ready-made and sympathetic national audience with enormous confidence in him, ready to make the sacrifices required in total war.[154]

No such open calls were made by the Nazi regime for sacrifices from the population. This was demonstrated in the opening months of the Second World War by the withdrawal of plans for labor mobilization after workers protested the impact that such a move would have on wages, working conditions and living standards. In a similar manner, as the war continued, other aspects of "business as usual" included the high level of consumer goods until 1943; the personal allowances paid to the wives of the military, which were so generous that most married women could not be persuaded to take up industrial work when that was belatedly sanctioned for women; and, finally, the incessant propaganda stressing that final victory was just around the corner, making sacrifice and effort appear less necessary. For Albert Speer, facing this twin psychological and economic deficit in national strategy on a daily basis, it was:

one of the oddities of this war that Hitler demanded far less from his people than Churchill and Roosevelt did from their respective nations. The discrepancy between the total mobilization of labor forces in democratic England and the casual treatment of this question in authoritarian Germany is proof of the regime's anxiety not to risk any shift in the popular mood. The German leaders were not disposed to make sacrifices themselves or to ask sacrifices of the people. They tried to keep the morale of the people in the best possible state by concessions .... Whereas Churchill promised his people only blood, sweat, and tears, all we heard during the various phases and various crisis of the war was Hitler's slogan: "The final victory is certain."[155]

The Military

Churchill and Hitler both viewed their early military experiences positively. For the Nazi leader writing six years after the First World War had ended, that conflict had been "a release from the painful feelings of my youth," a period in which he had been granted "the good fortune of being permitted to live."[156] And even during the next war, Hitler could still describe sending young men into the army, "whence they will return refreshed and cleansed."[157] Churchill's formative military experiences were from an earlier period. In his account of the 1897 Malakand Field Force campaign, he extolled the "healthy, open-air life" of the soldier, in which the "uncertainty and importance of the present, reduce the past and future to comparative insignificance .... And when all is over, memories remain, which few men do not hold precious."[158] Those memories were still alive when the British leader wrote a welcoming letter in November 1941 to Alanbrooke, his new CIGS, referring to his "old friendship for Ronnie and Victor, the companions of gay subaltern days and early wars."[159] The references to Brooke's two brothers were typical of the "old boy" military network, within which Churchill had always operated. He had met Victor in the 4th Hussars in 1895–96; and he and Ronnie Brooke had gone together through the fighting in the Boer War at Spion Kop, Vaal Krantz and the Tugela, and both had galloped into Ladysmith on the night of that town's liberation.[160]

Nevertheless, Churchill's view of the military leadership by 1940 was certainly not ringed by a romantic halo. There had been Jeffrey's gallant but misdirected leadership in 1897; Buller's almost incomprehensible procrastinations on the Tugela in the Boer War; the mishandling of the tanks as they evolved before the Battle of Cambrai in 1917; and, above all, the recurring slaughter of the First World War. "In truth," he wrote, "these high military experts all belong to the same school."[161] Equally important, as we have seen, Churchill's high-level civilian cabinet experiences in the Great War made him acutely aware of the limitations of the military when it came to the application of all elements of power to national strategy in total war. Policy must dominate strategy, he concluded, as opposed to Germany where from 1914–18 everything "had been sacrificed to the military view."[162]

Hitler emerged from the First World War with equally strong views about the German military elite – views he confided to his *Secret Book* of 1928.

> I did not learn about the war at a restaurant table reserved for regular customers, nor was I in this war one of those who had to give orders or to command. I was an ordinary soldier who was given orders for four-and-a-half years, and who, nevertheless, honorably and truly fulfilled his duty. But I thereby had the good fortune to know war as it is and not as one would like to see it.[163]

Once he assumed power the issue of policy dominating strategy was never in question. In August 1934, the Nazi leader, now President as well as Chancellor of Germany, extracted a personal oath of allegiance from the officers and men of the *Wehrmacht*. By 1938, when he became head of the armed forces, Hitler's successful use of psychological as well as other elements of national power in the so-called "March Crises" further convinced him of the limitations of his military advisers in terms of national strategy.

But Hitler was also a source of limitation in terms of national strategic support of Germany's military efforts in total war. In particular, his "worm's-eye view" of the First World War and his experiences in the subsequent *Kampfzeit* only confirmed his ideological and racist prejudices and his essentially ethnocentric, central European outlook. As a consequence, the Nazi leader consistently underestimated the strength of the United States and the Soviet Union and failed to develop and use allies effectively – a failure in which those allies often repaid Hitler in his own currency of surprise. In such circumstances, a common conspiracy against peace was merely the fortuitous crisscrossing of two lines of force, each in pursuit of different objectives. This was particularly true in the relationship between Germany and Japan, both of which, as one historian has noted, "practiced such secrecy and deception concerning their own objectives that even on those few occasions when their interests genuinely converged they were unable to coordinate their policies."[164]

For the widely traveled, infinitely more cosmopolitan Churchill, his experiences as a statesman in the First World War as well as his work on the biography of Marlborough in the inter-war years, only further demonstrated the vital importance of allies in the political–military scheme of grand strategy. Thus, there was his assiduous courting of President Roosevelt from a position of weakness throughout the 19 months before Pearl Harbor and his immediate support of the Soviet Union under the onslaught of Barbarossa despite a continuous and virulent antipathy to Bolshevism. As a result, the Grand Alliance actually comprised three distinct bilateral relationships. Two of them, the Anglo-American and the Anglo-Soviet, were consistently functioning political associations. The third, that between the Soviet Union and the United States, was significantly less developed, depending as it did on the emphasis attached to it by Roosevelt, who only began to assert himself at the 1942 Second Front negotiations. That divided framework provided more than enough room for Churchill in the early years of the war to seize the initiative in forming the Grand Alliance. All in all, that alliance was a stupendous accomplishment calling for skills similar to those demonstrated by the first Duke of Marlborough in forming an equally unlikely coalition against an earlier European despot.[165]

The creation and maintenance of such an alliance, particularly given Churchill's pivotal position, required extensive travel by him. Added to this were his many visits to battle fronts in North Africa and Europe, the contact with Commonwealth troops always a source of inspiration and rejuvenation for the British leader. In August 1942, for example, Churchill visited General Montgomery's forward positions at Alam Halfa. "I saw a great many soldiers that day," he could still recall with delight after the war, "who greeted me with guns and cheers."[166] Such contact only reinforced his determination that the troops not be involved in attritive, First World War-type blood baths, the memories of which, he admitted in his memoirs, "were not to be blotted out by time or reflection."[167] In the end, of the 1,825 days in which Churchill directed Britain's war effort, he spent approximately 407 of them abroad or in other parts of Britain – a testimony not only to the stability of his regime, but to his genuine interest in the welfare of his troops and to his incredible energy and stamina. "If disposal of all the Allied decorations were ... in my hands," General MacArthur commented on these travels, "my first act would be to award the Victoria Cross to Winston Churchill .... A flight of 10,000 miles through hostile and foreign skies may be the duty of young pilots, but for a statesman burdened with the world's cares, it is an act of inspiring gallantry and valour."[168]

By contrast, during the 2,069 days in which Hitler directed the war, the Nazi leader spent approximately 945 of them in a series of eastern front headquarters, 492 in Berlin and 366 in the area of his southern Bavarian retreat, the *Berghof*. It was a self-imposed restriction that Speer later perceived because of his post-war Spandau imprisonment as bearing a great resemblance to that of a prison, a perception with which Hitler would have concurred. "Here in the Wolfsschanze," he commented in 1942, "I feel like a prisoner in these dugouts, and my spirit can't escape."[169]

Equally important in comparison to Churchill was what Hitler's increasing isolation revealed about his relationship to the German troops whom he had greeted just before the war "as an old soldier with the feeling of comradeship."[170] When that conflict began, he appeared at the Reichstag in a simple grey uniform decorated only with his First World War Iron Cross, and announced that it would be his uniform until the war ended – a promise from which he never deviated. Nevertheless, after the Polish campaign, the German military, as General von Manstein described it, "never managed to get Hitler anywhere near a front line."[171] He continued to maintain, of course, that his "attachment and sympathy belong in the first place to the frontline soldier."[172] But there was an indifference to the fate of the German soldier even in victory that stood in sharp contrast to his laments for the fallen of the First World War. "Do you think it would have been good fortune for our troops if we had taken Poland without a fight?" he asked Speer. "... Victories without loss of blood are demoralizing."[173] And in defeat, the tendency was even more pro-nounced. "Paulus' report affected him not at all," General Zeitzler wrote after Stalingrad. "The figures of dead and wounded ... left him totally unmoved."[174]

Despite such basic differences, there were some striking similarities in the enthusiasm of both leaders for new military technology and intelligence, as well as in their tendencies to slip down from the political direction of the conflict to the theater strategic, operational and even tactical levels of war. In terms of technology, both statesmen were always aware of its connection with doctrine and surprise. Hitler was quick to grasp the implication of armor and *Blitzkrieg* – a combination which indulged his opportunistic and secretive nature and satisfied his love for speed, action and surprise. "That's what I need," he stated after his first glimpse of tank maneuvers. "That's what I want."[175] Most important, the combination allowed the Nazi leader to avoid the hated attrition of his experiences in the Great War and, thereby, the political, economic and psychological strains, which he recalled so acutely and fearfully, affecting the German people. In that same conflict, Churchill had played a major role in the development of the tank; and during the inter-war years he was in regular communication with Liddell Hart and other experts in the field. Nevertheless, as he admitted when the German *Blitzkrieg* smashed through the Allied lines in May 1940, "the idea of the line being broken even on a broad front, did not convey to my mind the appalling consequences that now flowed from it."[176]

Generally, however, Churchill brought a finely developed sense of what science and technology could contribute to warfare. On the romantic side, from his Victorian heritage, there was his belief in these elements as progressive entities that made him alive to their possibilities. On occasion this could lead to impulsive lapses such as his July 1943 request, under the code name Habbakuk, to examine the possibility of turning icebergs into floating air bases. In the end, the project never went beyond the memorandum stage because of a general skepticism, the nature of which could be gauged by one scientist's proposal to use the term "mili-habbakuk" as a new unit for measuring impracticability.[177] But normally, there was Churchill's realistic appreciation of how the civil military powers that he possessed as a twentieth century warlord could aid the application of science and technology in war though his tightly honed organization, his committee system and, above all, his encouragement without interference of the scientific community.

This combination of faith and pragmatism was illustrated by the British leader's total involvement throughout the war in the project to create artificial harbors. In 1917, he had proposed such a harbor in connection with a possible landing behind the German lines in Flanders. In the next war, the idea of Mulberry became the basic assumption for the 1944 cross-Channel invasion of Europe. But only because of the personal direction of Churchill, who kept a working model of the project next to the War Room in Whitehall. "Don't argue about it," he admonished the Combined Operations planners over a particularly difficult technical problem. "The difficulties will argue themselves."[178] That type of perseverance and faith produced a device that was decisive for the Allies. Strategically, Mulberry provided the planners with the freedom to select landing zones well away from the heavily fortified harbors on the Continent; psychologically, it gave the Allied leaders a degree of confidence, without which they might never have attempted what they perceived as an extremely hazardous cross-Channel operation. As it was, that "single brilliant technical device," as Speer described it, allowed the Allies to bypass the Atlantic Wall, thus rendering the German defense system irrelevant.[179]

Hitler also combined a grasp of detail with a gift for creative fantasy in military technological matters, frequently proposing, as Speer admitted, "convincing and usable innovations" ranging from the emplacement of sirens on Stuka dive bombers for psychological effect to the development of the highly successful 77mm anti-tank gun.[180] But the very dominance as a military decision-maker that allowed him to cut through bureaucratic red tape with such innovations ultimately led to the realities of Hitler's personality and organizational style. Thus, there was the hubristic interference combined with procrastination and lack of prioritization that hampered innovative projects such as the V weapons and the Me-262 jet fighter plane. All this, in turn, drew off resources from the German atomic bomb project, already hampered by the mass exodus of scientists because of the anti-Semitic policies and general totalitarian nature of the Nazi regime. Moreover, for those funds that were available, there was no coherence in their dispatch to research units that owed allegiance to several supervisory agencies, all of which were further fragmented by personal rivalries among the agency directors.[181]

In a similar manner, despite the fact that Hitler was an avid consumer of intelligence, the sycophancy and almost Darwinian bureaucratic infighting of his regime had a debilitating effect that perverted the meaning of intelligence by allowing information to be used by agencies and key personnel "to establish credibility in competition, not to discern 'truth.'"[182] All this was compounded by Hitler's curious mixture of insecurity and growing sense of infallibility. "As regards enemy intelligence," General Warlmont wrote of the Nazi leader in this regard, "he only accepted what suited him and often refused even to listen to unpalatable information."[183] On the other hand, Churchill's lifelong enthusiasm for intelligence was complemented by his superb leadership and organizational skills to form an efficient optimization for war – particularly noticeable in his encouragement and use of Ultra. For the British leader, intelligence meant power, not just in dealing with the enemy, but with the elements of the Clausewitzian trinity. As Minister of Defence, he knew what his chiefs did and could deal with them on equal terms. Ultra, in particular, strengthened his political–military position by providing him with inside information which he soaked into his prodigious memory and applied almost oppressively and often infuriatingly in combination with his passionate interest in the smallest details of military affairs. In terms of the government and the people, Churchill's possession of the latest intelligence greatly increased his central power in directing total war, which at times meant the management of that intelligence. "He claimed the right to approach the British people in his own way," the Director of Naval Intelligence noted of Churchill in early 1940; "he did not hesitate if it suited his conception of the interest of a nation at war to twist the truth or to paint a rosy picture that had no connection with veracity."[184]

Equally important, like Hitler, Churchill was fascinated by deception operations. Hitler's successes ranged from Barbarossa at the strategic level to the Battle of the Bulge at the operational level – all characterized by the personal involvement and direction as a military decision-maker that made him so effective at deception. Thus it was the Nazi leader who carried off the rare double bluff beginning in February 1941 by presenting the movement of German troops to the east as a deception to cover plans for the invasion of England, an example of "a gigantic military operation being effectively concealed from the very civil–military bureaucracy conducting it."[185] But Hitler's efforts never matched the centralized organization of the London Coordinating Section (LCS) established by Churchill to coordinate deception operations on a global basis. One of many successful results was the finely coordinated overarching deception framework for the cross-Channel operation containing 36 subordinate plans, including those of Fortitude, the deception most directly involved in the operation. Fortitude North called for containing German forces in Scandinavia by means of a phantom army ostentatiously assembled in Scotland and the threat of an Anglo-Russian invasion of northern Norway. There was some German skepticism concerning the joint invasion, but not enough to affect Fortitude North; and it was not until 16 June that Hitler ordered two divisions from that area to France – and even then only to the Pas de Calais. That, of course, was a tribute to the success of the most important deception plan, Fortitude South, which kept the other half of Rommel's Army Group B, the 15th Army, away from Normandy, spread roughly between the Scheldt and the Seine with the principal concentration at the Pas de Calais. And although Hitler had a last minute intuitive feeling about Normandy, his commanders in the west, including von Rundstedt and Rommel, remained convinced long after D-Day that the Pas de Calais was the key to Overlord.[186]

The central roles of Churchill and Hitler in deception plans was typical of their involvement in all facets of military operations. Both took combat seriously, perceiving themselves as part of great wartime leadership traditions, with Churchill finding sustinment and encouragement in the exploits of the elder and the younger Pitt, Hitler in those of Frederick Barbarossa and Frederick the Great. Both had a flair for some aspects of military operations, an instinctive understanding, as General Eisenhower noted of the British leader, of "the basic rules – a lack of dispersion, and using your mass, using surprise – all that kind of thing."[187] To this, the two leaders brought a love for the unconventional, symbolized by Churchill's enthusiasm for Ord Wingate and Hitler's use of Otto Skorzney, as well as a fascination with details, most pronounced in their preoccupation with maps. At the Munich conference, for instance, Hitler was so well briefed that he could draw boundaries over large scale maps.[188] And the director of Churchill's map room reported that it "was always a source of pleasure to the Prime Minister to mark in China-graph pencil in very considerable detail advances made by the various Divisions and Brigades."[189] Finally, both leaders consistently overestimated the efficiency of mass bombing and underestimated the value of logistics. In terms of the latter, Hitler's deficient conception of combat service support for Barbarossa is a classic example, as is Churchill's angry query of his Chief of Staff concerning the number of vehicles being added to what he considered the "interminable tail" in the Middle East. Ismay quoted in a reply that went unheeded a striking metaphor used by the British statesman before the turn of the century in *The River War*: "Victory is the beautiful bright-coloured flower: transport is the stem without which it would never have blossomed."[190]

Churchill's dealings with his military were very much like Chatham's, who chose Wolfe because he found expedients where other officers found difficulties. There had to be the offensive spirit, the desire to engage the enemy; and the British leader often quoted Nelson's Trafalgar memorandum in this regard: "No Captain can do very wrong if he places his ship alongside that of an enemy."[191] Nevertheless, Churchill was never so dominant, nor dared to be, as popular opinion imagined or his military advisers apprehensively believed; and he never overruled his chiefs when they were united in judgment. He did, however, submit them to gruelling examinations to test the firmness of that judgment so that no conceivable option was considered an impossibility until it had been thoroughly examined. The ultimate test, as Churchill realized, was that "in war you do not have to be nice – you only have to be right."[192] And ultimately, in this regard, like Elizabeth I and Chatham, he chose the right men. There were problems, many of his own making, and there was his lack of for the most part from military strategic considerations and increasingly enmeshed at the operational and tactical levels where, as General Warlemont described it, he could find some measure of comfort as the overall grand strategic scales turned against Germany.

> Commanders from the front, who were called out of the thick of fighting to Hitler's headquarters, were rarely given an opportunity to put in a word once they were in the presence of the dictator. He limited their responsibility to following his orders to the letter. Every independent command initiative at the front was suppressed, spied upon, and more and more frequently punished by dismissal or by even more severe and degrading methods.[204]

In the end, Hitler's approach to total war was like the play he had dictated to his sister when he was 15, which, in his estimation, "displayed a lofty, burning imagination." That notwithstanding, his sister touched on the basic flaw. "You know, Adolf," she said, "your play can't be acted."[205] And so it was with Hitler's strategy in which each success, as he pointed out in 1928, was "only the point of departure for a new struggle."[206] The Nazi leader liked to imagine a similarity between himself and both Bismarck and Frederick the Great, but he lacked the sense of limitation, of moderation that marked the career of Bismarck, who often quoted Goethe's injunction: "*In der Begrenzung zeigt sich der Meister.*"[207] And although Hitler spent his final days in the *Tiefbunker* before a candlelit portrait of "*der alte Fritz*" looking for inspiration and waiting for a miracle to deliver him, he never fully understood the strategic implications of Frederick's most fundamental dictum: "He who attempts to defend too much defends nothing."[208] Without a conception of an ultimate status quo, there could be no calculation of ends, ways and means – the essence of the strategic thought process And without that calculation, no interplay of the elements of Clausewitz's remarkable trinity could achieve success.

NOTES

1. Winston S. Churchill, *The Second World War. Vol. I. The Gathering Storm* (Boston: Houghton Mifflin, 1948), pp. 83–84. (After the initial citing of a volume, hereinafter referred to as *Second World War* with appropriate volume number.) Churchill's son, Randolph, was the apparent instigator. Hanfstaengel later wrote that the meeting was to occur at the Hotel Continental "(not the Regina Palace – Sir Winston's memory

plays him false) ...." Ernst Hanfstaengel, *Unheard Witness* (New York: J.B. Lippincott, 1957), p. 193. In 1937, the German Foreign Minister twice invited Churchill to visit Hitler. Churchill declined. "I would gladly have met Hitler with the authority of Britain behind me," he wrote after the Second World War.

> But as a private individual I should have placed myself and my country at a disadvantage. If I had agreed with the Dictator-host, I should have misled him. If I had disagreed, he would have been offended, and I should have been accused of spoiling Anglo-German relations. Therefore I declined, or rather let lapse, both invitations. All those Englishmen who visited the German Führer in these years were embarrassed or compromised. No one was more completely misled than Mr. Lloyd George, whose rapturous accounts of his conversations make odd reading today. There is no doubt that Hitler had a power of fascinating men, and the sense of force and authority is apt to assert itself unduly upon the tourist. Unless the terms are equal, it is better to keep away.

Winston S. Churchill, *The Second World War. Vol. III. The Grand Alliance* (Boston: Houghton Mifflin, 1950), pp. 249–50.

2. Carl von Clausewitz, *On War*, ed. and trans. Michael Howard and Peter Paret, rev. edn. (Princeton: P.U. Press, 1984), p. 607.
3. Ibid., pp. 592–3.
4. Ibid., p. 89.
5. Ibid., pp. 609–10.
6. Ibid., p. 89.
7. Michael I. Handel, *War, Strategy and Intelligence* (London: Frank Cass, 1989), p. 63; Michael Howard, *Clausewitz* (New York: Oxford University Press, 1986), pp. 3–4; and Martin van Creveld, "The Eternal Clausewitz," *Clausewitz and Modern Strategy*, ed. Michael I. Handel (London: Frank Cass, 1986), p. 36.
8. Handel, *War, Strategy and Intelligence*, p. 82. See also Dennis E. Showalter, "Total War for Limited Objectives: An Interpretation of German Grand Strategy," *Grand Strategies in War and Peace*, ed. Paul Kennedy (New Haven, CT: Yale University Press, 1991), pp. 110–11.
9. Gordon A. Craig, *The Politics of the Prussian Army 1640–1945* (New York: Oxford University Press, 1956), p. 107; Michael Howard, "The Armed Forces as a Political Problem," *Soldiers and Governments*, ed. Michael Howard (Westport, CT: Greenwood Press, 1978), p. 16.
10. Martin van Creveld, "Caesar's Ghost. Military History and the Wars of the Future," *The Washington Quarterly*, Winter 1980, p. 81. See also Michael Howard, "Men Against Fire: The Doctrine of the Offensive in 1914," *Makers of Modern Strategy*, ed. Peter Paret (Princeton: Princeton University Press, 1986), p. 521; and Handel, *War, Strategy and Intelligence*, pp. 21 and 64–8.
11. Howard, "Armed Forces as a Political Problem," p. 17. See also, Jack Snyder, "Civil Military Relations and the Cult of the Offensive, 1914–1984," *International Security*, Summer 1984, p. 109.
12. Theodore Ropp, *War in the Modern World* (New York: Collier Books, 1962), p. 229. Snyder, pp. 110–11, 130 and 132–33.
13. The French military elite made a mirror image of their disdain for reservists in their estimates of German strength. The German General Staff made extensive use of German reservists, however, and instead of the 68 German divisions which had been expected in the implementation of French Plan XVII, there were 83. Howard, "Armed Forces as a Political Problem," p. 17. Joffre's failure to use French reservists more fully in 1914 proved, as Douglas Porch has pointed out, "like going to war without your trousers on." "Arms and Alliances: French Grand Strategy and Policy in 1914 and 1940," *Grand Strategies in War and Peace*, p. 142. See also Snyder, pp. 108 and 133. It is true, of course, that had the French Army remained on the defensive instead of plunging into Alsace, it could have brought its full weight to bear on the German Army at the French frontier. Stephen Van Evera, "The Cult of the Offensive

41. Lord Moran, *Churchill. Taken from the Diaries of Lord Moran: The Struggle for Survival, 1940–1965* (Boston: Houghton Mifflin, 1966), p. 827.

42. Adolf Hitler, *Hitler's Secret Book*, trans. Salvator Attanasio (New York: Grove Press, 1961), p. 112. See also ibid, p. 169: "God is just as much on the side of ... the more determined, as well as ... on the side of those who are cleverer." That aim would not necessarily be understood in its own time, he noted in *Mein Kampf*, for "the greater a man's works for the future, the less the present can comprehend them ...." Adolf Hitler, *Mein Kampf*, trans. Ralph Manheim (Boston: Houghton Mifflin, 1943), p. 212.

43. Langer, p. 39.

44. Anthony Storr, "The Man," *Churchill Revised*, p. 238.

45. Manchester, p. 119, and Arthur Bryant, *The Turn of the Tide* (New York: Doubleday, 1957), pp. 12–13.

46. J. H. Plumb, "The Historian," *Churchill Revised*, p. 161. In the inter-war years, Churchill returned to the subject of Lloyd George's intuition: "He had the 'seeing eye.' He had that deep original instinct which peers through the surfaces of words and things – the vision which sees dimly but surely the other side of the brick wall or which follows the hunt two fields before the throng." Winston S. Churchill, *Great Contemporaries* (London: Thornton Butterworth, 1937), p. 280.

47. Herman Rauschning, *Hitler Speaks. A Series of Political Conversations with Adolf Hitler on His Real Aims* (London: Thornton Butterworth, 1939), p. 181.

48. Langer, p. 29. Alan Bullock, *Hitler, A Study in Tyranny* (New York: Harper & Row, 1962), p. 378. John Lukacs, *The Duel. The Struggle Between Churchill and Hitler, 10 May–31 July 1940* (New York: Ticknor and Fields, 1991), p. 35.

49. Pascal's *Pensées*. trans. W. F. Trotter, introduction by T. S. Eliot (New York: E. P. Dutton, 1958), p. 2.

50. Winston S. Churchill, *Thoughts and Adventures* (London: Odhams Press, 1949), p. 256.

51. *Secret Conversations*, 2–3 January 1942, p. 136.

52. Langer, pp. 33 and 159. *Mein Kampf*, p. 29.

53. Winston S. Churchill, *Savrola. A Tale of the Revolution in Laurania* (New York: Random House, 1956), p. 64.

54. Martin Gilbert, *Winston Churchill. Volume VI. Finest Hour. 1939–1941* (Boston: Houghton Mifflin, 1983), pp. 405–6.

55. Heinz Guderian, *Panzer Leader*, trans. Constantine Fitzgibbon (New York: E. P. Dutton, 1952), p. 443. Hitler's physical deterioration was also due, of course, to the results of the 20 July 1944 assassination attempt on him. See also Robert G. L. Waite's afterword to Langer, p. 235, and Lukacs, p. 44.

56. 26–27 February 1942. *Secret Conversations*, p. 276.

57. Anthony Storr, "The Man," *Churchill Revised*, pp. 230 and 232, and Manchester, p. 24. "When I was young," Churchill told Lord Moran in August 1944, "for two or three years the light faded out of the picture. I did my work. I sat in the House of Commons, but black depression settled on me." Moran, p. 179. The first Duke of Marlborough also suffered from depression. In the 1930s, Churchill described these symptoms in his biography of his illustrious ancestor whose "stress of soul and inward vexation were so great as to make him physically ill." After Blenheim in September 1704, Marlborough's "shining armour of serenity was heavy to bear." Winston S. Churchill, *Marlborough. His Life and Times. Vol. II. 1702–1706* (New York: Charles Scribner's Sons, 1933), p. 245, and *Vol. IV. 1704–1705* (New York: Charles Scribner's Sons, 1935), p. 144.

58. Winston S. Churchill, *The World Crisis 1915* (New York: Charles Scribner's Sons, 1929), p. 234.

59. Speer, p. 471. The other two whom Hitler *dutzt*: Christian Weber and Julius Streicher.

60. John Coleville, *The Fringes of Power. 10 Downing Street Diaries, 1939–1955* (New York: Norton, 1985), p. 179. Lukacs, pp. 165–6. Ronald Lewin, *Hitler's Mistakes* (New York: William Morrow, 1984), p. 26.

61. Speer, p. 47. See also *Secret Conversations*, 9 February 1942, p. 249: "There was a time when I could have wept for grief on having to leave Berchtesgaden."
62. *Thoughts and Adventures*, p. 231. See also Lukacs, p. 46.
63. *Secret Conversations*, 17 February 1942, p. 257, and Randolph S. Churchill, *Winston S. Churchill. Companion Volume I. Part II. 1896–1900* (Boston: Houghton Mifflin, 1967), p. 721. In describing the Hitler *putsch* of 1923, one police observer commented that the Nazi leader behaved "like a Red Indian." William Carr, *Hitler. A Study in Personality and Politics* (New York: St. Martin's Press, 1979), p. 170. An American General critical of Churchill likened the British leader in the Second World War to an Indian "intent upon obtaining the largest possible number of enemy scalps." Albert Wedemeyer, *Wedemeyer Reports* (New York: Henry Holt, 1958), p. 91.
64. James, "The Politician," p. 69. At the age of 13 at Harrow, for instance, Churchill learned 1000 lines of Macaulay, which included Horatius, from *The Lays of Ancient Rome. Roving Commission*, p. 18.
65. Ralph Manheim's "Translator's Note," *Mein Kampf*, pp. xii and xiv. Hitler dictated *Mein Kampf* primarily to Rudolf Hess and his *Secret Book* to Max Amann.
66. Esme Wingfield-Stratford, *Churchill. the Making of a Hero* (London: Gollancz, 1942), Frontispiece.
67. *Savrola*, p. 32. "To live in a dreamy quiet and philosophic calm in some beautiful garden far from the noise of man and with every diversion that art and intellect could suggest, was, he felt, a more agreeable picture. And yet he knew that he could not endure it." Ibid. See also Paul Fussell, *The Great War and Modern Memory* (New York: Oxford University Press, 1975), p. 24. Churchill could have been describing himself in his tribute in the Commons on the death of Lloyd George: "As a man of action, resource and creative energy he stood, when at his zenith, without a rival." Gilbert, Vol. VII, p. 1271.
68. Christopher Andrew, *Her Majesty's Secret Service* (New York: Viking, 1986), p. 494.
69. Hans Mend, *Adolf Hitler im Felde*, 2nd ed. (Dressen vor Muenchen: J.C. Hubers Verlag, 1934), p. 139.
70. Original emphasis. *Mein Kampf*, p. 236.
71. Rauschning, p. 224
72. Martin Gilbert, *Winston Churchill. Vol. IV. The Stricken World 1916–1922* (Boston: Houghton Mifflin, 1975), p. 50.
73. *Secret Conversations*, 20 January 1942, p. 187.
74. Erich Maria Remarque, *All Quiet On The Western Front* (Boston: Little, Brown, 1929), p. 290. Ironically, after assuming power, Hitler banned the book as too pacifistic.
75. "Renown awaits the Commander who first in this war restores artillery to its prime importance upon the battlefield, from which it has been ousted by heavily armored tanks." *Second World War. Vol. III*, p. 498. Manchester, p. 11. Lukacs, pp. 45–7. On wartime schedules, see Wilt, pp. 6 and 11. " Mr. Churchill was never amenable to regular routine." Lionel Hastings Ismay, *The Memoirs of General lord Ismay* (New York: Viking Press, 1960), p. 175. "Hitler "is unable to maintain any kind of working schedule. His hours are most irregular." Langer, p. 70.
76. Carter, p. 116. A lifelong trait. In 1958, Brendan Bracken commented that Churchill "would go to the stake for a friend." Moran, p. 796.
77. James, "The Politician," p. 116. For Violet Asquith, the gesture also demonstrated that Churchill was quite oblivious to the state of mind of the ordinary British citizens who, in her estimation, "expected from their King a Queen and not a hole-and-corner morganatic marriage." Carter, p. 9.
78. John Coleville, *The Churchillians* (London: Weidenfeld & Nicolson, 1981), pp. 10–11. "(J)udgment of men was not one of his strongest points." Ibid., p. 137.
79. S. W. Roskill, *Churchill and the Admirals* (New York: William Morrow, 1978), p. 177. "I have to consider my duty to the state," Churchill told Keyes, "which ranks above my personal friendship." Ibid., p. 176. See also Ibid, pp. 52 and 83.
80. Gilbert, Vol. VI, p. 1112.

81. Mrs Churchill originally tore up this note. Four days later, she pieced it together and gave it to Churchill. Ibid., pp. 588–9.
82. Winston S. Churchill, *The Second World War. Volume V. Closing the Ring.* (Boston: Houghton Mifflin, 1951), p. 85.
83. Bryant, p. 579.
84. Rauschning, pp. 214–15. See also Edward N. Peterson, *The Limits of Hitler's Power* (Princeton: Princeton University Press, 1969), p. 7, and Ian Kershaw, *The Nazi Dictatorship. Problems and Perspectives of Interpretation* (London: Arnold, 1985), pp. 73–4. On the roll of charismatic leadership in controlling factions of the Nazi party during the *Kampfzeit*, see Joseph Nyomarkay, *Charisma and Factionalism Within the Nazi Party* (Minneapolis: University of Minnesota Press, 1967). In all this, Hitler remained consistent to the idea of direct contact with the leader. "The best organization is not that which inserts the greatest, but that which inserts the smallest, intermediary apparatus between the leadership of a movement and its individual adherents." *Mein Kampf*, p. 346.
85. Joachim C. Fest, *The Face of the Third Reich. Portraits of the Nazi Leadership* (New York: Pantheon Books, 1970), p. 187.
86. Ibid., p. 189.
87. Otto Strasser, *Hitler and I* (Boston: Houghton Mifflin, 1940), p. 67.
88. Robert G. L. Waite, *The Psychopathic God. Adolf Hitler* (New York: Basic Books, 1977), p. 48. "When I first met him, his logic and sense of realities had impressed me, but as time went on he appeared to me to become more and more ... convinced of his own infallibility and greatness." Sir Neville Henderson, *Failure of a Mission* (New York: Putnam's, 1940), p. 177. A familiar cry from Hitler as the war proceeded was: "I was right at the time and no one wanted to believe me. Now I am right again." Speer, p. 233.
89. On Churchill's dexterity with bath faucets, see Gilbert, Vol. VII, p. 354. Churchill was often summoned from baths during emergencies such as the 1911 Sidney Street siege and the 1914 Dogger Bank episode. Eade, p. 75. In November 1944, Churchill and Eden were de Gaulle's guests at the Quai d'Orsay. Churchill was given a room with a gold bath. Eden recalled later going into Churchill's suite and hearing the Prime Minister's voice coming through the open bathroom door: "Come in, come in, that is if you can bear to see me in a gold bath when you have only a silver one." Gilbert, Vol. VII, p. 1057. See also Manchester, p. 12, and Lukacs, p. 44.
90. *Roving Commission*, p. 93.
91. Ismay, p. 116.
92. Original emphasis. Eade, p. 373.
93. D. C. Watt, "The Debate Over Hitler's Foreign Policy – Problems of Reality or Faux Problèmes?" *Deutsche Frage und europaisches Gleichgewicht*, eds. Klaus Hildebrand and Reiner Pommerin (Cologne: Boeheau, 1985), p. 159.
94. Speer, p. 83. On Hitler's ego-weakness, fear of criticism and sycophancy, see Norman Dixon, *On the Psychology of Military Incompetence* (New York: Basic Books, 1976), p. 322. Rauschning commented in 1934 on Hitler's younger days: "Already in that early period, he disliked hearing anything not calculated to strengthen his own convictions." Rauschning, p. 105. For his detailed discussion of sycophants, see Ibid, p. 204. Early in the *Kampfzeit* in his speeches Hitler learned to attack and diffuse the arguments that would be made to counter him: "*to strike the weapon of reply out of the enemy's hand myself.*" Original emphasis. *Mein Kampf*, p. 466.
95. *World Crisis 1916–1918*, Part 1, p. 193. "To have had an intense antagonism with an honored friend on a supreme issue," he wrote in the inter-war years, "without losing either his friendship or comprehension, has in it some enduring elements of comfort ...." *Great Contemporaries*, p. 106. But arguments could be exhausting with Churchill who looked on the process as something akin to a military art, constantly using such metaphors as "mustering" and "deploying" in his verbal encounters. Eade, p. 342. Timing was also essential, as Alanbrooke noted when he commented that asking "at the wrong moment was to court disaster. Once you had received a negative

reply it was almost impossible to get him to alter his verdict." Ronald Lewin, *Churchill as Warlord* (New York: Stein & Day, 1973), p. 155.

96. Ismay, p. 117. For more on how the different experiences of the two leaders contributed to their divergent attitudes toward unpleasant information, see David Kahn, *Hitler's Spies: German Military Intelligence in World War II* (New York: Macmillan, 1978), pp. 542–3.

97. Barton Whaley, *Codeword Barbarossa* (Cambridge: MIT Press, 1973), p. 132.

98. Speer, p. 100.

99. John Wheeler-Bennett, ed., *Action This Day* (New York: St. Martin's Press, 1969), pp. 122–3. Manchester, p. 36. Lukacs, p. 112.

100. Rauschning, pp. 20–1.

101. Winston S. Churchill, *The Story of the Malakand Field Force. An Episode of Frontier War* (New York and Bombay: Longmans, Green, 1901), p. 208.

102. *Thoughts and Adventures*, p. 169. See also Maurice Ashley, *Churchill as Historian* (New York: Charles Scribner's Sons, 1968) p. 143.

103. Churchill later commented: "My reference to Rommel passed off quite well at the moment. Later on I heard that some people had been offended. They could not feel that any virtue should be recognized in an enemy leader. This churlishness is a well-known streak in human nature, but contrary to the spirit in which a war is won or a lasting peace established." Winston S. Churchill, *The Second World War. Volume IV. The Hinge of Fate* (Boston: Houghton Mifflin, 1950), p. 67. Churchill was not always consistent in this view. In October 1940, for example, a British admiral sent a message to the Admiralty paying tribute to the gallantry displayed by Italian destroyers in a recent engagement in the Mediterranean. "This kind of kid glove stuff," Churchill replied, referring to the ongoing Battle of Britain, "infuriates the people who are going through their present ordeal at home." Roskill, p. 169. Rommel returned the compliment. In the winter of 1942–43, he remarked to a confidant: "Winston Churchill is the only man who can save Europe." Lewin, *Churchill*, p. 167.

104. *Second World War, Vol. I*, p. 52.

105. Winston S. Churchill, *The Second World War. Volume II. Their Finest Hour* (Boston: Houghton Mifflin, 1949), p. 121. The meetings with Mussolini in the 1920s were pleasant, and Churchill had only "with difficulty escaped the highest [Italian] decoration." Ibid. See also Fraser J. Harbutt, *The Iron Curtain. Churchill, America and the Origins of the Cold War* (New York: Oxford University Press, 1986), p. 29, and Lukacs, p. 68.

106. *Great Contemporaries*, p. 265. But see also ibid., p. 41: "Something may be said of dictatorships, in periods of change and storm; but in these cases the dictator rises in time relation to the whole moving throng of events. He rides the whirlwind because he is part of it. He is the monstrous child of emergency. He may well possess the force and quality to dominate the minds of millions and sway the course of history. He should pass with the crisis."

107. Ashley, p. 33. For Churchill's unsuccessful attempts in 1912 as First Lord of the Admiralty to convince George V to name one of the four capital ships of 1912–13 after Cromwell, a regicide, see R. Churchill, Vol. II., pp. 628–31. "His Majesty is the heir of all the glories of the nation," Churchill wrote at the time, "and there is no chapter of English history from which he should feel himself divided." Ibid., p. 629.

108. Lukacs, p. 179.

109. Bullock, p. 772. See also *Secret Conversations*, 27 March 1942, p. 299: "From Churchill one may finally expect that in a moment of lucidity – it's not impossible – he'll realize that the Empire's going inescapably to its ruin, if the war lasts another two or three years."

110. Winston S. Churchill, *The World Crisis 1916–1918*, Part II (London: Thornton Butterworth, 1927), p. 544.

111. Wheeler-Bennett, p. 83. Isaiah Berlin, *Mr. Churchill in 1940* (Boston: Houghton Mifflin, nd), pp. 17–18.

112. Gilbert, Vol. VI., p. 81. In 1940, Churchill referred to Hitler as this "wicked man, the

repository and embodiment of many forms of soul-destroying hatreds, this monstrous product of former wrongs and shame." Ibid., p. 779. "You know," Churchill told one of his Private Secretaries, "I may seem to be very fierce, but I am fierce with only one man – Hitler." Wheeler-Bennett, p. 140.

113. C. P. Snow, *Variety of Men* (New York: Charles Scribner's Sons, 1967), p. 125.
114. Storr, p. 259.
115. Moran, p. 290.
116. Berlin, p. 12. See also ibid., p. 10.
117. R. Churchill, *Winston S. Churchill. Companion Vol. I, Part II*, p. 839. Lukacs, p. 50.
118. Lukacs, p. 47. Gertrude Himmelfarb, *Victorian Minds* (New York: Knopf, 1968), p. 268.
119. Trevor-Roper, "The Mind of Hitler," *Secret Conversations*, pp. xxvii–xxix. On historical facts, see ibid., p. 185: "I have learnt a great deal from Marxism, as I do not hesitate to admit." The Franco-Prussian War provided Germany "a position of infinite esteem in Europe." See also *Secret Book*, p. 47.
120. Waite, p. 323. Joachim Fest, "On Remembering Adolf Hitler," *Encounter*, Vol. XLI, No. 4 (Oct. 1973), p. 24.
121. Langer, p. 69. "This is the foundation of Hitler's character," Langer wrote of Hitler's life up to the end of the First World War. "Whatever he tended to do afterwards is only a superstructure, and superstructure can be no firmer than the foundations on which it rests." Ibid., p. 126. There was "no development, no maturing in Hitler's character and personality." Sebastian Haffner, *The Meaning of Hitler*, trans. Ewald Osers (London: Weidenfeld & Nicolson, 1979), pp. 6–7. "Even a man of thirty will have much to learn in the course of his life, but this will only be a supplement." *Mein Kampf*, p. 67.
122. Both quotes in Fest, *Faces*, p. 163. "It never entered Bismarck's head to lay down a political course tactically and theoretically for all time." *Mein Kampf*, p. 656. Trevor-Roper, "Mind of Hitler," pp. xxv–xxvi.
123. *Mein Kampf*, p. 234. On Hitler as both an ideologue and as a leader with a pragmatic talent for exploiting opportunities, see Alan Bullock, "Hitler and the Origins of the Second World War," *The Origins of the Second World War*, ed. E. M. Robertson, (London: Macmillan, 1971), pp. 192–3.
124. "And the greatest miracle of all ... is perhaps due solely to the experience of 1923 that we were able to sail around the rock which faces any revolution." Norman H. Baynes, ed., *The Speeches of Adolf Hitler*, Vol. I (New York: Howard Fertig, 1969), p. 157.
125. David Irving, *The Rise and Fall of the Luftwaffe* (London: Futura Publications, 1976), p. 39. "I am willing to sign anything. I will do anything to facilitate the success of my policy." Rauschning, p. 109. Later, Hitler extended the concept of the coup to the war. "Our present struggle is merely a continuation, on the international level, of the struggle we waged on the national level." *Secret Conversations*, 2 November 1941, p. 88. See also ibid., 19 November 1941, p. 110.
126. Rauschning, p. 280.
127. Fest, "On Remembering Adolf Hitler," p. 19.
128. Gilbert, Vol. VI., p. 844.
129. Lukacs, pp. 156–7 and 39.
130. Ibid., p. 221.
131. Gordon A. Craig, "The Political Leader as Strategists," *Makers of Modern Strategy*, pp. 481–509.
132. *Savrola*, p. 240.
133. C. P. Snow, *Science and Government* (Cambridge, MA: Harvard University Press, 1961), p. 4.
134. Eade, pp. 79–80. Churchill used his wartime authority to make sure that the Chamber was rebuilt almost exactly as it had been before, R. Churchill, Vol. II, p. 4. Churchill, of course, had spent almost his entire adult life in the Commons, and had long since polished his oratorical skills in that House. Nevertheless, he practiced each speech as if it were his first. One day while Churchill was soaking in the bath, his valet responded

to what he thought was his master's call. "I wasn't talking to you, Norman," Churchill told him. "I was addressing the House of Commons." R. W. Thompson, *The Yankee Marlborough* (London: Allen & Unwin, 1963), p. 247.

135. Wheeler-Bennett, p. 150. Churchill later pointed out that "an efficient and a successful administration manifests itself equally in small as in great matters." Ibid., p. 151.
136. Gilbert, Vol. VI., p. 659.
137. Wheeler-Bennett, p. 235.
138. Waite, p. 90.
139. General Ismay brought to his job years of invaluable experience on the Committee of Imperial Defence. He was a hard worker of unchallenged reliability, a consummate bureaucrat who could elicit decisions and compromises without antagonizing – a staff officer, in the words of one of the Prime Minister's Private Secretaries, "to whom Churchill owed more and admitted that he owed more than to anybody else, military or civilian in the whole of the war." Colville, *Churchillians*, p. 124. Lord Moran termed Ismay "a perfect oil-can." Moran, p. 121.
140. Ismay, p. 162; Lewin, *Churchill*, pp. 60–2; and *Second World War*, Vol. III, p. 122.
141. All quotes from *Mein Kampf*, pp. 192, 197 and 195.
142. Gilbert, Vol. VI, p. 665.
143. Ibid., p. 840. Churchill's experiences with the British people throughout the war only reinforced this impression. In June 1945 while presiding over a meeting of the Crossbow Committee to discuss the V-1 menace, the Prime Minister talked of the Home Front in terms of a flying bomb that had killed more than 60 people the day before, many of them serving officers, during a service at the Guards Chapel. "He was at his best," one participant noted in his diary, "and said the matter had to be put robustly to the populace, that their tribulations were part of the battle in France, and that they should be very glad to share in the soldiers' danger." Ibid., Vol. VII, p. 810. A few days later in a letter to Stalin, Churchill reconfirmed his faith in those who manned the Home Front. "You may safely disregard all the German rubbish about the results of their flying bomb," he wrote, "... The people are proud to share in a small way the perils of our own soldiers." Ibid., p. 835.
144. *Mein Kampf*, pp. 212 and 250.
145. Ibid., p. 473. See also pp. 456–7.
146. *Secret Book*, pp. 95–6. "The impossibility for justifying the necessity for enduring the war helped to bring about its unfortunate outcome." Ibid, p. 77.
147. *Mein Kampf*, p. 229.
148. H. R. Trevor-Roper, *Last Days of Hitler* (London: Macmillan, 1947), p. 92.
149. Original emphasis. *Second World War*, Vol. II, p. 266.
150. Moran, p. 13.
151. Gilbert, Vol. VIII, p. 391.
152. Ismay, pp. 185–6. "I was completely undermined, and wept," Churchill later recounted. *Second World War*, Vol. II, pp. 307–8.
153. Gilbert, Vol. VI, pp. 602–3.
154. Churchill, of course, managed the news, particularly the fairly constant flow of information concerning war-related disasters in the first two years of his leadership. On 17 June 1940, for instance, the *Lancastria* was bombed during an evacuation of 5,000 soldiers from St. Nazaire. Almost 3,000 soldiers were killed. "When the news came to me in the great Cabinet Room during the afternoon," Churchill recorded in his memoirs, "I forbade its publication saying 'the newspapers have quite enough disaster for today at least.'" *Second World War*, Vol. II, p. 194. Churchill forgot to lift the ban until July 1940. Gilbert, Vol VI, p. 685. See also ibid., pp. 564–5.
155. Speer, p. 214. See also Karl Hardach, *The Political Economy of Germany in the Twentieth Century* (Berkeley: University of California, 1980), pp. 76–9.
156. *Mein Kampf*, p. 161.
157. *Secret Conversations*, p. 548.
158. *Malakand Field Force*, pp. 252–3.
159. Gilbert, Vol. VI, p. 1235.

160. *Second World War*, Vol. II, Footnote 5, p. 265.
161. Eade, p. 29.
162. *Thoughts and Adventures*, p. 109.
163. *Secret Book*, p. 184.
164. Johanna Meskie, *Hitler and Japan: The Hollow Alliance* (New York: Atherton Press, 1966), pp. 3–4.
165. Harbutt, p. 35. Churchill's view of history taught him nothing about the Far East, and he was almost as ignorant of the realities in that region as Hitler was in terms of the United States. This was graphically illustrated by his failure to grasp the deficiencies in defense arrangements for Singapore until disaster struck and by his various proposals throughout the war for invasions of Java and Sumatra. Moreover, despite an abstract love of Empire, the fact that he never understood the Far East as well as other areas meant that his feelings for New Zealand and Australia were not so deep as for other Imperial members. It was all very well, for instance, to take the broad view in 1942 that it did not matter what the Japanese did since their actions had brought the United States into the war. But that view was of small consolation to the Australians. Wheeler-Bennett, p. 203. In discussing the fall of Singapore in his memoirs, Churchill admitted that "it had never entered into my head that no circle of detached forts of a permanent character protected the rear of the famous fortress. I cannot understand how it was I did not know this." *Second World War*, Vol. IV, p. 49.
166. *Second World War*, Vol. IV, p. 464.
167. Ibid., Vol. V, p. 38.
168. Gilbert, Vol. VII., p. 217. Churchill's favorite song from Gilbert and Sullivan was from the *Mikado* which began: "A wandering minstrel I ...." Ibid., p. 1227, and Moran, p. 282. Churchill's travels also included vacations and recuperation at such locations as Florida, Cairo and Marrakesh. Wilt, p. 11.
169. 26–27 February 1942. *Secret Conversations*, p. 276. Speer, p. 302. Wilt, pp. 5–6. Hitler was referring to the Wolf's Lair, his headquarters in East Prussia in the vicinity of Rastenberg. On the locations of Hitler's wartime field headquarters, see Annex 16, Office of the Chief of Military History, Manuscript T–101. (All manuscripts from OCMH hereafter referred to as MS followed by the lettered manuscript series, the number of the manuscript within that series, and the page numbers in the manuscript.)
170. 4 June 1939, Baynes, Vol. II, p. 1666.
171. Erich von Manstein, *Lost Victories* (Novato, CA: Presidio Press, 1982), p. 21; Lukacs, p. 17.
172. 4 April 1942. *Secret Conversations*, p. 322.
173. Speer, p. 166. See also *Secret Conversations*, 19–20 August 1941, p. 24: "If I am reproached with having sacrificed a hundred or two thousand men by reason of the war, I can answer that, thanks to what I have done, the German nation has gained, up to the present, more than two million five hundred thousand human beings."
174. W. Richardson and S. Freidin, *The Fatal Decisions* (Manchester: World Distributors, 1965), p. 178.
175. Guderian, pp. 29–30.
176. *Second World War*, Vol. II, p. 443.
177. Churchill chose the code name from the biblical text: "Behold ye among the heathen and regard, and wonder marvellously: for I will work a work in your days which ye will not believe, though it be told you." Gilbert, Vol. VII, p. 446. "Alone among politicians, he valued science and technology as something approaching their true worth." Jones, p. 107.
178. Alfred Stanford, *Force Mulberry* (New York: William Morrow, 1951), p. 39, and Guy Hartcup, *Code Name Mulberry* (New York: Hippocrene Books, 1977), p. 28.
179. Hartcup, p. 141.
180. Speer, p. 232.
181. Of those German scientists who left Nazi Germany for the United States, five were Nobel Prize winners and six were subsequently to win the prize. Lewin, *Hitler*, p. 56. See also ibid., p. 88.
182. Geyer, p. 340.

183. Walter Warlimont, *Inside Hitler's Headquarters 1939–1945*, trans. R. H. Barry (New York: Praeger, 1964), p. 244. In referring to the military intelligence organization, Warlimont pointed out that "there is not the slightest proof that a different organizational structure would have ... modified any of Hitler's decisions." MS T-101 K1, p. 114. This was particularly true at the strategic level where, as David Kahn has pointed out, "admission of error in his racism or anti-semitism or sense of geopolitical mission would have undermined not only his political power but his very personality." Kahn, p. 538.
184. Donald McLachlan, *Room 39. Naval Intelligence in Action 1939–45* (London: Weidenfeld & Nicolson, 1968), p. 125.
185. Whaley, pp. 41 and 147.
186. Gilbert, Vol. VII, pp. 700–1, 812, 833 and 837.
187. Nelson, p. 21. Wilt, p. 2.
188. Sir Ivone Kirkpatrick, *The Inner Circle* (New York: St. Martin's Press, 1959), p. 118.
189. Gilbert, Vol. VII, p. 463.
190. Ismay, p. 16. Lukacs, p. 212.
191. Ismay, p. 210.
192. A. L. Rowse, *The Later Churchills* (London: Macmillan., Ltd., 1958), p. 492. This type of adversarial approach to decision-making could be exhausting, particularly with Churchill who thrived on the combative give-and-take of spirited argument. At one point in June 1940, for instance, as Churchill was questioning Alanbrooke by telephone concerning the situation in France, he insisted that the general comply with his wishes for troop dispositions. For almost half an hour, Alanbrooke resisted the arguments in an increasingly heated manner. "At last," he noted in his diary, "when I was in an exhausted condition, he said: 'All right, I agree with you.'" Bryant, p. 136.
193. Rowse, p. 455.
194. In fact, Part III of the Directive was entitled "Tactical Employment of the Above Force." *Second World War*, Vol. II, p. 430.
195. John Connell, *Auchinleck* (London: Cassell, 1959), p. 309.
196. A. J. P. Taylor, "The Statesman," *Churchill Revised*, p. 43.
197. Bryant, p. 412. In another incident, Churchill was "completely startled" when Eisenhower informed him that a cable the Prime Minister was planning to send to Alexander would cause Eisenhower to resign if he received it from his American commanders. The message, Eisenhower explained, "looks as though you, who are three thousand miles away ... feel that you are more competent to judge on the readiness of this army to attack, its morale, its equipment, its strength, its positions, training and everything, than is General Alexander. This I don't think is possible." Churchill's reply was almost plaintive. "As a military man and as the Chief Minister of His Majesty," he queried, "don't I have the right to ask questions?" Nelson, p. 27.
198. Mathew Cooper, *The German Army 1933–1945* (New York: Stein and Day, 1978), p. 3.
199. *Secret Conversations*, 2 November 1941, p. 90.
200. Lewin, *Hitler*, p. 139.
201. Guderian, p. 378.
202. In terms of Wavell, General Dill advised Churchill to "Back him or sack him." The British leader replied: "It is not so simple as that. Lloyd George did not trust Haig in the last war – yet he could not sack him." Kennedy, p. 119. On the OKW, see MS T-101 K1, p. 83 and Michael Howard, *Studies in War and Peace* (New York: Viking Press), p. 116.
203. Franz Halder, *Hitler as War Lord*, trans. Paul Findley (London: Putnam, 1950), p. 49.
204. MS T-101 K1, p. 82. Warlimont also pointed out that "decisions down to the shifting of units of battalion and even lesser strength had to be submitted to Hitler practically every day." Ibid., p. 81. See also Gerhard Engel, "Adolf Hitler as Supreme Commander of the Armed Forces," Annex 15, MS T-101, pp. 9–10 & 14.
205. *Secret Conversations*, 8–9 January 1942, p. 158.
206. *Secret Book*, p. 43.
207. Waite, pp. 308–9.
208. Jay Luvaas, ed. and trans., *Frederick the Great and the Art of War* (New York: Free Press, 1966), p. 120.

2
Churchill:
The Making of a Grand Strategist

"Churchill is the most bloodthirsty of amateur strategists that history has ever known," Adolf Hitler stated in a 1941 speech. "He is as bad a politician as a soldier and as bad a soldier as a politician."[1] Hitler was wrong on all counts. Winston Churchill was a competent, experienced and enthusiastic soldier who served as an officer in four wars, beginning as a subaltern on India's northwest frontier and ending as a battalion commander on the Western Front in the First World War. And while he could not match Hitler's horrific life as a runner in the trenches for most of that war, the future British leader had experienced intense, close-quarters combat first hand in many campaigns, in many lands before his twenty-fifth birthday.

In terms of his political career, there is a tendency to focus on Churchill's years in the wilderness of the 1930s. But that period was slight compared to his time spent gainfully as a successful politician. Between 1905 and 1922 with only a two-year interruption, for example, he held high offices ranging from Home Secretary and First Lord of the Admiralty to Secretary of State for War and Colonial Secretary. The results were as diverse as they were successful, including much needed prison and naval reforms as well as new initiatives concerning the pacification of Ireland and the organization of mandated nations in the Middle East.

It was the combination of Churchill's experiences as a soldier and as a politician that gave lie to the Nazi leader's estimation of his British counterpart as an "amateur strategist." For it was this combination which ultimately allowed Churchill to master grand strategy. That mastery did not occur overnight; nor was it the result of reading such great strategic thinkers as Sun Tzu and Clausewitz. It was, instead, the result of a long apprenticeship in military and public affairs.

Over the decades during that apprenticeship, Churchill also earned his living as a professional writer and historian. The result was a series of books and articles that described with vivid and visceral immediacy many of the historic events in which he had personally participated. These works provide a valuable collection of Churchill's reflections on Britain's recent and more distant past. Equally important, they provide a means to trace the evolution of the future British leader's thoughts on strategy.

The purpose of this chapter is to demonstrate by means of these writings how Churchill's approach to grand strategy was formed. Through these works, it is possible to follow his first tentative strategic steps as he dealt with the rapidly changing nature of warfare at the turn of the century. That development caused him to broaden his military viewpoint beyond the purely tactical realm. Finally, as Britain passed through the 1914–18 crucible, the change in Churchill's perspective concerning military power was also complemented by an appreciation of the use in war of all the instruments of national power – the essence of grand strategy.

The Building Blocks – The Vertical Dimension

The First World War demonstrated repeatedly that a single battle was no longer sufficient to achieve a strategic victory, that in fact an engagement or a battle would normally not determine the outcome of a campaign, much less a war. The frustration at this turn of events was captured by a character in F. Scott Fitzgerald's novel *Tender is the Night* when he visited the Somme Valley after that war. "See that little stream," he said, "we could walk to it – a whole empire walking very slowly, dying in front and pushing forward behind. And another empire walked very slowly backwards a few inches a day leaving the dead like a million bloody rugs."

Clausewitz had foreseen this trend early in the previous century. For him, the higher commander must create something that was more than the sum of its individual tactical parts. "By looking on each engagement as part of a series," he wrote, "at least insofar as events are predictable, the commander is always on the high road to his goal."[2]

That high road became increasingly complex in the second half of the nineteenth century. By that time, Königgrätz and Sedan notwithstanding, a series of developments had made it increasingly difficult for nation-states to achieve strategic outcomes by means of a single decisive battle. To begin with, there was the dramatic increase in populations that allowed large nations to deploy more than one field army, each capable of simultaneously conducting a campaign in its own right. As the century drew to a close, this size was compounded by the growth in Europe of a complex and sophisticated alliance system that facilitated the formation of huge multinational armies that could fight on many fronts, extending a theater of war to encompass an entire continent. Finally, there were technological

innovations, ranging from breechloading weapons to smokeless pow-
der which, in conjunction with these other factors, meant that concentration of armies on small, limited battlefields was no longer
feasible.[3]

It was in this milieu that the First World War was fought; and it was
that conflict which demonstrated the inadequacy of classical strategy
to deal with the intricacies of modern warfare. Napoleon had defined
that strategy as the "art of making use of time and space."[4] But the
dimensions of the two variables, as we have seen, had been stretched
and rendered more complex by demographics as well as geopolitical
and technological factors. And that very complexity, augmented by
the lack of decisiveness at the tactical level, impeded the continuum of
war outlined in Clausewitz's definition of strategy as "the use of the
engagement for the purposes of war."[5] Only when the continuum was
enlarged, as the Great War demonstrated, was it possible to restore
warfighting coherence in modern conflict. And that, in turn, required
the classical concept of strategy to be positioned at a midpoint, an
operational level, designed to orchestrate individual tactical engagements and battles in order to achieve strategic results. In the aftermath of the First World War, the Soviets incorporated this new
perspective of the continuum of war into their military doctrine.
"Tactics," a faculty member at the Frunze Academy wrote in 1927,
"make the steps from which operational leaps are assembled; strategy
points out the path."[6]

The United States was slower to explore this continuum. In 1982
and 1986, the US Army incorporated the three "broad divisions of
activity in ... conducting war" into that organization's basic manual
on operations.[7] And in 1990, the US Joint Chiefs of Staff made the
three-level continuum of war (Figure 1) official for the Armed
Services:

> Operational Level of War is the level of war at which campaigns
> and major operations are planned, conducted, and sustained to
> accomplish strategic objectives .... Activities at this level link
> tactics and strategy .... These activities imply a broader dimen-
> sion of time or space than do tactics; they provide the means by
> which tactical successes are exploited to achieve strategic objec-
> tives.[8]

The Early Years

The young Churchill, of course, did not consider this vertical continuum, despite an abiding interest in all things military. Instead, his
formative years in the Indian summer of the Victorian era left him

with a reverence for the great captains of the past whose decisive victories in battle had led to the scarlet splash on the world map that marked the British Empire. As a young boy, for instance, he was introduced to the majestic prose of Macaulay's *History of England*. Later, he acquired his own works of that author and, as he described it, "voyaged with full sail in a strong wind" as he "revelled" in

Figure 1

Macaulay's essays on such great leaders as Chatham, Frederick the Great and Clive.[9] Added to these were the works of George Alfred Henty, who two years after Churchill's birth published the first of his 80 novels and serials, many of which dealt with English and imperial history. Whether it was with Clive in India (1884), Wolfe at Quebec (1887) or adventures in the Punjab (1894) and Afghanistan (1902), young Victorians like Churchill could relive vicariously every British triumph throughout the Empire. In 1898, the year Churchill observed Kitchener's victory over the Mahdi at Omdurman, Henty's annual sales were estimated to be as many as 250,000.[10]

At Harrow, the normally indifferent student could always muster an infectious enthusiasm for military activities. In 1889, Churchill described to his mother a "grand sham" battle at Aldershot conducted by the Rifle Corps from the various public schools in which his force of 3,500 students, two batteries of guns and a cavalry regiment was defeated by an attacking student force of 8,000.[11] And later that year, he focused on the Japanese defeat of the Chinese at Pyongyang in the Sino-Japanese War. "I take the greatest interest in the operations," he wrote to his mother, "both of the fleets and armies. Anything so brilliant as the night attack of Pung Yang is hard to find in modern war. The reports ... show that the Japanese concentration was so accurately timed and their assault so skillfully delivered that the celestials had 'no show' at all."[12]

At Sandhurst, Churchill's curriculum initially kept him grounded at the lowest level of war – "all ... very elementary, and our minds were not allowed to roam in working hours beyond a subaltern's range of vision."[13] Nevertheless, he managed to order a number of books through his father's bookseller dealing with the American Civil, Franco-Prussian and Russo-Turkish Wars, "which were then our latest and best specimens of wars. I soon had a small military library," he wrote many years later, "which invested the regular instruction with some sort of background."[14] More importantly, he was invited at various times to dine at the nearby Staff College where, as he described it, he could at least broaden his tactical horizons.

> Here the study was of divisions, army corps and even whole armies; of bases, of supplies, and lines of communications and railway strategy. This was thrilling. It did seem such a pity that it all had to be make-believe, and that the age of wars between civilized nations had come to an end forever. If it had only been 100 years earlier what splendid times we should have had! Fancy being nineteen in 1793 with more than twenty years of war against Napoleon in front of one![15]

The ironic tone of Churchill's description, written while he was still reacting to the slaughters of the Great War, should not obscure the solid military education he received at Sandhurst, where he graduated twentieth out of 130 and excelled in tactics, fortifications and riding.[16] "He would talk about the battle of Cannae," General Eisenhower commented years later in this regard, "just as well as could a professional soldier."[17] That, of course, was because Churchill was a professional soldier off and on for five years after graduating from Sandhurst, personally passing through four different regiments and three different wars in that twilight of the Victorian era. In the first two of these wars, there was nothing that would draw Churchill to the vertical continuum of war. Certainly, the minor engagements that he observed as part of Sir Bindon Blood's Malakand Field Force on India's northwest frontier in 1897 fitted the lowest tactical parameters of most Victorian wars. And the following year, there could be no doubt as to the decisiveness of the battle of Omdurman in terms of the Sudan campaign, when Lord Kitchener's forces, by means of Maxim guns, naval and high velocity artillery shells, and Dum Dum rounds, slaughtered between 10,000 and 12,000 dervish followers of the Mahdi with a loss of 48 dead.

There was, however, no such decisiveness in the Boer War. In South Africa, the British were not dealing with the Pathan and Omdurman tribesmen. This time it was the Boers with a panoply of modern weapons ranging from machine guns, which shredded the dense ranks of the Queen's army, to distant artillery known as Long Toms, which were emplaced far beyond the reach of the British cavalry, rapidly firing 40 pound, 4.7 inch shrapnel shells that dismembered men in the attack or in static positions. Added to this were the sandbagged entrenchments and the barbed wire. As British casualties mounted at such battles as Spion Kop and Vaal Krantz, regimental histories began to record phrases that would become set-pieces for the total wars of the twentieth century. Battles became "enshrined forever" in history; engagements were "imperishable" and "immortal."[18]

Those changes were not lost on Churchill, who along with Ghandi served at the battlefields along the Tugela River. "Colenso, Spion Kop, Vaal Krantz, and the third day at Pieters were not inspiring memories," he wrote. At the battle of Pieters, he watched as British units were repeatedly cut down by "the hideous whispering Death" from Mauser bullets. And Spion Kop left an indelible impression concerning the effects of artillery shrapnel on a 2,000-man British brigade crowded into a space "about as large as Trafalgar" on the bare top of the kop – "scenes ... among the strongest and most terrible I have ever witnessed."[19] Moreover, those scenes had been produced by far less than a battery of howitzers. "Yet in a European war," Churchill concluded, "there would have been ... three or four batteries. I do not see how troops can be handled in masses in such conditions."[20] He returned to this theme in 1906 at the German Army maneuvers in Silesia where he watched "with astonishment" the dense columns of German troops attacking entrenched forces who "burned blank cartridges in unceasing fusillade."

> I had carried away from the South African veldt a very lively and modern sense of what rifle bullets could do. On the effects of the fire of large number of guns we could only use our imagination. But where the power of the magazine rifle was concerned we felt sure we possessed a practical experience denied to the leaders of

these trampling hosts . . . . Whatever else this might amount to, it
did not form contact with reality at any point. Besides South
Africa I had also vividly in my mind the Battle of Omdurman,
where we had shot down quite easily, with hardly any loss, more
than 11,000 Dervishes in formations much less dense, and at
ranges far greater than those which were now on every side
exhibited to our gaze. We had said to ourselves after Omdurman, "This is the end of these sort of spectacles. There will never
be such fools in the world again."[21]

The effect of the new technology on Churchill's perception of war at
the time of the South African conflict should not be overstated.
During the Second World War, Churchill's physician noted in this
regard that "the P.M. always goes back to the Boer War when he is in
good humour. That was before war degenerated. It was great fun
galloping about."[22] Certainly, there was a tendency at the time for
Churchill to gloss over the evolving nature of warfare. At Diamond
Hill on 14 June 1900, for instance, there was an almost palpable sense
of relief when the British reverted to a cavalry charge, "a fine gallant
manoeuvre, executed with a spring and an elasticity wonderful and
admirable ... in troops who have been engaged ... in continual
fighting with an elusive enemy."[23] As for the new technologies,
Churchill also had a warning firmly grounded in the nineteenth
century. "Battles now-a-days are fought mainly with firearms," he
wrote, "but no troops ... can enjoy the full advantage of their
successes if they exclude the possibilities of cold steel and are not
prepared to maintain what they have won, if necessary with their
fists."[24]
Nevertheless, there was also a sense of change that pervaded most
of Churchill's writings on the period. By 1900, his dispatches to the
Morning Post, while not neglecting the tactical aspects of what he
observed, were sprinkled with insightful glimpses up the continuum of
war. In January of that year, for instance, he noted that "it is
impossible not to admire the Boer strategy. From the beginning they
have aimed at two main objectives: to exclude the war from their own
territories, and to confine it to rocky and broken regions suited to
their tactics."[25] There was also a realization that the vast spaces of
those regions as well as the new technology of warfare in the hands of a
trained, well-armed, entrenched enemy, enjoying the advantages of
interior lines, made it impossible for one battle in a campaign to
achieve decisive strategic results. Nowhere was that more evident to
Churchill than in the inept, ponderous and dilatory campaign for the

relief of Ladysmith by General Redvers Buller, in which no attempt was made to mold the scattered minor tactical parts into anything resembling an operational whole. That disconnect in terms of the continuum of war applied also to the campaign objective.

> Whoever selected Ladysmith as a military centre must sleep uneasily at night .... Tactically Ladysmith may be strongly defensible, but for strategic purposes it is absolutely worthless. It is worse. It is a regular trap .... Not only do the surrounding hills keep the garrison in, they also form a formidable barrier to the advance of a relieving force.[26]

On-The-Job Training

The First World War provided Churchill with a continuing education on the vertical continuum of war. At the highest military level, as he pointed out after that conflict, the "entry of Great Britain into war ... was strategically impressive. Her large Fleets vanished into the mists at one end of the island. Her small Army hurried out of the country at the other."[27] As First Lord of the Admiralty, Churchill was personally involved in the first decision. On 26 July 1914, he ordered the fleet, assembled for review at Portland, not to disperse, in view of the increasingly tense international situation. That order was one of the decisive acts of the war, for while free from the provocation inherent in any army mobilization, it placed the British Navy automatically in control of the sea, particularly after the unnoticed dispersion of the fleet on 29 July to its war station at Scapa Flow. From that location in the Orkney Isles, the Grand Fleet controlled the passage between northern Britain and Norway and began the invisible pressure on Germany's arteries until, in those same waters in November 1918, the German fleet surrendered to a force which it had only briefly glimpsed in over four years of a naval twilight war.[28]

As for the army, its arrival during the Marne campaign, Churchill added, "reached in the nick of time the vital post on the flank of the French line. Had all our action been upon this level, we should to-day be living in an easier world."[29] But it was not to be so simple. The Marne campaign was a German attempt to achieve a decisive victory in the manner of Austerlitz or Königgrätz at the military strategic level. The operational speed required for the Schlieffen Plan to work, however, could not be matched by the immense and complex German armies whose pace was still governed by the speed of the foot soldier. Moreover, the scale of the huge operation overwhelmed the still primitive telegraph and radio communications, and this, combined with the inevitable friction at all levels of the war, caused the Marne campaign to end in an operational and thus military strategic stalemate.[30]

Soon the entire war seemed to be locked in that stalemate – a situation, Churchill realized, made all the more terrible in total war.

> Wealth, science, civilization, patriotism, steam transport and world credit, enabled the whole strength of every belligerent to be continually applied to the war .... But at the same time that Europe had been fastened into this frightful bondage, the art of war had fallen into an almost similar helplessness. No means of procuring a swift decision presented itself to the strategy of the commanders, or existed on the battlefields of the armies.[31]

How far that situation had moved operational art from the earlier decisiveness of classical strategy was summarized by Churchill after the war:

> Compared with Cannae, Blenheim or Austerlitz, the vast world-battle ... is a slow-motion picture. We sit in calm, airy, silent rooms opening upon sunlit and embowered lawns, not a sound except of summer and of husbandry disturbs the peace; but seven million men, any ten thousand of whom could have annihilated the ancient armies, are in ceaseless battle from the Alps to the Ocean. And this does not last for an hour, or for two or three hours .... Evidently the tests are of a different kind; it is certainly too soon to say that they are of a higher order.[32]

The Continuum of War

Churchill's recognition of the vertical continuum of war was evident in his analysis of two key operations conducted by General von Ludendorff at the beginning and at the end of the First World War. The first was the Tannenberg–Masurian Lakes campaign in August and September 1914 on the Eastern Front where there was, he noted, "the opportunity for manoeuvre, and for that kind of tactics or battlefield strategy."[33] At the tactical level of that campaign, Churchill later expressed his appreciation of the German expertise in his biography of the first Duke of Marlborough, written in the 1930s. In particular, there was Oudenarde, the deft and decisive battle fought by his illustrious ancestor in the 1708 campaign, which with its "looseness and flexibility of all the formations" and "movement of the Allies, foreshadowing Tannenberg, present us with a specimen of modern war which has no fellow in the rest of the eighteenth century."[34] But it was at a higher level of warfare that Churchill reserved his greatest admiration for Marlborough, whose operational artistry "applied with the highest technical skill, and with cool judgment in the measuring and turning of events, exactly harmonizes with Napoleon's processes, and may very well have suggested some of them." It was this artistry, he concluded in the inter-war years, that allowed Marlborough to move in time and space beyond one tactical encounter, that "enabled him to make a second or a further move, *foreseen in all its values from the beginning*, to which there could be no effective resistance."[35]

In a similar manner, Churchill appreciated the operational opportunities as the Russian forces moved westward in August 1914 by the Masurian Lakes in the eastern theater of operations. "Here too on a smaller front," he commented in this regard, "the Germans had a war on two fronts."[36] The task for the German commander, he realized, was to orchestrate his forces at the tactical level to achieve operational results that would stop the westward flow of Russian troops at the theater strategic level, while also allowing him, in a worst-case situation, to form a continuous strategic fighting front behind the Vistula. It was, Churchill concluded, a "situation at once delicate and momentous, requiring the highest qualities, but offering also the most brilliant opportunities to a Commander-in-Chief! The task was one in which Marlborough, Frederick the Great, Napoleon or the Lee–Stonewall Jackson combination would have revelled . . . ."[37]

Operational synchronization is a difficult process because it requires nothing less than "the arrangement of battlefield activities in time, space and purpose to produce maximum relative combat power at the decisive point."[38] The key word in the definition is "produce," which takes synchronization beyond just the adjustment of activities to one another – the essence of coordination. It also means that the process will involve more than bringing forces and fires together at a point in time and space, as is normally envisaged when concentration takes place. Synchronization at the operational level, in fact, may often be necessary between activities far removed from each other in either time or space, or both. Nevertheless, as the US Army operations manual points out, "these activities are synchronized if their combined *consequences* are felt at the decisive time and place."[39]

That process formed the basis for Churchill's analysis of Ludendorff's opening campaign in 1914. "With that sorry wisdom that judges after the event," he wrote, "one may ask why the Russian strategic plan ever contemplated an advance of two separate armies, with all the advantages it gave the Germans with their breakwater of lakes and fortifications and their network of railways."[40] Those advantages were put to good use when the German cavalry screen in front of the First Russian Army caused that army commander to believe that he still faced the bulk of the German Eighth Army. Using this screen, the German commanders moved two corps to the south against the Russian Second Army, already engaged against a corps from the Eighth Army. At that encounter, coordination between units and the ultimate concentration of forces and fires achieved the decisive German victory. But, as Churchill fully realized, the synchronization process by the German commanders had begun days before in the north with a series of activities, whose combined results led to the final campaign victory. "The double battles of the Eighth German Army under Hindenburg and Ludendorff against the superior armies of Samsonov and Rennenkampf," he concluded, "are not only a military classic but an epitome of the art of war."[41]

Churchill provided no such commendation to Ludendorff for his massive spring offensive in 1918. "War ... should be a succession of climaxes ... toward which everything tends and from which permanent decisions are obtained," he pointed out at the time of the offensive. "These climaxes," he added, "have usually been called battles."[42] But without overarching military strategic guidance there could be no permanent decisions. And this was the case with the spring offensive, in which Ludendorff elected to follow the tactical line of least resistance by attacking where breakthroughs were easiest. That development was not lost on Churchill. "Five divisions engaged out of an army of seven may fight a battle," he wrote in this regard. "But the same operation in an army of seventy divisions ... sinks to the rank of petty combat. A succession of such combats augments the losses without raising the scale of events."[43]

The results of the Ludendorff offensive at this lower tactical scale were spectacular, particularly by the First World War standards, occasioning British Field Marshal Haig's famous "backs to the wall" order. At the operational level, however, the campaign degenerated into a series of uncoordinated and unproductive thrusts. "Of the ... great battles which had been fought," Churchill wrote of the campaign, "the first three ... had failed to achieve any one of the progressively diminishing strategic results at which they had aimed. The fourth ... was ... very spectacular but without strategic consequence."[44]

None of that was helped by Ludendorff's decision at one point in the campaign to reinforce failure on his stalled right flank with his limited operational resources, instead of exploiting the extraordinary and unexpected tactical successes of the 18th Army on his left. But ultimately, as Churchill realized, it was Ludendorff's choice to ignore the continuum of war which was decisive. By mid-summer of 1918, the tactical results of his offensive had been more than reversed; and the Quartermaster General was well on his way to bringing down the Second German Empire. "What then had been gained?" Churchill asked after the war.

> The Germans had reoccupied their old battlefields and the regions they had so cruelly devastated .... Once again they entered into possession of these grisly trophies. No fertile province, no wealthy cities, no river or mountain barrier, no new untapped resources were their reward. Only the crater-fields extending abominably wherever the eye could turn, the old trenches, the vast graveyards, the skeletons, the blasted trees and the pulverized villages, ... the Dead Sea fruits of the mightiest military conception and the most terrific onslaught which the annals of war record.[45]

The Break in the Continuum

Decisiveness at the operational level of war remained a problem throughout most of the First World War. Part of that problem, Churchill believed, lay in the key operational variables of space and time. "In the West the armies were too big for the country," he observed of the former; "in the East the country was too big for the armies."[46] Even in the East, all that was achieved militarily was to make a continuous front mobile. It was not that the Germans lacked the operational wherewithal. "The number of trains which can be moved north and south on the German side of the frontier is at least three times the comparable Russian figure," Churchill noted in a memorandum on the situation in June 1915. "This superiority of lateral communication applied to an 800-mile front has also enabled the Germans to deliver offensive strokes of the most formidable character."[47] But operational concentration was not the answer if space in the vast eastern theater could be traded for time, if "a retirement of 100 to 200 miles enables the Russians to recover their strength, and deprives the enemy of his advantage."[48]

In addition, there was also the problem of coordinating the combined forces of the Central Powers into a strategic whole at the theater of operations level – a fact noted by Churchill as he examined the "strategic barrenness" of the 1915 German winter campaign on the Eastern Front after the war, while quoting approvingly from Hindenburg's post-war description of that campaign: "In spite of the great tactical success ... we had failed ... strategically. We had once more managed practically to destroy one of the Russian armies, but fresh enemy forces had immediately come up to take its place, drawn from other fronts to which they had not been pinned down."[49]

On the Western Front, the inability to achieve decisive operational results in the troglodyte world of the trenches made an indelible, lifelong impression on Churchill. "Before the war," he wrote, "it seemed incredible that such terrors and slaughters, even if they began, could last more than a few months. After the first two years it was difficult to believe that they would ever end."[50] A major problem, as Churchill saw it after leaving his Admiralty post as First Lord, concerned the capability to conquer sufficient space at the operational level to achieve anything approaching strategic outcomes. "Although attacks prepared by immense concentrations of artillery have been locally successful in causing alterations of the line," he wrote to the Asquith cabinet in June 1915, "the effort required is so great and the advance so small, that the attack and advance, however organized and nourished, are exhausted before penetration deep enough and wide enough to produce a strategic effect has been made." The result was that the line would be "merely bent" at particular points on the tactical spectrum that "do not ... compromise other parts." In the end, he concluded, despite ferocious tactical combat,

> no strategic results are obtained in France and Flanders ... from making, at an inordinate cost, an advance of 3 or 4 miles. For beyond the ground captured so dearly lies all the breadth of Flanders before even the Rhine is reached, and before the artillery of the attack can move forward and re-register, a new line of entrenchments not less strong than the old has been prepared by the enemy.[51]

Closely allied with space at the operational level, and even more important as far as Churchill was concerned, was the factor of time – so critical in synchronizing and sequencing events into a larger whole. At the operational level, for example, there might be a series of tactical victories which could produce a larger, equally favorable outcome if exploited. "But none of these consequential advantages," Churchill wrote, "will be gained if the time taken ... is so long that the enemy can make new dispositions." When that happened, the attacker would be "confronted with a new situation, a different problem," which in turn would result in "operations consisting of detached episodes extending over months and divided by intervals during which a series of entirely new situations are created." In such circumstances, without an overarching operational whole, attrition between relatively evenly balanced forces would ensure that there were no military strategic decisions. "It is not a question of wearing down the enemy's reserves," Churchill concluded, "but of wearing them down so rapidly that recovery and replacement of shattered divisions is impossible."[52]

The Maneuver Solution

Within a few months of the Marne campaign, as the conflict settled into a familiar pattern of stalemated attrition, Churchill began to examine the possibilities of strategic maneuver. Germany's strategic position was the key to this approach, as he pointed out years later in his study of Marlborough. "The kingdoms of France and Spain were in a central position in 1702 similar to that of Germany and Austria in 1914," he wrote. "They had the advantage of interior lines and could ... throw their weight now against this opponent, now against that."[53] There were, however, disadvantages to this position if the theaters of operation were not properly managed. "There are two enemies and two theaters," Churchill pointed out:

> the task of the commander is to choose in which he will prevail. To choose either is to suffer grievously in the neglected theater. To choose both is to lose in both. The commander has for his guides the most honoured principles of war and the most homely maxims of life .... It is the application of these simple rules to the facts that constitutes the difficulties and the torment. A score of good reasons can be given not only for either course, but also for the compromises which ruin them. But the path to safety nearly always lies in rejecting the compromises.[54]

Those types of compromises, Churchill believed, prevented the Germans throughout the war from achieving operational successes sufficient to change the strategic balance in any of the theaters of operation. The most egregious example for him was in the opening days of the war when a campaign in one such theater adversely affected a campaign in another theater of operations that was just on the point of achieving an operational success with important strategic consequences. On 20 August 1914, after the initial encounter with the Russians advancing westward in East Prussia, the German commander sent an alarming message to Moltke and the Central Command and began a phased withdrawal to the Vistula. In the West at that point, everything appeared to be proceeding smoothly. As a consequence, Moltke agreed to send six corps from the western forces, two of which were reserve and could thus be sent immediately. "Thus the wheeling wing of the Schlieffen plan," Churchill concluded, "was weakened at its most critical moment by the withdrawal of the two corps which would otherwise in a fortnight have filled the fatal gap at the Marne."[55]

Again and again as he looked back on the Great War, Churchill focused on this type of oscillation between the two major points in the theater of war. It was a matter, he believed, of failing to determine when a theater of operations could be decisive. In 1914, for example, after Falkenhayn had replaced Moltke, the new commander became absorbed in the western "race to the sea" and would not send reinforcements to the Austrians reeling backwards from their impact in the southeast with Russian and Serbian forces. As a consequence, four corps were withdrawn from the Hindenburg–Ludendorff combination's Eighth Army in the northeast and sent to the south as the Ninth Army to buttress the Austrian north flank on the Silesian frontier. It was just at that point, as their campaign forces were being reduced, that those eastern warlords, Churchill noted sympathetically, "believed that with six or eight additional army corps they could destroy quite swiftly the military power of Russia .... After that everybody could turn ... and ... finish with the West."[56] Ironically, by the following winter, the reputations of the two eastern commanders had increased to such an extent that Falkenhayn was forced to send troops from the west to the east. "The four corps which he had longed to hurl into a new offensive in the West had been wrested from him," Churchill concluded. "They had marched and fought in the Winter Battle, gaining new cheap laurels for his dangerous rivals, but producing as he had predicted no decisive strategic result."[57]

There were lessons in all this, Churchill believed, for the forces of the Triple Entente. Faced with operational indecisiveness in one theater of operations, he began to enlarge his perspective. "He never ceased to think of the war as a whole," he later wrote of Marlborough in 1703. "To him the wide scene of strife and struggle ... was but one."[58] And so it was with Churchill. "The essence of the war problem was not changed by its enormous scale," he wrote in 1915.

> The line of the Central Powers from the North Sea to the Aegean and stretching loosely beyond even to the Suez Canal was, after all, in principle not different from the line of a small army entrenched across an isthmus, with each flank resting upon water. As long as France was treated as a self-contained theatre,

a complete deadlock existed, and the front of the German invaders could neither be pierced nor turned. But once the view was extended to the whole scene of the war, and that vast war conceived as if it were a single battle, and once the sea power of Britain was brought into play, turning movements of a most far-reaching character were open to the Allies. These turning movements were so gigantic and complex that they amounted to whole wars in themselves.[59]

For the young First Lord early in 1915, the key to such an enlarged perspective was to determine where decisive operational and thus strategic results could be achieved. "The Decisive theatre," he wrote in this regard, "is the theatre where a vital decision may be obtained at any given time. The Main theatre is that in which the main armies or fleets are stationed." He added later, in a far ranging memorandum to the War Cabinet, that the main theater beginning in late 1914 "ceased to be for the time being the decisive theatre."[60] But to recognize the problem was not necessarily to solve it. The creation of a new theater of operations did not always lend decisiveness to the strategic whole – a fact, as Churchill knew, which had bedeviled Marlborough who had been "forced to acquiesce for years in a lamentable drain of troops and money from his own forces to regions where nothing decisive could be gained."[61] It was in this light that Churchill also considered Allenby's successful campaigns in Palestine.

No praise is too high for these brilliant and frugal operations, which will long serve as a model in theatres ... in which manoeuvre is possible. Nevertheless their results did not simplify the general problem. On the contrary, by opening up a competing interest which could not influence the main decision, they even complicated it. The very serious drain of men, munitions and transport which flowed unceasingly to the Palestine Expedition ought to have been arrested by action far swifter in character and far larger in scale. Brevity and finality, not less at this period than throughout the war, were the true tests .... Prolonged and expanding operations in distant unrelated theatres, whether they languished as at Salonica, or crackled briskly and brightly forward under Allenby in Palestine, were not to be reconciled with a wise war policy.[62]

In both of those instances, of course, Churchill was still reacting to the failure of the 1915 Gallipoli campaign in a theater of operations where "the true strategic direction could have been armed with tactical force."[63] There, he continued to believe throughout his life, the continuum of war could have stretched upward beyond the strategic objectives in the theater of operations and decisively affected the coalition's military strategy for the entire theater of war. "If we are successful," he wrote at the time to the War Cabinet, "results of the greatest magnitude will follow, and ... dominate the whole character of the Great War and throw all other events into the shade."[64] But all that depended on success at the tactical and operational levels of war in the theater and that, Churchill was convinced, could only have been achieved by speed and the concomitant element of surprise – all of which were lost as the campaign evolved. "Time was the dominant factor," he wrote of the situation after the initial landings at Gallipoli.

> The extraordinary mobility and unexpectedness of amphibious power can, as has been shown, only be exerted in strict relation to limited periods of time. The surprise, the rapidity, and the intensity of the attack are all dependent on the state of the enemy's preparations at a given moment. Every movement undertaken on one side can be matched by a counter movement on the other. Force and time in this kind of operation amount to almost the same thing, and each can to a very large extent be expressed in terms of the other. A week lost was about the same as a division. Three divisions in February could have occupied the Gallipoli Peninsula with little fighting. Five could have captured it after March 18. Seven were insufficient at the end of April.[65]

The failure of the Dardanelles campaign closed the most promising phase of the war for Churchill. "There was nothing left on land now," he wrote, "but the war of exhaustion .... No more strategy, very little tactics."[66] The result was continued and fruitless bloodletting on the Western Front in 1915, impelled by the offensive spirit of the military leaders from both sides. "Neither of them," Churchill wrote of Joffre and Falkenhayn as they prepared for new attacks on each other that year, "... had ever sufficiently realized the blunt truth – quite obvious to common soldiers – that bullets kill men."[67] The Somme campaign the following year, he noted in a memorandum at the time, showed a similar indifference, weakening the Allies, "while the actual battle fronts were not appreciably altered, and ... no strategic advantage of any kind had been gained."[68] Once again, there were the military leaders "unequal to the prodigious scale of events," such as General Haig. "It needs some hardihood," he wrote of Haig's biographer after the war concerning the Somme, "... to write: 'The events of July 1 ... amply justified the tactical methods employed.'"[69] Years after the war, it was impossible for Churchill to maintain his objectivity as he looked back on Haig's persistence in attempting to bridge the continuum of war with masses of human beings. For him the British commander would always be

> a great surgeon before the days of anaesthetics; versed in every detail of such science as was known to him: sure of himself, steady of poise, knife in hand, intent upon the operation; entirely removed in his professional capacity from the agony of the patient, the anguish of relations .... He would operate without excitement, or he would depart without being affronted; and if the patient died, he would not reproach himself.[70]

The pattern did not change in 1917, a year in which the "obstinate offensives" continued to be "pursued regardless of loss of life until at length ... the spirit of the British army in France was nearly quenched under the mud of Flanders and the fire of the German machine-guns."[71] The problem, as Churchill saw it, was a lack of operational coordination. The Germans "only had to face ... disconnected attacks by the British." As a consequence, he concluded, "although each military episode, taken by itself, wore the aspect of a fine success, with captures of ground and guns and prisoners, in reality we were consuming our strength without any adequate result."[72]

That result would only come from an operational "succession of climaxes ... toward which everything tends, and from which permanent decisions are obtained. The climaxes have usually been called battles." The major difficulty in all this, Churchill came to believe, was that the British by 1917 were confusing operational intensity and decisiveness with casualties, the latter, in reality, only reflecting tactical disjointedness. "All the great operations of 1916 and 1917," he wrote, "although so prolonged as to cause very heavy casualties, have involved the simultaneous employment only of comparatively small forces on comparatively small fronts. The armies have been fighting in instalments." Consequently, for Churchill the war in the west had dwindled down to tactical siege operations "on a gigantic scale which however bloody and prolonged cannot yield a decisive result."

> Thus, when a great battle is raging on the British front, six or eight British divisions are fighting desperately, half a dozen others are waiting to sustain them, the rest of the front is calm; twenty British divisions are remaining quietly in their trenches doing their daily routine, another are training behind the lines; 20,000 men are at school, 10,000 are playing football, 100,000 are on leave.[73]

The Defensive Solution

As he looked back in revulsion at the indecisive bloodletting of the
Allied offensives against the entrenched German defenses, Churchill
recalled a "sense of grappling with a monster of seemingly un-
fathomable resources and tireless strength, invulnerable – since
slaughter even on the greatest scale was no deterrent."[74] That had not
always been the case, he realized later, as he examined the life of his
ancestor and discovered the shock that the sanguinary battle of
Malplaquet in 1709 had had on the "intricate polite society of the Old
World."[75] During that battle, the French commander conducted
defensive operations that "extracted from the Allies a murderous toll
of life," as he maneuvered back and forth within his entrenchments.
Despite Marlborough's ultimate victory, Churchill noted that "not
one of the allied generals, if he could have gone back upon the past,
would have fought the battle, and none of them ever fought such a
battle again."[76] That, of course, was not the case with the Allied
commanders in the First World War who, as Churchill constantly and
emotionally pointed out after that conflict, sent their men struggling

> forward through the mire and filth of the trenches, across the
> corpse-strewn crater fields, amid the flaring, crashing, blasting
> barrages and murderous machine-gun fire .... The battlefields
> ... were the graveyards of Kitchener's Army. The flower of that
> generous manhood ... was shorn away forever .... Uncon-
> querable except by death, which they had conquered, they have
> set up a monument of native virtue which will command the
> wonder, the reverence and the gratitude of our island people as
> long as we endure as a nation among men.[77]

The major problem in all this, as Churchill realized early on, was
that the offensive at the higher levels of war was not a viable option.
The opportunity for such solutions had faded at the Marne when,
although the Germans had reached their "culminating point," the
Allies were equally exhausted and could not unsheath Clausewitz's
"flashing sword of vengeance." The situation was not unique. "Military history," as he wrote shortly after the Marne campaign, "shows
many examples of commanders marching swiftly into an enemy's
country and seizing some key position of defensive strength against
which the enemy is afterwards forced to dash himself. Thus are
combined the advantages of a strategic offensive with those of a
tactical defensive."[78] Added to this in the Great War was the combination of technology and entrenchments which insured that "the

power of the defensive is as 3 or 4 to 1. We are therefore in the unsatisfactory position," Churchill concluded, "of having lost our ground before the defensive under modern conditions was understood, and having to retake it when the defensive has been developed into a fine art."[79]

For Churchill, as the war proceeded, it was obvious that the defensive posture of the Germans on the Western Front was the key to their continual success. "It is certain surveying the war as a whole," he wrote, "that the Germans were strengthened relatively by every Allied offensive ... launched against them, until the summer of 1918."[80] The situation changed that year, however, because of the great German attack in the west. "It was their offensive, not ours, that consummated their ruin," Churchill observed. "They were worn down not by Joffre, Nivelle and Haig but by Ludendorff."[81] Moreover, without that last operational offensive gasp, the German military situation would have remained relatively favorable. "Had they not squandered their strength in Ludendorff's supreme offensive in 1918," he concluded, "there was no reason why they should not have maintained their front in France practically unaltered during the whole of the year, and retreated at their leisure during the winter no farther than the Meuse."[82]

The answer for the Allies, Churchill maintained as early as 1915, was an "active defense" in the west, combining defensive operations at the lowest level with operational offensive assaults. "If our whole strategy and tactics had been directed to that end," he asked, "would not the final victory have been sooner won?"[83] The key to the "active defense," Churchill believed, was the deliberate weakening of various sections of the line in order to invite German attacks at the tactical level. Once the enemy had pushed in great pockets at different points in the yielding line, the Allies would "strike with independent counter-offensive on the largest scale and with deeply planned railways, not at his fortified trench line, but at the flanks of a moving, quivering line of battle!"[84] This type of thinking led Churchill in 1916 to assert that the French should have sacrificed ground at Verdun to gain a "greater manoeuvering latitude."[85] And this conceptual understanding of the interplay of defense and offense up and down the continuum of war allowed Churchill to provide Lloyd George during the German offensive of 1918 an explanation of Clausewitz's culminating point that the Prussian philosopher of war might have articulated himself. On 24 March 1918, the British Prime Minister took Churchill aside and asked why, if the Commonwealth soldiers could not hold the line they had fortified so carefully, they should be able to hold positions further to the rear with troops who had already been defeated. "I answered," Churchill replied,

> that every offensive lost its force as it proceeded. It was like throwing a bucket of water over the floor. It first rushed forward, then soaked forward, and finally stopped altogether until another bucket could be brought. After thirty or forty miles there would certainly come a considerable breathing space when the front could be reconstituted if every effort were made.[86]

Force Multipliers

The stalemate of the First World War drew Churchill to a variety of means by which decisive linkage could be restored to the continuum of war (See Figure 2). The idea of such force multipliers, he discovered in his inter-war studies of his ancestor, were very much a part of the first Duke of Marlborough's military philosophy. "As clever at piercing the hidden designs of his enemy as in beating him on the field of battle," Churchill wrote admiringly of Marlborough, "he united the cunning of the fox to the force of the lion."[87] And so it was with Churchill. He became a firm believer in deception operations at all levels of war from his experiences in the First World War as well as from his later studies of Marlborough's campaigns. And to an appreciation of tactical intelligence and all that could reveal of enemy

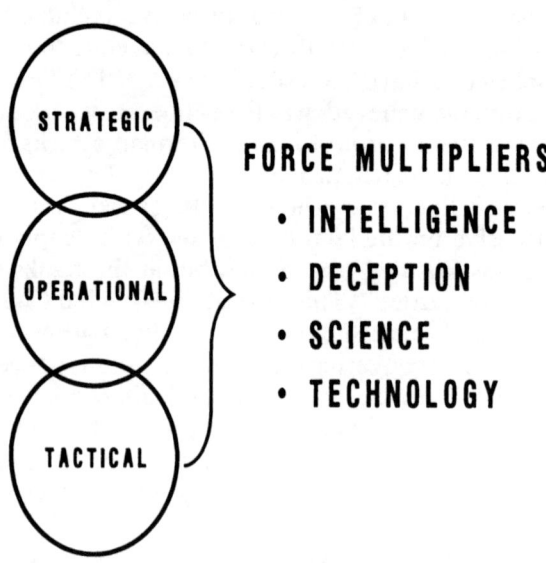

Figure 2

capabilities, he added in the Great War an understanding of what operational and strategic intelligence could mean in terms of enemy intentions. The advent of wireless communication, in this regard, opened up an entirely new field of signal intelligence (Sigint), the impact of which Churchill, as First Lord of the Admiralty, grasped immediately.

The science and technology that spawned Sigint also caused the attritive stalemate on the Western Front. Ironically, part of the cure for that stalemate lay in the cause – a fact that Churchill realized from the first bloody encounters in the trenches. As a result, he devoted much of his efforts throughout the war to creating the means for technological surprise. It was not an easy task. "In nearly every great war there is some new mechanical feature introduced the early understanding of which confers important advantages," Churchill noted in his study of Marlborough. "Military opinion is naturally rigid. Men held in the grip of discipline, moving perilously from fact to fact, are nearly always opposed to new ideas."[88] But those ideas must be explored, he concluded, if the vertical continuum of war was to be restored.

> There is required for the composition of a great commander not only massive common sense and reasoning power, not only imagination, but also an element of legerdemain, an original and sinister touch, which leaves the enemy puzzled as well as beaten. It is because military leaders are credited with gifts of this order which enable them to ensure victory and save slaughter that their profession is held in such high honour. For if their art were nothing more than a dreary process of exchanging lives, and counting heads at the end, they would rank much lower in the scale of human esteem.[89]

Intelligence

Early British codebreaking success in the First World War began Churchill's lifelong enthusiasm for Sigint. In November 1914, he issued a memorandum, for which he devised a new formula, "Exclusively Secret," directing the study of all decoded intercepts "in order to penetrate the German mind and movements."[90] By the end of that year, a small staff of cryptographers were implementing that charter from Room 40 of the Admiralty Old Building, and as their numbers grew, from a series of adjoining rooms, but still known by the innocuous collective title as "Room 40 O.B." Churchill's enthusiasm for this new organization was quickly justified. In January 1915, Admiral von Pohl submitted a memo to William II recommending

German submarine attacks on merchant shipping and the dispatch of airships to attack Britain. "So excellent was our Intelligence Service," Churchill wrote later, "that reports of what was passing in the minds of the German Naval Staff reached us even before Admiral von Pohl's memorandum had been laid before the Emperor."[91]

At the strategic level, so long as Room 40 could continue this type of successful decryption effort, there would always be a forewarning of any German move into the North Sea. This meant, in turn, that it was no longer necessary for the Admiralty to keep the Grand Fleet in a constant state of complete readiness, nor were the sweeps by that fleet of the North Sea required on a continual basis. The implications for the fleet as well as for the British people were not lost on the First Lord.

> Without the cryptographers' department . . . the whole course of the naval war would have been different. The British Fleet could not have remained continuously at sea without speedily wearing down its men and machinery. Unless it had remained almost continuously at sea, the Germans would have been able to bombard two or three times all our East Coast towns . . . and returned each time safely, or at least without superior attack, to their home bases . . . . The nation would have been forced to realize that the ruin of its East Coast towns was as much their part of the trial and burden as the destruction of so many Provinces to France.[92]

At the operational level, Sigint could be equally effective as Churchill demonstrated after the war in his analysis of the opening German eastern campaign in that conflict. On 24 August 1914, Ludendorff hesitated to dispatch two of his corps from their positions opposite Rennenkampf's advancing First Army in the north to attack the flank of Samsonov's Second Army in the south at Tannenberg. At that key moment, Churchill wrote, the Russian "radio now with bland simplicity proclaimed to the world in two uncoded messages exactly what Rennenkampf and Samsonov would do or not do on the 25th and 26th. The German wireless station . . . listened to these amazing disclosures" which "told them . . . that Rennenkampf's army could not take part in Samsonov's battles."[93]

After the war, as he worked on the biography of his ancestor, Churchill reinforced his appreciation of operational intelligence. That appreciation was particularly apparent in his description of the 1704 campaign, as Marlborough's "Scarlet Caterpillar" crawled across Europe from the North Sea to southern Germany. During the march, the British leader was provided information by an agent on the entire French campaign plan. The plan had been taken from the cabinet of the War Minister in Paris, deciphered, and sent to Marlborough at his camp by long and circuitous routes through France and Germany. For Marlborough, Churchill pointed out, the plan "only confirmed what his occult common sense had divined. But it must have been nonetheless very reassuring."[94] So must have been the continual flow of information on his French opponent, Marshal Tallard, provided by Marlborough's elaborate Secret Service, on which the normally stingy British commander consistently lavished funds. "Even more remarkable was Marlborough's own Intelligence," Churchill added in describing the Blenheim campaign, "for on July 3 he already knew almost exactly the number of battalions and squadrons which the King had so secretly assigned to Tallard only ten days before at Versailles. A message could hardly have covered the distance quicker."[95]

Deception

Deception was always a favorite means for Churchill to facilitate the vertical linkage in the continuum of war. Early in the First World War, he was instrumental in such an operation at the tactical level – an operation that had decisive military operational and strategic results. On 26 August 1914, as First Lord, he dispatched a brigade of marines to Ostend "in the hopes that it would attract the enemy and give him the impression that larger forces would follow by sea."[96] To the small force about to embark, Churchill pointed out that the object was "to create a diversion . . . and to threaten the western flank of the German southward advance. It should therefore be ostentatious."[97] That same day, in order to provide further publicity, he announced in the Commons that a British force had begun landing at Ostend. By 5 September, the momentous day in the Battle of the Marne when French troops from Paris were moving to strike at the flank of the German First Army, that army believed that its rear was seriously menaced, primarily due to the Ostend expedition. How much effect that belief had on the German command at that crucial stage is not definitely known. Nevertheless, the 5 September message that the "English are disembarking fresh troops continuously on the Belgian coast," must have placed insidiously enervating pressures on the German commanders as the Battle of the Marne reached its climax.[98]

   In the inter-war years, as he immersed himself in Marlborough's eighteenth century campaigns, Churchill lingered in great detail over every deception operation conducted by his ancestor. At Elixem in 1705, Marlborough's deception plan, which included the construction of eleven bridges that were never to be used, was not only designed to fool the French but his cautious Dutch allies as well. On the morning of 18 July, the Duke's forces overwhelmed the outnumbered French units at Elixem, breaching the Lines of Brabant, the great defensive system that ran from Antwerp to Namur, while the preponderance of the French army, reacting to Marlborough's deception, was far to the south. "There is no moment in war more thrilling," Churchill wrote of that operation's successful denouement, "than a surprise attack at dawn."[99] In a similar manner, there was Marlborough's feint before Tournai in 1709 where the "surprise was complete and the fortress was caught with barely five thousand men."[100] And in the 1711 campaign, marked by "the artifices and strategems which he used," Churchill detailed a series of operational level deceptions implemented by Marlborough's forces as they "traversed those broad undulations between the Vimy Ridge and Arras which two centuries later were to be dyed with British and Canadian blood."[101]

But it was the Blenheim campaign of 1704 that truly reinforced Churchill's appreciation of deception as a force multiplier. In August and September of 1932, Churchill toured his ancestor's battlefields on the Continent, and was struck by the success of the overall deception operation that governed the Grand Alliance march into Bavaria. On 25 September, he wrote to his cousin, the ninth Duke of Marlborough, about the first Duke's departure from Coblenz on the march that would end at Blenheim, emphasizing that:

> none of the hostile watching armies ready to spring, not even our army, was sure where it was going to. They still thought it was a campaign in Alsace; but no, a fortnight later the long scarlet columns swing off to the Danube .... This marvellous march was distinguished ... for its absolute secrecy and mystery – no one knew, not the Queen, not Sarah, not the English government, except Eugene.[102]

Eugene was, of course, Prince Eugene of Savoy, Britain's great ally who was to play a major role in Marlborough's deception plan at Blenheim. Before the battle, Eugene gave Tallard the impression through numerous spies and deserters that he was moving back to his old position on the Lines of Stollhofen, thinly held against the French General Villeroy by the corps of the Prince of Anhalt. He then marched his forces ostentatiously north in the direction of the Lines. On 27 July, he reached Tübingen, then vanished from the French view among the desolate hills of Swabia. Villeroy was convinced that Eugene was still close to the lines and showed no sign of movement toward Bavaria. As a consequence, Churchill concluded, "Villeroy, gaping at the half-vacant lines of Stollhofen, need no longer be considered as a factor in the fateful decisions impending upon the Danube."[103]

Marlborough coordinated every aspect of that deception. There was thus no sudden surprising arrival of Eugene to rescue the coalition commander at Blenheim in the nick of time, as had sometimes been reported until Churchill set the record straight in his biography. In fact, as Churchill convincingly demonstrated, Marlborough had a superabundance of information from his own Secret Service and confirmed it where necessary by Eugene's field reports. "The accuracy of his information about the enemy," he wrote, "and also the speed with which it reached him is remarkable. He knew ... exactly what had happened ... and where Tallard was baking his bread and would march."[104] The key to all that was Eugene, moving secretly with his forces to join those of Marlborough. "Eugene knew that, whatever might miscarry behind him," Churchill concluded, "... he must arrive on the Danube somewhere between Ulm and Donauwoerth at the same time that Tallard joined the Elector. Marlborough in all his conduct counted on him to do this, and his own arrangements made the juncture sure and certain."[105]

With all the coalition forces gathered together, Marlborough engaged in another deception operation on 12 August by planting four deserters in the French camp. As Churchill described it, each deserter told the same story: Marlborough had arrived with his troops; but the entire allied army was going to retire under a bold display toward Noerdlingen on the morning of 13 August. That information appeared to be confirmed by reports from the French cavalry scouts, who had watched the dust clouds above the allied baggage columns which Marlborough had sent off on a false march, and by planted rumors that came in from the countryside. Marshal Tallard and his staff agreed as a result that they should not attack so strong an army, but more important, that the allies themselves would not attack. At 7 a.m. on 13 August, just before Marlborough's attack on the French forces, Tallard wrote a letter to Louis XIV, describing how the enemy forces had begun to assemble before daybreak at 2 a.m. and were now drawn up at the end of their camp. "Rumour in the countryside," he concluded confidently, "expects them at Noerdlingen."[106]

Science and Technology

Churchill's natural interest in scientific gadgetry deepened as the First World War progressed. Increasingly, he saw science and technology as a means to break the military deadlock, much as innovations ranging from the stirruped cavalry horse to the long bow and the Maxim gun had enabled armies to achieve surprise and win unexpected victories in the past. "Machines save lives," he asserted in the March 1917 Army Estimates debate, "machine-power is a substitute for man-power." Unless new devices were developed, he continued, "I do not see how we are to avoid being thrown back on those dismal processes of waste and slaughter which are called attrition."[107] At the end of the year, Churchill reflected on Allied progress with "new devices" and the fact that the Germans appeared bent on resuming the offensive. "Let them traipse across the crater fields," he wrote. "Let them rejoice in the occasional capture of placeless names and sterile ridges; & let us dart here & there armed with science and surprise ...."[108]

Churchill's attitude about the role that scientific and technological surprise would play in the continuum of war was demonstrated by his involvement in the evolution of the tank in the First World War. In the beginning, his restless imagination concentrated on protecting his fledgling Royal Naval Air Service with an armored machine possessing large driving wheels and rollers in front to crush both barbed wire and trenches. It was not until 1915, because of the failure of those experiments coupled with the persuasive advice of several army officers, that Churchill advocated a tank with a caterpillar system which could advance into German lines, "smashing away all the obstructions and sweeping the trenches with their machine-gun fire."[109]

Nevertheless, Churchill quickly learned as First Lord of the Admiralty that the process of inventing was only part of the problem in the complex power hierarchy of total war. In 1915, for instance, he tied the lack of progress in tank development directly to government indifference. "The problem," he wrote, "of crossing two or three hundred yards of open ground and of traversing ... barbed wire in the face of rifles and machine guns ... ought not to be beyond the range of modern science if sufficient authority ... backed the investigation. The absence of any satisfactory method cannot be supplied by the bare breasts of gallant men."[110] And after the war, Churchill returned to the subject when he credited the armor pioneers in the British officer corps with seizing the idea of the tank and even presenting specific proposals before the War Office. "These officers," he noted, "had not however the executive authority which alone could ensure progress and their efforts were brought to nothing by the obstruction of some of their superiors. They were unfortunate in not being able to command the resources necessary for action, or to convince those who had the power to act."[111]

With authority must come coordination extending from the lab to the trenches. "A hiatus exists between inventors who know what they could invent, if they only knew what was wanted," Churchill wrote in 1916, "and the soldiers who know, or ought to know, what they want and would ask for it if they only knew how much science could do for them."[112] That belief was reinforced by his experience with the Stokes gun, a hand-held mortar whose design was based on a front line need for immediately responsive, short-range indirect fire to be used in attacks on trenches at close quarters. "All the ideas on which this scheme rests," he wrote after the demonstration, "have come from officers who have been themselves constantly engaged in trench warfare. In order to give a fair chance to such a method of attack, it is necessary that it should not be attempted until it can be applied on a very large scale."[113]

With that last injunction, Churchill returned to the vertical continuum of war and the key element of operational surprise. In any armor attack, he emphasized as early as December 1915, tactical surprise should not be squandered when the means were lacking to mesh the results into an operational whole. Less than a year later at the Somme, however, 35 tanks were dispersed in small, ineffective groups along the entire front of the Fourth Army as it attacked. Lloyd George informed Churchill just before the assault. "I was so shocked at the proposal to expose this tremendous secret to the enemy upon such a petty scale," Churchill recalled, "... that I sought an interview with Mr. Asquith."[114] But to no avail. The attack was a limited success which, in Churchill's opinion,

> recklessly revealed to the enemy a secret that might have produced allied victory in 1917. The immense advantage of novelty and surprise was thus squandered while the number of the tanks was small, while their condition was experimental and their crews almost untrained. This priceless conception ... was revealed to the Germans for the mere petty purpose of taking a few ruined villages .... The enemy was familiarized with them by their piecemeal use.[115]

It was not until the Battle of Cambrai on 29 November 1917 that massed armor combined with sufficient tactical surprise to achieve a breakthrough in the German lines. "All the requisite conditions were at last accorded," Churchill wrote. "The tanks were to operate on ground not yet ploughed up by artillery, against a front not yet prepared to meet an offensive. Above all, Surprise! The tanks were themselves to open the attack."[116] That development had occurred because of the technological progress in other areas such as the science of gunnery which meant, in Churchill's words, that "indispensable preparation" would no longer destroy "indispensable surprise."[117] By the fall of 1917, as a consequence, artillery did not require preliminary registration to be on target, and the British were able to open accurate, preplanned fire at H-Hour just as the British tanks moved forward in the attack in the first half light of the dawn.

Despite technological advances, however, Cambrai still represented for Churchill only a single tactical event separated by a wide and disjointed gulf from the higher levels of war. "If British and French war leaders had possessed ... the vision and comprehension which is expected from the honoured chiefs of great armies," he wrote, "there is no reason why ... three or four concerted battles like Cambrai could not have been fought simultaneously in the spring of 1917 .... Then indeed the roll forward of the whole army might have been achieved and the hideous deadlock broken."[118] Those leaders, in Churchill's judgment, had been captured by technology instead of harnessing it for novel, imaginative use that restored operational surprise and thus operational maneuver to warfare. Nevertheless, as he also realized, the quantity and scope of the technological innovations that were provided to the Allies by 1918 ultimately allowed even Marshals Haig and Foch to move to the higher levels of war's continuum.

> Both were now provided with offensive weapons, which the military science of neither would have conceived .... The Goddess of Surprise had at last returned to the Western Front. Thus both ... were vindicated in the end. They were throughout consistently true to their professional theories, and when in the fifth campaign of the war the facts began for the first time to fit the theories, they reaped their just reward.[119]

The Building Blocks – The Horizontal Dimension

The US Department of Defense *Dictionary of Military and Associated Terms* (JCS) defines national strategy as the "art and science of developing and using the political, economic, and psychological powers of a nation, together with its armed forces, *during peace and war*, to secure national objectives."[120] The inclusion of peace as well as war is understandable, given the emphasis in the nuclear age on what

Clausewitz termed "preparation for war." Nevertheless, it is useful to focus on national strategy purely in a wartime setting, for conflict ultimately tests the ability of a nation and its leaders to make the calculated relationship of means and ends that is the essence of successful strategy.

That type of national calculation is described by the term "Grand Strategy," the role of which, as Liddell Hart has pointed out, "is to coordinate and direct all the resources of a nation, or band of nations, toward the attainment of the political objective of the war . . . ." This aspect, he concludes, causes a grand strategy to look "beyond the war to the subsequent peace. It should not only combine the various instruments, but so regulate their use as to avoid damage to the future state of peace . . . ."[121]

Figure 3

Grand Strategy, then, is the use in wartime of all the instruments of national power. Those instruments can be conveniently broken down on a horizontal plane into the categories described in the JCS definition of national strategy: political, economic, psychological and military (Figure 3).

The linchpin in this horizontal design is the military instrument of power at the national strategic level – the apex, as we have seen, of the vertical continuum of war. Normally, no matter how sophisticated the interaction of all elements of national power during wartime, military force plays the ultimate and decisive role. Moreover, the various levels of war will not only influence the national military strategy, but can directly affect at the national policy level the other instruments of power that support the overall grand design. Compounding this already intricate picture is the fact that force multipliers at all levels of the vertical continuum are often inextricably entwined with the political, economic and psychological aspects of grand strategy (Figure 4).[122]

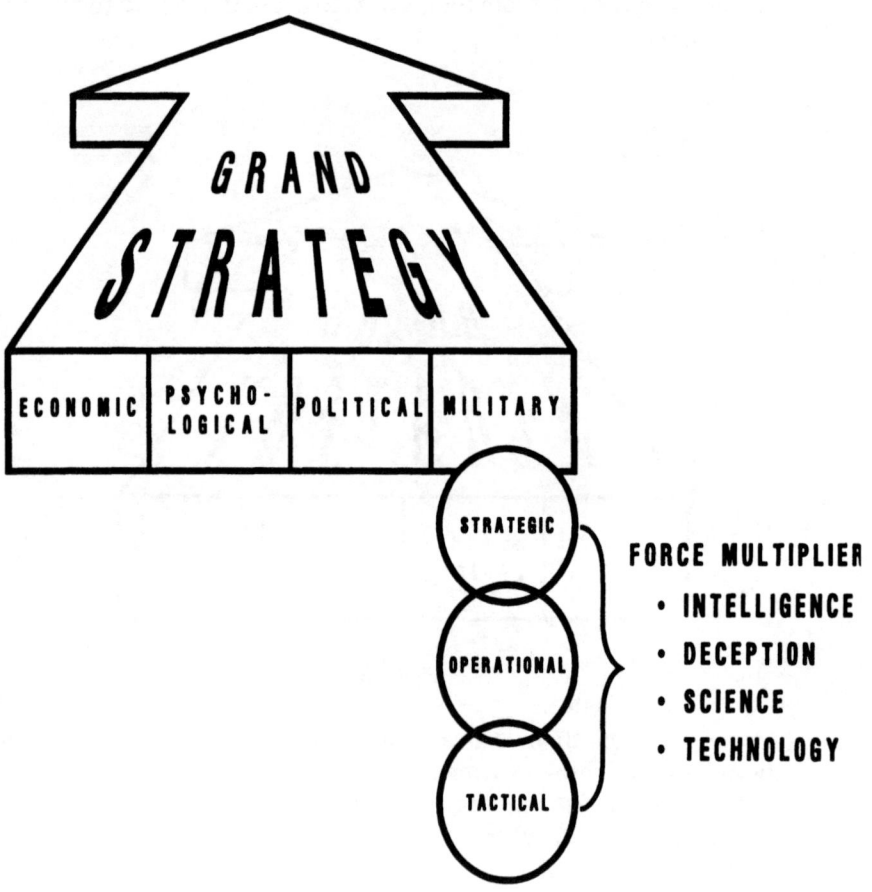

Figure 4

In his early encounters with Victorian wars, Churchill showed some awareness of these elements that make up grand strategy. To be aware, however, was not necessarily to approve of the intrusion of non-military matters. "Of course nowadays every budding war is spoiled and nipped by some wily diplomat," he wrote to his mother from India in 1895; and a few years later with the Malakand Field Force on India's northwest frontier, he did not hide his contempt for the political officers assigned to that unit.[123] Nevertheless, in his account of the 1898 Sudan campaign, the young Victorian showed a marked appreciation of the political and psychological interaction with the military aspects of that operation launched by the Conservative government: "The diplomatist said 'it is to please the Triple Alliance.' The politician said 'It is to triumph over the Radicals.' The polite person said 'It is to restore, the Khedive's rule over the Sudan.' The man in the street ... said 'It is to avenge General Gordon.'"[124]

That type of grand strategic interaction was further brought home to Churchill by the increased involvement of the British public in the Boer War at the turn of the century. To begin with, there was the national outcry at the "Black Week" of December 1899, during which occurred the defeats of Lord Methuen at Magersfontein, General Gatacre at Stormberg, and Sir Redvers Buller at Colenso – the latter with the loss of eleven artillery pieces without breaking the Boer investment of Kimberley and Ladysmith. And conversely, there was the national exultation on the night of 18 May 1900 when the siege of Mafeking was raised and all London rioted with joy. That night, in the interval before the last act of *Lohengrin* at Covent Garden, the Prince and Princess of Wales stood with the audience and sang the national anthem. But the humorist, Saki, put the victory in perspective in one of his brief sketches and thus introduced a new word to the English language.

> Mother, may I go and maffick,
> Tear around and hinder traffick?[125]

For Churchill, these aspects of the Boer War demonstrated that the vertical military continuum was no longer sufficient in itself to meet the demands of modern conflict. In a prescient speech in May 1901, he pointed out that "a European war cannot be anything but a cruel, heart-rending struggle, which, if we are ever to enjoy the bitter fruits of victory, must demand, perhaps for several years, the whole manhood of the nation, the entire suspension of peaceful industries, and the concentration to one end of every vital energy in the country." Earlier wars on the margin of the Empire were one thing, he concluded,

but now, when mighty populations are impelled against each
other, each individual severally embittered and inflamed – when
the resources of science and civilization sweep away everything
that might mitigate their fury, a European war can only end in
the ruin of the vanquished and the scarcely less fatal dislocation
and exhaustion of the conquerors. Democracy is more vindictive
than cabinets. The wars of peoples will be more terrible than the
wars of kings.[126]

In the First World War, Churchill confirmed this aspect of a larger
wartime strategy. "It is established that henceforth whole populations
will take part in war," he wrote, "all doing their utmost, all subjected
to the fury of the enemy." Out of that conflict emerged for the young
statesman certain sombre facts, "solid, inexorable, like the shapes of
mountains from drifting mist," which directly concerned the role of
grand strategy. "There are many kinds of manoeuvers in war, some
only of which take place upon the battlefield," he concluded. "There
are manoeuvers far to the flank or rear. There are manoeuvers in
time, in diplomacy, in mechanics, in psychology; all of which are
removed from the battlefield, but react often decisively upon it."[127]

It was a broader picture of wartime strategy that would have been
understood by Marlborough, Churchill believed, who "was not only
Commander-in-Chief of the English and Dutch armies, but very
largely a Prime Minister as well." As a consequence, he concluded in
the inter-war years, his ancestor "could ... feel towards the whole
problem a responsibility different from that of the leaders of individual armies, however large."[128] It was in this context that Churchill
came to realize that the vertical military continuum was only one
aspect of total war. A massive defeat like Tannenberg at the operational and theater strategic level, for instance, did not mean the fall of
a government with sufficiently strong nerves and untapped resources.
Clear decisions were no longer possible on that continuum alone.
Even campaigns and major operations were no more than competitions in mutual attrition in which "the strength being eroded had to be
measured in terms not simply of military units but of national manpower, economic productivity, and ultimately the social stability of
the belligerent powers."[129]

The Political Connection

In the 1930s, looking back on Marlborough's career, Churchill neatly
summarized much of his thoughts on the interaction of the military
and political instruments of power in grand strategy. "His life was a

ceaseless triple struggle," he wrote of his ancestor, "first to preserve the political foundation in England which would enable her to dominate the continental war; secondly, to procure effective military action from the crowd of discordant, jealous, and often incompetent or lukewarm allies; and thirdly – and this was the easiest part – to beat the French in the field."[130] That the political foundations of his nation were inextricably linked with the fortunes of war at all levels was by that time a truism for Churchill who had been thrust from office as First Lord by the operational defeats on the Gallipoli Peninsula in 1915. "At home," he could thus note sympathetically of Marlborough's situation two centuries before, "... the wolves were always growling."[131] In 1703, as an example, there were the violent attacks by the High Tory Party on the Duke's conduct of the war. The Whig Party, Churchill concluded, "had hitherto most loyally voted the supplies and sustained the policy of a grand land war; but they expected results. Without victories and solid gains they saw themselves stultified and pilloried in the party fight."[132]

The answer in that instance lay in the victorious Blenheim campaign of 1704. Yet even that outcome was colored by politics. After the storming of the Shellenberg in that campaign where Marlborough pressed the central attack, in Churchill's words, "with a disregard of human life unusual in these prolonged and stately wars," the Tories were not alone in asking at home: "What was the sense of capturing a hill in the heart of Germany at such heavy loss? Were there not many such hills?"[133]

Marlborough's Blenheim campaign was the result of Bavaria's joining France and Spain in 1702, the strategic equivalent, Churchill believed, of Turkey's coalition with the Central Powers in 1914, since Bavaria's defection separated the Austrian Empire from the West much in the way Turkey's hostility cut Russia off from the Allies in the First World War. The role of a government in recruiting allies, Churchill realized, was a vital part of grand strategy. "The manoeuver which brings an ally into the field," he wrote, "is as serviceable as that which wins a great battle."[134] That was, of course, particularly true of the US entry into the war, an event captured in microcosm by his description at the Chemin des Dames in May 1918 of the "impression made upon the hard-pressed French by this seemingly inexhaustible flood of gleaming youth in its first maturity of health and vigour."[135]

Even the failure to recruit such allies by either side, Churchill realized, could have momentous results. There were, for instance, the disadvantages to the Allies of the Dutch neutrality throughout the war which kept the Rhine open for Germany and closed the Scheldt to

Antwerp. In that case, Churchill pointed out, "a neutral Holland was of far more use to Germany than a hostile, a conquered, or even an allied Holland."[136] But neutrality was not normally such a boon to the Central Powers whose efforts to recruit allies, like those of the Entente, depended heavily on the fortunes of war at every level of the vertical continuum. The twin defeats in September 1914, for instance, of the Germans at the Marne and the Austrians around Lemberg decisively influenced several neutrals. "Roumania had actually decided at the beginning of September," Churchill observed, "to make proposals to the Central Powers . . . . But under the decisions of the battlefields in France and Galicia the offer was suppressed . . . . Bulgaria, the spectator of the Austrian repulse by Serbia, wrapped herself in impenetrable reserve."[137] This linkage of the vertical continuum to the political part of the horizontal policy plane, Churchill also emphasized, proved to be equally strong for the Western allies.

> After war had been declared, diplomacy counted little with neutrals. They were no longer concerned with what was said or promised. The questions they asked themselves were, What was going to happen, and who was going to win? They were not prepared to accept British assurances upon either point. We were astonished to find that many of these neutrals seemed to doubt that Great Britain would certainly be victorious. One pitied their obliquity. But they persisted in it. The Foreign Office talked well; but it was like talking to the void.[138]

Once allies were recruited, Churchill realized how hard it was to keep them together. The problems with coalition warfare in the First World War were part of his personal experiences. Those impressions were reinforced in the inter-war years as he examined Marlborough's efforts to coalesce the disparate Dutch, British and German forces that formed the core of the Grand Alliance against Louis XIV's France in the War of the Spanish Succession. "It was never in his power to give orders which covered the whole field of the war," Churchill observed in this regard, "and in many quarters ... his command was disputed."[139] Added to this were the centrifugal tendencies of sovereign states with their own national interests. "The history of all coalitions," he concluded at the end of the study of his ancestor's Grand Alliance, "is a tale of the reciprocal complaints of allies."[140]

And those complaints, Churchill knew, could reach serious proportions if coalition progress throughout the vertical military continuum was not forthcoming. "Fear and hatred of French ascendancy," he wrote of the situation in 1703, "would not hold the Alliance together beyond the hour when hope of beating France departed .... The Grand Alliance quivered at this moment in every part of its fragile organization. Marlborough saw that without some enormous new upholding force it must come clattering down."[141] The answer once again was the Blenheim campaign, in which Marlborough's operational art produced political–military results on a grand strategic scale. "The wide plain, bathed in the morning sunlight," Churchill wrote of the Blenheim battlefield, "was covered with hostile squadrons and battalions, already close at hand and steadily marching on."

> But behind this magnificent array, if the count could have discerned them, were the shapes of great causes and the destinies of many powerful nations. Europe protested against the military domination of a single Power. The Holy Roman Empire pleaded for another century of life. The ancient rights of the Papacy against Gallicanism and the ascendancy of a Universal over a National church – despite the mistaken partisanship of the reigning Pope – were, in fact, fatefully at stake. The Dutch Republic sought to preserve its independence, and Prussia its kingdom rank. And from across the seas in England the Protestant succession, Parliamentary government, and the future of the British Empire advanced with confident tread. All these had brought their cases before the dread tribunal now set up in this Danube plain.[142]

The Psychological and Economic Connections

The First World War required a total mobilization for which there was no example to guide the British government. It was a total war in the sense that its rigors were not just confined to front line combatants, but to entire populations on what came to be called the "Home Front," no matter how distant from the battle area. For the people of Britain, as the war progressed, the conflict brought every aspect of their lives under an ever-increasing control and regimentation ranging from freedom of action and speech to employment and even diet.

Churchill was intimately involved in many of the major issues that arose from the regimentation of the Home Front. As early as 1915, he was urging a more discriminate acceptance of volunteers in terms of civilian employment in war essential industries; and during the conscription crisis of 1916, he fought against the compromise solution which "neither secured the numbers of men that would be needed, nor did it meet the new fierce demand for equalization of sacrifice."[143] But his head-on encounter with industrial unrest, after joining Lloyd George's government in 1917 to head the Ministry of Munitions, brought home to Churchill the importance of the psychological and economic aspects of grand strategy. Typically, he answered many of the grievances with pragmatic, common sense solutions that demonstrated to the workers their importance in the total effort. That effort, he told over 4,000 workers at the Ponders End munitions factory in October 1917, "was not a war only of armies, or even mainly of armies. It was a war of whole nations . . . ."[144]

The key to the Home Front, Churchill came to realize, was national will molded by strong leadership and propaganda into an unswerving belief in the worthiness of the war effort. In Churchill's only novel, *Savrola*, the protagonist is asked in this regard how he knows he will ultimately triumph. "Because we have got might on our side, as well as moral ascendancy," he replies.[145] In the inter-war years, Churchill returned in his study of Marlborough to the dominant influence this psychological aspect of grand strategy could have on the vertical continuum of war, even entitling one chapter of that biography, "The Home Front." And in his analysis of the closing years of the War of the Spanish Succession, he admitted that even his ancestor could be wrong concerning that influence. In the discussions in that war about a possible conflict in Spain, Churchill concluded in terms that presaged the limited wars of a later era:

> Marlborough himself considered that a single campaign would suffice. It may well be that he greatly underrated the resisting power of a nation, and thought of it in terms of professional armies. He might have fallen into the same trap as was a hundred years later to ruin Napoleon. There was always the possibility which Bolingbroke ... was many years later to describe: "That armies of twenty or thirty thousand men might walk about the country til Doomsday ... without effect."[146]

This type of will meant even the most crushing military defeats in the field would not be necessarily decisive – particularly if they could be counterbalanced by positive news, as occurred for the Germans in the fall of 1914. "The dazzling victories in the East," Churchill observed, "came just at that moment when the German people became aware of the fact that they had been repulsed from Paris .... They nursed and warmed themselves with the good tidings that Hindenburg had smashed the Russians."[147]

On the other hand, if the national will was weakened or lacking, the most trifling military defeat at the tactical or operational level could be decisive. In 1918, for example, the two battles of Salonica and Vittorio Veneto, both almost negligible by earlier standards, brought about the collapse of the exhausted Central Powers. In that regard, the subsequent "undefeated" claims of the German military feeding into the "stab in the back" myth were appeals to anachronistic criteria that were as irrelevant as they were untrue. The vertical continuum of war could no longer be divorced from its parent society. "Once the national will to war had been exhausted," Michael Howard has pointed out, "that great reserve of enthusiasm and patriotism and endurance built up over a century of careful training and squeezed to the last drop by relentless war propaganda, the military instruments of that will were as useless as empty suits of armour."[148]

Nowhere was this phenomenon better illustrated than in the 1918 Ludendorff offensive in which the psychological and economic instruments worked in adverse synergism against Germany. "Here then was the wearing down," Churchill wrote,

> which coming at the moment when the German national spirit was enfeebled by its exertions during four years and by the cumulative effects of the blockade, led to the German retreat on the Western Front; to the failure to make an effective withdrawal to the Antwerp–Meuse line with all the bargaining possibilities that this afforded; and to the sudden final collapse of German resistance in November 1918.[149]

In all that, the inexorable economic pressure of the British blockage was particularly effective, a fact realized by Churchill who criticized Falkenhayn for turning west in 1916 at Verdun to attack the strongest enemy at the strongest point. "It was a decision that not only underestimated the allied defenses in France, but took altogether too narrow and too purely military view of the general position of Germany and her allies." The vital need, he observed, was for Germany to break the blockade which with its "vast process of starvation not only in food but in materials indispensable to modern armies was remorselessly and unceasingly at work." Since that was impossible, there should be a drive to the east and southeast where Germany "could find the feeding grounds and breathing room ... without which her military strength however impressive was but a wasting security."[150] In this way, Churchill concluded, Falkenhayn could

> have gained the vast food and fuel regions which stretched from Galicia to the Caspian Sea. He would thus have broken the naval

blockage by continental conquest, and gained from the land much that the British Navy denied upon the sea. Instead, in approved professional spirit he chose to gnaw the iron hills of Verdun and their steel defenders. Thus were the Allies delivered from the penalties which their strategic follies in 1915 had deserved, and the equipoise of the war preserved for another bloody year.[151]

Falkenhayn's decision to attack Verdun was based to a large extent on psychological factors – a "simple solution," in Churchill's words, "for world-wide problems" in which whether Verdun "was taken or not, the French Army would be ruined and the French nation sickened of war."[152] In a military sense, Churchill knew that Falkenhayn was right. "At its highest," he observed, "the capture of Verdun would have been a military convenience to the Germans, and in a lesser degree an inconvenience to the French."[153] The problem, however, was that the "psychological conceptions which had led Falkenhayn to select Verdun as the point of the German attack became mingled in the tactical sphere."[154] As a result, "anything less than absolute victory would count as a failure for Germany .... Only one result, and that the most difficult, could achieve his purpose. A hundred variations would meet the modest requirements of his antagonists."[155] Thus it was, Churchill concluded, even though the "Germans had performed prodigies, ... the world sustained only one impression, namely that the French held Verdun; and that impression was grievous to the German cause."[156]

In a similar manner, the Battle of Jutland that same year demonstrated for Churchill how entwined the psychological and economic instruments of power were with the military instrument at all levels of war. Those considerations made the possible outcome of the battle a very uneven proposition for the former First Lord. On the one hand, a decisive British defeat could be catastrophic, which is why Churchill considered the commander of the British Grand Fleet to be "the only man on either side who could lose the war in an afternoon."[157]

> The trade and food-supply of the British islands would have been paralysed. Our armies on the Continent would have been cut from their base by superior naval force. All the transportation of the Allies would have been jeopardized and hampered. The United States could not have intervened in the war. Starvation and invasion would have descended upon the British people. Ruin utter and final would have overwhelmed the Allied cause.[158]

On the other hand, Churchill concluded concerning a decisive German defeat, the "psychological effect upon the German nation cannot be estimated."[159] In fact, the German High Seas Fleet won a minor tactical victory at Jutland. For Germany, the strategic situation remained at least as bad as it had previously, and arguably worse since the German commander only allowed that fleet in the future to put to sea in an even more grudging and sporadic fashion than before. It was, as far as Churchill was concerned, a situation that already obtained and one which should not have been risked on one roll of the dice at the middle and lower levels of the vertical military continuum. "Although the battle squadrons of the Grand Fleet have been denied ... decisive battle," he wrote of the British forces, "yet from the beginning they have enjoyed all the fruits of a complete victory. If Germany had never built a Dreadnought, ... the control and authority of the British Navy could not have been more effective .... There was no need for the British to seek battle at all."[160]

In 1917, the national will of the Russian Empire collapsed in Revolution, a psychological blow to the Entente that had an impact on all instruments of grand strategy, to include the naval encirclement of Germany. Churchill was aware of this connection, pointing out to the War Cabinet that "our policy of blockage on which the Navy have hitherto relied can no longer be counted upon to produce decisive results now that the Germans have got enormous portions of Russia at their disposal."[161] The counterbalance to this development in the east was the entrance of the United States into the war, the moral and psychological consequences of which, he concluded, marked the turning point in the conflict.

> The war had lasted nearly three years; all the original combatants were at extreme tension; on both sides the dangers of the front were matched by other dangers far behind the throbbing lines of contact. Russia has succumbed to these new dangers; Austria is breaking up; Turkey and Bulgaria are wearing thin; Germany herself is forced even in full battle to concede far-reaching Constitutional rights and franchise to her people; France is desperate; Italy is about to pass within an ace of destruction; and even in stolid Britain there is a different light in the eyes of men. Suddenly a nation of one hundred and twenty million unfurls her standard on which is already the stronger side; suddenly the most numerous democracy in the world, long posing as a judge, is hurled, nay, hurls itself into the conflict. The loss of Russia was forgotten in this new reinforcement. Defeatist

movements were strangled on the one side and on the other inflamed. Far and wide through every warring nation spread these two opposite impressions – "The whole world is against us" – "The whole world is on our side."[162]

The Horizontal–Vertical Interface

The issue of overall centralized government direction of grand strategy was for Churchill the key problem that had emerged from the complications of modern war. To begin with, there was the decision-making process. In peacetime, under a popular and democratic form of government, he acknowledged, that process could be decentralized to some extent with compromise being "very often not merely necessary but actually beneficial."[163] But in war, particularly total war, decision-making on the grand strategic plane had to be centralized. There could be no prevarication on the part of political leaders, such as that which he had witnessed by Arthur James Balfour at the Supreme Council of the Allies at Versailles during the First World War. Balfour spoke for ten minutes; and when he finished, Clemenceau abruptly inquired of him: "Pour ou contre?"[164] Looking back on those experiences, Churchill concluded that the grand strategic decision-making process in modern war was and must be entirely different from that of national strategy in peacetime.

> There is no place for compromise in War. That invaluable process only means that soldiers are shot because their leaders in Council and camp are unable to resolve. In War the clouds never blow over, they gather unceasingly and fall in thunderbolts. Things do not get better by being let alone. Unless they are adjusted, they explode with shattering detonation. Clear leadership, violent action, rigid decisions one way or the other, form the only path not only of victory, but of safety and even of mercy. The State cannot afford indecision or hesitation at the executive centre.[165]

Grand strategic policy was the result of decisions at that "executive centre," and as Churchill had learned during Britain's initial skirmish in South Africa with modern war, that type of policy must determine military strategy. "It is not enough," he wrote to Joseph Chamberlain in 1901, "for the Government to say 'we have handed the war over to the military: they must settle it: all we can do is to supply them as they require!' I protest against the view. Nothing can relieve the Government of their responsibility."[166]

That belief was reinforced by Churchill's pre-war problems as First

Lord with the high level naval bureaucracy, which prevented the formation of a naval general staff and prevented or slowed many of his cherished projects. But it was the Admiralty's resistance to the convoy system, in which the "means of salvation were forced upon them from outside," that in all probability confirmed Churchill's views.[167] The adoption of that system, he noted, was the result of a struggle

> between the amateur politicians, thrown by democratic Parliamentary institutions to the head of affairs ... and the competent, trained, experienced experts of the Admiralty .... The astonishing fact is that the politicians were right ... upon a technical professional question ostensibly quite outside their sphere, and the Admiralty authorities were wrong upon what was, after all, the heart and centre of their own peculiar job.[168]

As he became more immersed in the total conflict of the First World War, Churchill became increasingly concerned that grand strategic policy must dominate military strategy at the highest leadership. "The distinction between politics and strategy," he wrote, "diminishes as the point of view is raised. At the summit true politics and strategy are one."[169] But the civilian apparatus for control at that level in Britain during the First World War was lacking. After the Dardanelles campaign, Churchill pointed out that "no one had the power to give clear brutal orders which would command unquestioning respect. Power was widely disseminated among the many important personages who in this period formed the governing instrument."[170] And to Asquith, he wrote concerning Lord Kitchener in October 1915: "The experiment of putting a great soldier at the head of the War Office in time of war has not been advantageous. In the result we have neither a Minister responsible to Parliament nor a General making a plan."[171] But the key was responsible power. "I will never accept political responsibility," Churchill concluded after the Dardanelles debacle, "without recognized regular power."[172]

Even the more highly controlled and centralized government under Lloyd George, who took office in 1916, was not adequate to the task as far as Churchill was concerned. To begin with, there was not a tight executive core, ensuring that policy evolved from the equivalent of what in Marlborough's time was "the eye and brain and soul of a single man, which from hour to hour are making subconsciously all the unweighable adjustments, no doubt with many errors, but with ultimate accuracy."[173] Moreover, by that time the dispute between the military, or "brass hats," and the politicians, or "frocks," was public knowledge. In fact, the people were becoming more and more influential in the dispute since the general public impression was that the military leaders must be right on matters of war. "The feeble or presumptuous politician," Churchill wrote in this regard,

> is portrayed cowering in his office, intent in the crash of the world on Party intrigues or personal glorification, fearful of responsibility, incapable of aught save shallow phrase-making. To him enters the calm, noble, resolute figure of the great Commander by land or sea, resplendent in uniform, glittering with decorations, irradiated with the lustre of the hero, shod with the science and armed with the panoply of war. This stately figure, devoid of the slightest thought of self, offers his clear farsighted guidance and counsel for vehement action or artifice or wise delay. But his advice is rejected; his sound plans put aside; his courageous initiative baffled by political chatterboxes and incompetents. As well ... might a great surgeon, about to operate with sure science and the study of a lifetime upon a desperate case, have his arm jogged or his hand impeded, or even his lancet snatched from him, by some agitated relation of the patient.[174]

That impression, Churchill pointed out, "was not entirely in accordance with the facts, and facts, especially in war are stubborn things."[175] The basic fact, as he increasingly realized, was that the broad issues of grand strategy were so complex and far reaching that the coordination of ends, means and ways for the war effort could only be accomplished at the highest policy level. "The General no doubt was an expert on how to move his troops," he observed, "and the Admiral upon how to fight his ships .... But outside this technical aspect they were helpless and misleading arbiters in problems in whose solution the aid of the Statesman, the financier, the manufacturer, the inventor, the psychologist, was equally required."[176]

Reacting to all this after the First World War, Churchill wrote in exasperation that there "was no supreme authority in London as in Berlin .... It was only one man's opinion against another."[177] Despite the note of envy, Churchill would not have had Britain's adversarial grand strategic policy direction evolve in any other way. For the alternative, as he knew full well, was the dominance of that policy by military strategy – a prescription for unmitigated disaster in both the Triple Entente and the Triple Alliance. In France, for instance, the Minister of War had attempted to dissuade General Neville from his disastrous campaign in 1917. "The German retreat, the outbreak of the Russian revolution, the certain and imminent entry of the United

States into the war against Germany," Churchill wrote, " – surely these had introduced some modification into the problem." But Neville would not alter his plan, and Churchill summed up that effort after the war by quoting a French general's analysis of the Neville campaign: "Prisoners yes, guns yes, a narrow band of territory or perhaps 10 or 12 kilometres; but at an outrageous cost, and without strategic results."[178]

In Germany by that time, there was the Hindenburg–Ludendorff dictatorship which not only dominated the war machine, but "increasingly absorbed to themselves the main political authority in Germany."[179] In particular, there was Ludendorff, a man, Churchill pointed out, who "loved his country, but he loved his task more."[180] He was, in short, "a man of the German General Staff. This military priesthood was ... the dominating and drawing power of Germany ... through the fifty-two months of the war."[181] As a consequence, Churchill concluded, grand strategic decision-making in the Second German Empire dealt only with:

> the high confederation of the General Staff; only men who know what they are talking about; only men who talk the same technical language; only men who are thinking of war propositions in war terms to the exclusion of all other considerations! Quite a small gathering, a rigidly limited few, competent experts in blinkers, their eyes riveted on the job, their own job, with supreme knowledge in their sphere and little inkling that other or larger spheres existed.[182]

Reflecting on this development after the war, Churchill pointed to three cardinal mistakes, all stemming from military ascendancy in Germany over grand strategy. "The invasion of Belgium and the unlimited U-boat war were both resorted to on expert dictation as the only means of victory," Churchill wrote, citing two of the mistakes. "They proved the direct cause of ruin .... Nothing could have deprived Germany of victory in the first year of war except the invasion of Belgium; nothing could have denied it to her in its last year except her unlimited submarine campaign."[183] The war plan of the General Staff determined grand strategy in the opening days of the war. "The violation of Luxembourg and Belgium by the German armies marching upon France," Churchill concluded, "will stare through the centuries from the pages of History."[184]

For Churchill, the involvement of Hindenburg and Ludendorff in the U-boat decision was typical of what occurred when the narrow military framework of strategy, over dependent "upon a purely mechanical device," dominated the broader implications that grand strategic policy must consider. "They looked too little," he concluded, "to the tremendous psychological reactions upon the Allies, upon the whole world, above all upon their own people, which must follow the apparition of a fresh, mighty antagonist among the forces against Germany."[185] The German declaration of unlimited U-boat war, Churchill noted, was only one of three major events in 1917, the others being the intervention of the United States and the Russian revolution. The order in which these events were placed, he concluded, proved to be decisive, leading to more of the "terrible ifs" of the First World War.

> If the Russian revolution had occurred in January instead of in March, or if, alternatively, the Germans had waited to declare unlimited U-boat war and consequently no intervention of the United States. If the Allies had been left to face the collapse of Russia without being sustained by the intervention of the United States, it seems certain that France could not have survived the year, and the war would have ended in a Peace by negotiation or, in other words, a German victory. Had Russia lasted two months less, had Germany refrained for two months more, the whole course of events would have been revolutionized. In this sequence we discern the footprints of Destiny. Either Russian endurance or German impatience was required to secure the entry of the United States, and both were forthcoming.[186]

This type of indifference to grand strategy led to the Ludendorff offensive, the third of what Churchill considered the great German mistakes, with its negative implosive effect on the Home Front, which in turn destroyed an excellent possibility for Germany of a negotiated peace. "The passion for revenge ran high, and stern was the temper of the Allies," Churchill wrote; "but retribution, however justified, would not in the face of real peace offers have been in itself a sufficient incentive to lead the great war-wearied nations into another year of frightful waste and slaughter."[187] But Ludendorff, "altogether lacking that supreme combination of the King-Warrior-Statesman," was not interested in such considerations, being "captivated" instead, Churchill concluded, by "the scale and mechanism of the enterprise . . . ."

> These were the calculations on which he had spent his life. This was the quintessence of all he had learnt and wrought. Here were intense, precise, tangible propositions. The larger arguments about making peace with the Allies while time remained,

and of compromising on both sides in the West at the expense of caitiff Russia, seemed quite unimportant. The practical warnings addressed to him in the winter by the ablest German industrialists upon the danger of continuing the war were brushed aside. All this was to him merely a vague, pale, tenuous mist, in the centre of which lay his own gigantic red-hot cannonball. To fire that shot, to pull that spring, and press that button, to let loose those mighty pent-up energies, must have seemed an end in itself.[188]

Grand Strategy and the Search for Decisiveness

The vertical impasse in the continuum of the First World War impelled Churchill to work within the larger framework of grand strategy. That framework, in turn, provided some measure of comfort to the young statesman who increasingly saw victory in the cumulative effect that all instruments of Allied power could bring on the Central Powers. "The old wars were decided by their episodes rather than by their tendencies," he explained to the Commons while resigning over the Dardanelles disaster.

> In this war the tendencies are far more important than the episodes. Without winning any sensational victories we may win this war. We may win it even during a continuance of extremely disappointing and vexatious events. It is not necessary for us in order to win the war to push the German lines back over all the territory they have absorbed, or to pierce them. While the German lines extend far beyond her frontiers, and while her flag flies over conquered capitals and subjugated provinces, while all the appearances of military success attend her arms, Germany may be defeated more fatally in the second or third year of the war than if the Allied Armies had entered Berlin in the first.[189]

Victory, then, could come not as a result of Napoleonic orchestration of battles in campaigns of maneuver to achieve strategic ends, but rather as the result of the collapse of entire nations strained beyond endurance. It was in a sense a return to the time of Marlborough when the grand strategic objectives had centered on the exhaustion of the enemy's economic resources. The attrition accomplished in that eighteenth century *Ermattungsstrategie*, however, had been accomplished primarily by avoiding battles in campaigns that did not focus on the destruction of enemy forces but sought instead, as Churchill noted in his study of Marlborough, "no prize that was not geographic."[190] In the First World War, on the other hand, the attrition was accomplished by provoking battle; and the basic linchpin of that attrition became, as the conflict continued, the production of human casualties.

Seen in this light, even achievements at the lowest end of the continuum of war could feed into the larger grand strategic whole. "Every year ... the close of the campaign has seen the enemy's front, however dented, yet unbroken," Churchill wrote in a 1917 memorandum to the War Cabinet.

> But this in itself is by no means conclusive; for the effects of our efforts upon the enemy have been cumulative, the exhaustion of his manpower and the deterioration of his morale have been progressive .... Therefore it may well be ... that the assertions and hopes that have proved unjustified in four successive campaigns might be vindicated in the end.[191]

After the war, he returned to the subject in his analysis of the Paschendaele campaign, so costly for the British in 1917. "The losses and anxieties inflicted upon the enemy," he concluded, "must not be underrated. Ludendorff's admissions are upon record."[192]

There was, however, a great deal of pragmatic rationale in all of this – an attempt to make the best of what Churchill considered a horrible situation. For him, there would always be better ways for activities on the military continuum to influence the "tendencies" of grand strategy, at least in the initial stages of exhausting the enemy. This had been the lesson before the turn of the century of Alfred Thayer Mahan whose emphasis on the influence of sea power in history in his best seller of that title had apparently influenced Churchill in his early advocacy as a young statesman of the so-called Maritime School of strategy. Certainly by 1905, Churchill was opposed to continental involvement by the British army, referring to that organization in the same year in an argument against the supplemental army budget as "those gorgeous & gilded functionaries with brass hats and ornamental duties who multiply so luxuriously on the plains of Aldershot & Salisbury."[193] In the coming years, however, that attitude changed completely due at least partially to the influence of Sir Julian Corbett, one of the few strategic theorists Churchill ever studied.

Unlike Mahan, who postulated strategic principles and then looked for historical examples, Corbett studied history in order to formulate such principles. From these studies, he came to believe that modern war must be viewed in the light of an all-encompassing grand strategy – "a complex sum of naval, military, political, financial, and moral factors."[194] Within this perspective, Mahan's doctrine concerning command of the sea was too simplistic for Corbett. In his 1900 study of British naval history since Elizabethan times, Corbett wrote that what "the period teaches us is the limitation of maritime power." Jointness was the answer because land and naval power were mutually dependent upon each other in the support of grand strategy; and since this was the case, the histories of the two services should not be studied separately. "The real importance of maritime power," the British strategist concluded, "is its influence on military operations."[195]

Whether Churchill actually read *The Successors of Drake* cannot be determined; but the need for joint operations was a constant theme in most of Corbett's works, including the influential *Some Principles of Maritime Strategy*. In 1911, the year in which he became First Lord of the Admiralty, Churchill referred to both Mahan and Corbett in lamenting the lack of strategic works produced by the British navy. "The standard work on Sea Power was written by an American Admiral," he pointed out. "The best accounts of British sea fighting and naval strategy were compiled by an English civilian."[196] And that same year in his prescient memorandum outlining the course of any future war, the First Lord outlined the vital role army forces would play on the continent in such a conflict.[197]

Equally influential was Corbett's interpretation of Francis Bacon's maxim: "He that commands the sea is at great liberty and may take as much or as little of the war as he will."[198] For Corbett, this meant the selection of a theater of operations where the application of limited power could achieve the unlimited grand strategic objective of total victory by aiding the larger operations of continental forces. This peripheral approach with joint army–navy amphibious forces found a ready audience in Churchill, anxious in the first new year of the First World War to restore a decisive role in grand strategy to the vertical military continuum. "Julian Corbett writes one of the best books in our language upon political and military strategy," Lord Esher noted in his diary during the year of Gallipoli. "All sorts of lessons, some of inestimable value, may be gleaned from it. No one except Winston, who matters just now has ever read it."[199]

To Churchill, the Dardanelles–Gallipoli campaign would always be the great lost opportunity for a true intersection of the vertical continuum with all the instruments of power in the horizontal grand strategic plane. But it required leaving "the great armies scowling at each other in the trenches and the great navies hating each other in strict routine from widely separated harbours," and breaking in "upon this new weak opponent" and beating "him down by land and sea."[200] If all that had been successful, he was certain that Constantinople would have fallen and thus "the only ally the Teutonic Empires had gained would be irretrievably broken." Militarily, that would have opened up Germany's most vulnerable flank in Austria to a deadly Allied thrust with the great port as its base. Moreover, it would have allowed the flow of arms and equipment to Russia. "Even more serious," Churchill added, "would be the political consequences." The Allies "could offer to Italy, Greece and Romania, all three already trembling on the verge of joining them, ample and highly-coveted rewards. They could act upon Bulgaria both by the threat of isolation amid a hostile Balkan Peninsula, and by potent bribes."[201] It was all, he remarked bitterly on the ill-fated campaign after the war, "again very simple; again very difficult."

> The politicians are attracted, the Generals and Admirals mutter "To break away from a first-class war, the sort of war that only comes once in a hundred years, for an amphibious strategic-political manoeuvre of this kind is nothing less than unprofessional." Divided councils, half-hearted measures, grudged resources, makeshift plans, no real control or guidance.[202]

Despite, or perhaps because of, his experience in the Dardanelles campaign, Churchill reaffirmed his belief in joint military operations in support of grand strategy as he immersed himself in the 1930s in his work on Marlborough. During Duke John's era, he pointed out in this regard, "the Tories *obstinately* championed the policy that if we were drawn into a war, we should go as little to the Continent, send as few troops as possible," and "fight as near to the coast as possible."[203] That approach, Churchill believed, could be as injurious to the grand strategic balance of power in Marlborough's time as it had recently been in the Great War. The English political hierarchy, he emphasized in his opening volume of *Marlborough*, was "as sure in 1688 that Belgium must not be conquered by the greatest military power on the Continent as were all parties and classes in the British Empire in August 1914."[204] This did not mean, however, that he had lost his belief in the role that peripheral military operations in other theaters could have in the larger context of grand strategy. In fact, it was this belief that caused Churchill to make the somewhat forced comparison of Marlborough's 1704 campaign into Germany to the ill-starred Dardanelles venture.

> The strategic results of Bavaria joining France and Spain in 1702 resemble curiously in many points those that followed the accession of Turkey to Germany and Austria in 1914. The

enemy in his central position had gained a state which lay across the circuitous communications of the allies. The defection of Bavaria separated the large, loosely knit, ill-equipped, but none the less indispensable mass of the Empire from the rest of the confederacy, in the same way as the hostility of Turkey cut Russia off from the allies in the Great War. The isolation and forcing of the Empire into a separate peace in 1704 seemed as certainly fatal to the allied cause as the same events in Russia would have been in 1915 .... On both occasions grave differences of opinion prevailed which aggravated the difficulties of decisive action. But there was also a great contrast. The allies of 1914 could, if they so resolved, strike down Turkey with ease and swiftness by a naval or amphibious operation. Their forbears in 1704 could only reach Bavaria by a long and hazardous march across Europe and amid its moving armies.[205]

Nevertheless, there was a dominant theme, growing stronger in each succeeding volume of *Marlborough*, which emphasized that grand strategic decisions could only be achieved ultimately in northwest Europe. After Blenheim, Churchill noted approvingly, Marlborough planned "the decisive invasion of France" along the Moselle by establishing "the strongest possible forces in winter quarters for an advance toward Paris in the spring."[206] But the great war leader lacked cooperation from his German allies and was soon compelled to return with his army to the Netherlands rather than pursue the decisive attack through Lorraine against the enemy capital.

A similar opportunity occurred for Marlborough after his victory over the French in the spring of 1706 at Ramillies when all of the Spanish Netherlands submitted to the Grand Alliance. Once again a major force from that coalition was poised in the north in the major theater to strike into the heart of France. This time, however, there was to be a diversionary amphibious operation at Toulon that could act as a supplement in southern France to the major theater by creating "the root of an immense rodent growth in the bowels of France, leading to a fatal collapse either on the northern or the southern front, or perhaps on both. Here was the way to achieve the full purpose of the Allies and finish the war." It was a plan, Churchill concluded, which "lay in that high region of strategy where all the forces are measured and all the impulsions understood."[207] Unfortunately, it was also a plan that was never implemented because of problems within the Grand Alliance.

Ultimately, as Churchill demonstrated in his later volumes on

Marlborough, "a war of the circumference against the centre" must strike toward the middle in order for the military instrument to have a full effect on the other instruments of grand strategic power. The Whigs in England at the time of Marlborough, he noted approvingly in this regard, "sought, with the largest army that could be maintained, to bring the war to an end by a thrust at the heart of France, the supreme military antagonist, arguing that thereafter all the rest would be added to them."[208] Moreover, the encircling ring could only be closed so far before other pressures in the grand strategy of coalition warfare pressed in. "Was the war to drag on in costly, bloody gnawings around the frontiers of France," Churchill asked concerning the conflict over the Spanish succession, "until perhaps it died down in disastrous futility, until the Alliance, reforged on the anvil of Blenheim, broke again to pieces? For a thrust at the heart, the chance, the means, the time, and – might he not feel? – the man had now come."[209]

Once again an opportunity for "a thrust at the heart" was provided by a great allied victory over the French – at Oudenarde in July 1708. After that battle, Churchill wrote, his ancestor devised "his greatest strategic design. The whole combined army should invade France, ignoring the frontier fortresses and abandoning all land communication with Holland." It would require seizing the sea base of Abbeville northeast of Paris by a British force descending upon it from the Isle of Wight. British and Dutch sea power would be used to ferry masses of stores, weapons and equipment to the sea base and thereafter maintain a constant flow of logistical support. The bulk of the Grand Alliance forces would then bypass the French frontier fortresses and march directly to Abbeville. From there, Churchill summed up the operation, "Marlborough ... would march on Paris through unravaged country at the head of a hundred thousand men, and bring the war to a swift and decisive close."[210]

But the War of the Spanish Succession was not destined for such an early end. Once again Marlborough's plan for the invasion of France was thwarted by his allies – "one of the cardinal points of the war," in Churchill's words. "The lesser processes to which he was confined," he wrote of Marlborough after the invasion plan was abandoned, "although yielding immediate necessary supplies, did not procure the strategic result." And that result, he emphasized, lay squarely on the horizontal axis of grand strategy. "Great battles would have been fought in the heart of France," Churchill concluded, "and victory would have provided in 1708 that triumphant peace which after so much further bloodshed the Grand Alliance was still to seek in vain."[211]

Conflict Termination

Churchill had no more illusions about the effect of the Great War on the victors than he had about its effect on the vanquished. "The shadow of victory is disillusion," he wrote. "The reaction from extreme effort is prostration. The aftermath even of successful war is long and bitter."[212] That disillusion was evident throughout the inter-war period, in which he often returned in his writings and speeches to the attritive slaughter of the trenches where combat had been "reduced to a business like the stockyards of Chicago."[213] And in 1922, his notes for an election speech began:

> What a disappointment the Twentieth Century has been
> How terrible & how melancholy
> is long series of disastrous events
> which have darkened its first 20 years.[214]

It was natural in those circumstances that Churchill's thoughts would return to the "palmy days" of his youth. "Winston waxed very eloquent on the subject of the old world & the new," one friend noted in her diary in January 1920, "taking up arms in defence of the former."[215] But there was no immediate comfort or direction for him from the past in terms of the unprecedented cataclysm through which Britain had passed. "The Muse of History to whom we all so confidently appeal has become a Sphinx," he wrote. "A sad half mocking smile flickers on her stone war-scarred lineaments."[216] In a long and bitter discourse touched with longing, Churchill lashed out at the era of his youth. "How those Victorians busied themselves and contended about minor things!" he wrote. "... They never had to face as we have done, and still do, the possibility of national ruin. Their main foundations were never shaken."[217] That was not a past, he believed, that could help in the cold grey dawn of grand strategy in modern war. Commenting on the effect of the new era on John Morely, the statesman whom he considered a quintessential Victorian, Churchill found that Morely "was dwelling in a world which was far removed from the awful reality."

> At such juncture his historic sense was no guide; it was indeed an impediment. It was vain to look back to the Crimean War, to the wars of 1866 and 1870, and to suppose that any of the political reactions which had attended their declaration or course would repeat themselves now. We were in the presence of events without their equal or forerunner in the whole experience of mankind.[218]

Despite these developments, Churchill's fundamental belief in historical continuity was only momentarily suspended by the First World War. Already in 1929 he was referring to history "as a guide to present difficulties," to Katherine Asquith. "How strange it is," he wrote, "that the past is so little understood and so quickly forgotten. We live in the most thoughtless of ages. Every day headlines and short views."[219] And in the following decade as he moved back and forth in the late seventeenth- and early eighteenth-century life of his ancestor, Churchill began to appreciate the totality of the grand strategic stakes in the more limited warfare of Marlborough's time. The battle of Blenheim, for instance, had changed the political axis of the world in that period. If that battle had been lost, he concluded, the "collapse of the Grand Alliance and the hegemony of France in Europe must have brought with them so profound a disintegration of English political society that for perhaps a century at least vassalage under a French-inspired king might well have been our fate."[220] Even attritive warfare had its characteristics of continuity. In fact, as Churchill concluded, the "spectacle of one of the battlefields of Marlborough, Frederick, or Napoleon was . . . incomparably more gruesome than any equal sector of the recent fronts in France or Flanders."[221] In that light, the First World War became for Churchill part of the grand historical pattern of Britain.

> For the fourth time in four successive centuries she had headed and sustained the resistance of Europe to a military tyranny . . . . Spain, the French Monarchy, the French Empire and the German Empire . . . . During 400 years England had withstood them all by war and policy, and all had been defeated and driven out. To that list of mighty sovereigns and supreme military Lords which already included Philip II, Louis XIV and Napoleon, there could now be added the name of William II of Germany.[222]

His renewed faith in historical continuity caused Churchill in the last decade of the inter-war years to examine his ancestor's problems concerning conflict termination within the broad grand strategic object of winning the peace. The fundamental tenet in that process, he believed, was to negotiate from strength. "We must admire the dual process . . . of earnestly seeking peace while . . . preparing for war on an ever greater scale," he wrote of Marlborough's efforts in 1706. "Nearly always Governments which seek peace flag in their war efforts, and Governments which make the most vigorous war preparations take little interest in peace." The solution, he realized, was to bring both efforts together for it was "only by the double and, as it might seem, contradictory exertion that a good result can usually be procured."[223]

There were many instances in which such an effort almost brought permanent peace to the War of the Spanish Succession. "Marlborough's exertions for five months to have large forces at his disposal during the negotiations had succeeded beyond his hopes," Churchill noted of the 1709 preparations. "'All the facts,' wrote Eugene to the Emperor, 'go to show that . . . we can . . . if we wish obtain everything we ask for. We have only to hold together and preserve good understanding among ourselves.'"[224] That, of course, was easier said than done in coalition warfare. Already in 1704, despite the fact that Marlborough "had rescued the Empire from ruin and the Grand Alliance from collapse, the fruits of victory were largely cast away by the jealousies of the allies."[225] Louis XIV, of course, was aware of that divisiveness, and in the spring of 1707 made an all-out effort to arrange peace. Churchill was in accord with Duke John, however, in believing the French monarch's type of conflict termination could be fatal in coalition warfare. "Between equals and similars," he concluded,

> there always is much to be said for peace even though a drawn war; but to a wide numerous disconnected coalition, faced by a homogeneous military nation . . ., a drawn war embodied in a treaty spelt permanent defeat. One man still carrying with him the British island in its most remarkable efflorescence of genius and energy, stood against this kind of accommodation. Marlborough, harassed and hampered upon every side, remained unexhausted and all-compelling.[226]

But without all the strands of grand strategy at his command, the military instrument of power was ultimately useless to Marlborough in his efforts at conflict termination. There was, for instance, the problem of war aims. In 1703, in order to lure Portugal into the Grand Alliance, the Allies had moved away from the original goal of partitioning the Spanish Empire between the Bourbons and the Habsburgs. After Blenheim, the original war aims would probably have been acceptable to Louis XIV whose "object henceforward was only to find a convenient and dignified exit from the arena in which he had so long stalked triumphant."[227]

Negotiations continued to flounder in the future, however, even under the bludgeoning of such battles as Ramillies and Malpaquet, on the Allied insistence that the entire Spanish Empire be peacefully transferred to the Habsburg claimant. Thus it was, Churchill noted, that the costly Allied victory at Malpaquet in the autumn of 1709 "cast a lurid reproach upon the failure to make peace in the spring." Moreover, whatever disputes there might be about the consequences of that battle, he concluded, "one fact was certain: peace was no nearer than in June."[228] In the end, the Tory government negotiated a separate peace with France behind Marlborough's back, followed by the end of 1711 with his humiliating dismissal by Queen Anne. Looking back over the centuries at the entire sorry episode, Churchill wrote: "Here in foretaste we may read the bitter story of how in the eighteenth century England won the war and lost the peace."[229]

Churchill was referring at the time, of course, to what he considered the disastrous Versailles Treaty, the results of which were brought home daily to him in the "Locust Years" of the 1930s. The fundamental defect of that treaty for the British statesman was its violation of his lifelong belief in victorious magnanimity. As one who had been bullied and misused as a child, and as a public school product, Churchill had always had an instinctive sympathy for the underdog. Consequently, he never painted the enemies in his Imperial conflicts in stark black, displaying instead a fairness and generosity unusual even by Victorian chivalric standards. " 'Never despise your enemy' is an old lesson," he wrote of the 1897 Indian campaign, "but it has to be learnt afresh, year after year, by every nation that is warlike and brave."[230]

That outlook was reinforced by his experience as a captive in the Boer War, in which his captors treated him with fairness and in many cases with generosity, "a great surprise."[231] In later years, Churchill was drawn to General Smuts for many reasons, not the least of which was his perception of the South African as a gallant foe who had become a loyal and devoted subject of the King. The British statesman had always taken great pride in helping to establish a just and generous peace after the Boer War, and Smuts was living proof for him that magnanimity and good sense in victory would achieve best the ultimate goal of grand strategy.[232]

But magnanimity in the realm of grand strategic conflict termination must also be accompanied by a firm dose of realism. For Churchill, escalation of Allied war aims had been as disastrous for the Entente as it had been for Marlborough's Grand Alliance. In Marlborough's time, that escalation had taken the traditional form of territorial aggrandizement. In the First World War, the process had acquired ideological and moral overtones for the Allies – particularly damaging, Churchill pointed out, in the form of Woodrow Wilson and his 14 Points, of which he later approvingly quoted Clemenceau's judgment: "Même le bon Dieu n'avait que dix."[233] This type of escalation had brought vast populations into the most important aspect of grand strategy, but not, Churchill emphasized, with propitious results. "The peoples, transported by their sufferings and by the mass teachings with which they had been inspired," he wrote, "stood around in scores of millions to demand that retribution should be exacted to the full. Woe betide the leaders now perched on their dizzy pinnacles of triumph if they cast away at the conference table what the soldiers had won on a hundred blood-soaked battlefields."[234]

Epilogue – The Art of the Possible

Figure 4 (see page 90) represents a rational confluence of the horizontal plane of grand strategy with the vertical continuum of war. But that confluence could be skewed, as Churchill realized after the First World War, a conflict in which "events passed very largely outside the scope of conscious choice."[235] War was a period for him in which "action is circumscribed within practical limits. There are only a certain number of alternatives open ... in a world of reality where theories are constantly being corrected and curved by experiment. Resultant facts," he concluded, "accumulate and govern to a very large extent the next decision."[236] In such an environment, warfare in both dimensions of the paradigm was the art of the possible. "We cannot make war as we ought," Churchill quoted Lord Kitchener; "we can only make it as we can."[237]

This seemed hardly evident when Churchill assumed power in May 1940 and immediately proclaimed that the overriding grand strategic goal was victory at any price. Members of his cabinet early in the war raised the subject of negotiations with the Germans, and the Foreign Office held out against repudiating the Munich settlement until 1942. But Churchill was resolute. "I have only one aim in life," he remarked at the time, "the defeat of Hitler, and this makes things very simple for me."[238] This simplicity not only eased the Prime Minister's burden, but it suited his temperament as well. For although more fertile in imagination and ideas than most leaders, he remained fundamentally a straightforward person, always ready to strike out for goals that he could see.

Nevertheless, strategy in any form remains the calculated relationship of means to ends; and given the state of Britain's resources, Churchill's war aim seems in retrospect much less rational than that of Hitler's at the time. But Churchill, unlike the Nazi leader, was operating in both dimensions of strategy. On the horizontal plane, the British leader centralized his power by the assumption, with the King's blessing but without parliamentary mandate, of the post of Minister of Defence, a position unknown during the First World War. Churchill was careful not to define the powers of the new post too precisely. But, in this combined capacity, as Minister of Defence and Prime Minister, he was able to supervise all the power instruments in Britain's grand strategy, both as the ultimate political authority and as the specific warlord directing defense policy. Thus, in terms of grand strategy, Churchill became in 1940, as he had written of Marlborough, "the central link on which everything was fastened . . . . It is not until we reach Napoleon, the Emperor-statesman-captain, that we see the threefold combination of functions – military, political and diplomatic – which was Marlborough's sphere, applied again upon a Continental scale."[239]

Finally at the helm, where "politics and strategy are one," Churchill immediately began to orchestrate the instruments of grand strategy. On the economic and psychological side, there was the country's mobilization and the British leader's embodiment of the British people in the *annus mirabilis* of 1940. "Come then: let us to the task, to the battle, to the toil – each to our part, each to our station," he exhorted that year. "Fill the armies, rule the air, pour out the munitions, strangle the U-boats, sweep the mines, plough the land, build the ships, guard the streets, succor the wounded, uplift the downcast, and honor the dead."[240] And politically, perhaps Churchill's supreme achievement was the formation of a triumphant Grand Alliance, in the manner of his great ancestor. For the United States, in this regard, there was his assiduous courting of President Roosevelt from a position of weakness throughout the 19 months prior to Pearl Harbor – a process one historian assessed as "feline, adroit, and far-sighted."[241] And for the Soviet Union, despite his consistent and virulent antipathy to Bolshevism, there was Churchill's immediate support of that country under the onslaught of Barbarossa and all that portended positively on the horizontal scale – particularly when compared to Hitler's gratuitous declaration of war upon the United States after Pearl Harbor.

Against this grand strategic background Churchill applied his meager military resources. Nowhere were those means more vital than in the Battle of Britain. In the midst of that battle, the Russian ambassador called on the British leader and inquired as to his overall strategy. "My general strategy at present," Churchill replied drawing on his cigar, "is to last out the next three months."[242] But the Battle of

Britain, coming in the wake of Dunkirk and the Battle of France, was also a boost not only to the national will of the Home Front, but to Britain's relationship with its potential ally across the Atlantic, engaged at the time in a presidential campaign. For as Churchill well knew, the United States would not side with a losing cause; and the image of strength and defiance orchestrated by the Prime Minister in the summer and fall of 1940 was intended for the American as well as the British people.

It was in this context in the early years of the war that Churchill pursued the military art of the possible. The Battle of the Atlantic, a struggle against German submarines to keep open Britain's economic lifeline, was only won at great cost and at an excruciatingly ponderous pace. The strategic bombing of Germany was as much concerned with bolstering British morale as it was with the psychological and economic impact it could have across the Rhine. Perhaps more. "You need not argue the value of bombing Germany," Churchill wrote the Air Staff in this regard early in 1942, "because I have my own opinion about that, namely, that it is not decisive, but better than doing nothing."[243] That urge to do something, to take action, was also fundamental to Churchill's character. "It was his paramount duty to make the Second Army fight *somewhere*," he had written of the Austrian commander in 1914 who was not able to reach Galicia for the central battle.[244] And so it was with the Eighth Army in North Africa, the only area in the European theater of war until the invasion of Sicily where British landpower could grapple with the Axis.

Such limited application of military power still had to be handled very carefully within the framework of grand strategy in the early years of the war. For as Churchill knew from his experience in the First World War, military options at the operational and tactical levels on the vertical continuum of war, even in a peripheral theater of operations, could have major impact along the horizontal plane of that framework, particularly in the realm of national will and the acquisition of allies as well as in his own ability to remain in power. It was this perspective that at least partially explains many of the British leader's more controversial decisions early in the war, ranging from the destruction of the French fleet at Oran to the priority given to the Middle East in Britain's strategy over the Far East. It was this perspective that also, in part, drew the Prime Minister down to the operational and sometimes even tactical level in his interface with his field commanders throughout much of the desert campaigns.

Finally, the lack of resources on the vertical continuum, and the concomitant need for that continuum to contribute fully in the horizontal dimension of strategy, reinforced the British leader's natural enthusiasm and interest in force multipliers. As a consequence, he was a major force in the use and development of means to improve intelligence and counterintelligence, the most graphic examples being Ultra and the Double Cross system. In a similar manner, Churchill was instrumental in encouraging and centralizing deception activities at all levels of war, the most important outcome of which was the strategic "bodyguard of lies" in the form of the Fortitude operation that protected the cross-Channel invasion. Added to this was the Prime Minister's abiding interest in technological surprise as demonstrated at all points up and down the vertical continuum during the Second World War, from the strategic "Mulberry" harbors, to the "Window" clutter at the operational level, down to the tactical "sticky bomb."

By December 1941 with the help of the Japanese surprise attack at Pearl Harbor, Churchill had consolidated the other instruments of power sufficiently to be able to ward off a series of disasters on the vertical continuum that began after the Arcadia Conference that month and continued for six more months until Rommel was halted at the first battle of Alamein. By the end of February 1942, Singapore had been captured, General Auchinleck had been pushed back to Gazala, and the *Scharnhorst* and *Gneisenau* had traversed the Channel with humiliating ease. On 25 June 1942, after Auchinleck's defeat, Churchill faced a motion in the Commons that "this House, while paying tribute to the heroism and endurance of the Armed Forces of the Crown ... has no confidence in the central direction of the war."[245] On 1 July, the debate began on the Vote of Censure, the thrust of which was to eliminate Churchill's position of Minister of Defence. That same day, German forces reached El Alamein, 130 miles inside Egypt, 80 miles from Cairo, and in the Crimea captured Sebastopol.

Churchill would not turn away from the institution he revered. At the same time, however, harkening back to the First World War and his experiences thus far in the current conflict, he would not approach leadership in terms of grand strategy without possession of the means necessary to work effectively on that horizontal plane. "I am your servant," he told the Commons on 2 July, "and you have the right to dismiss me when you please. What you have no right to do is to ask me to bear responsibilities without the power of effective action."[246] At the conclusion of the debate, only 25 members supported the vote, a consoling figure to the historically-minded Prime Minister in more than one way, since it duplicated the number of votes against the way the younger Pitt was conducting the war in 1799. Equally important was a message from across the Atlantic. "Good for you," Roosevelt telegraphed. "Action of House of Commons today delighted me."[247]

The message was a reminder of Churchill's great triumph in forging the Grand Alliance. For the first 19 months of his tenure, he operated without that full coalition. And for the next 22 months of the war, he played a dominant role in it. That dominance notwithstanding, however, the grand strategic patterns of the alliance during that period were molded by circumstances, necessity, trial and error and, above all, compromises among its leaders. Those compromises were never more in evidence than with the question of a cross-Channel invasion. The United States wanted that operation to begin as early as 1942. Churchill did not question the ultimate and necessary decisiveness of a thrust into the heart of northwest Europe. It was, after all, what Marlborough had advocated so many centuries before at Abbeville; and no one worked harder than the British leader in the early days of the coalition to make the operation an ultimate reality.[248]

But grand strategy remained for Churchill the art of the possible, and he became convinced that the strategic resources were lacking for the American timetable. In the summer of 1942, he persuaded President Roosevelt, also very much aware of the vertical–horizontal interface, that Allied operations that year should be in French North Africa, the only proposal "which, either at the time or in retrospect, seemed to make strategic sense."[249] That those operations would preclude a 1943 cross-Channel invasion became apparent as they continued well into that year. That this was a fortunate by-product of the 1942 Roosevelt–Churchill agreement, which had bypassed the American Chiefs of Staff, has also become apparent over the years. "A cross-channel offensive in 1942," Ronald Lewin has noted in this regard, "would have been a guaranteed, and in 1943 an almost certain, failure."[250]

But if the pragmatic caution and realism of Churchill was an invaluable and necessary element in the early efforts at combined strategic planning, the stubborn perseverance of the American military leaders in the closing years of the war ensured the ultimate implementation of the Allied grand strategy. For by mid-to-late 1943, Churchill's enthusiasm for an attack on northwest Europe had disappeared, and he increasingly regarded the Mediterranean not as a subsidiary theater, but a primary one where successful operations were themselves the ultimate justification for that primacy. It was in that theater that he saw the remaining possibility for enhancing British military prestige as well as for indulging in his continuing passion to direct the war. Certainly, there was also the lure of the Eastern

Mediterranean from the earlier war. But there were also the memories from that war of northwest Europe, the source of the attritive blood baths, which he admitted in his memoirs "were not to be blotted out by time or reflection."[251] In the end, like the First World War, the decisive campaign had to be fought in France, and the dominant and consistent American pressure made that possible.[252]

That dominance had been progressively more evident at Casablanca, Washington and Quebec. By 1943, both Roosevelt and Stalin were operating from a position of increasing power and influence, while that of Churchill's was in decline. That vulnerability was suddenly and humiliatingly exposed in November of that year at the Teheran Conference, in which the American president actively sided both publicly and privately with the Soviet leader against Churchill on several key strategic issues. "I realized at Teheran for the first time," Churchill later recalled, "what a small nation we are. There I sat with the great Russian bear on one side of me, with paws outstretched, and on the other side the great American buffalo, and between the two sat the poor little English donkey who was the only one, the only one of the three, who knew the right way home."[253]

The "right way home" for Churchill was the path that would lead to the ultimate goal of grand strategy, the winning of the peace. All his experience over the years had demonstrated to him that in support of that ultimate goal there must be a flow from the vertical military continuum into the dominant horizontal plane. But to the British leader that flow appeared to be reversed in the West once the Allies under American direction were on the Continent. There, in the absence of clear and consistent political guidance from Washington, General Eisenhower increasingly made decisions primarily on the basis of military considerations, despite, as Churchill realized, the growing political character of the war. The final result in the West in the last year of the European war was "a conventional war of concentration, a technical military game."[254] "In Washington especially," Churchill later lashed out in his memoirs, "larger and wider views should have prevailed."[255]

Whether the outcome would have been different even had Britain retained its dominant position in the Grand Alliance is problematical. Part of the reason was that the objective of unconditional surrender had been woven into the fabric of Allied grand strategy. Whatever positive warfighting aspects that objective may have possessed, it did not prove to be a viable peace goal, masking instead the divergent Allied interests while providing no basis for bringing those interests closer together. As a result, the only common denominator for the

Allied coalition by 1945 was the defeat of Germany. But as those partners moved closer together and Germany's defeat seemed imminent, the Grand Alliance began to unravel. In the end, Maurice Matloff has argued in this regard, "the war outran the strategists and the statesmen."[256] For Churchill, who combined those occupations, it was a situation he had addressed in the preceding decade when describing the "great decline in Marlborough's personal power . . . ."

> He had since 1700 woven together a Grand Alliance and carried it forward by management, tact, and great victorious battles to mastery. At every stage he had had to hold in check divergent and competing aims. The fear of being defeated and destroyed had joined the Allies together. Now his own victories had destroyed that fear. Thus at the moment when his work should have given him the greatest authority, and when that authority might have been most beneficently exercised, he found himself alone.[257]

None of the events in the last year of the Second World War, however, could diminish Churchill's colossal achievement in the realm of grand strategy. The British leader's focus on the "tendencies" in that horizontal dimension, as he had learned so many years before, proved ultimately to be successful, particularly because of Hitler's fundamental disregard for that same overarching aspect of modern war. The Nazi leader's failure of statecraft in the horizontal plane made defeat in the end a virtual certainty even as the tactical, operational and even theater strategic German victories piled up in the early stages of the war. Those victories on the vertical continuum could only help Germany win the war if they decisively affected Allied strength in the horizontal dimension. But that only occurred early in the war when the Nazis overran countries like Poland or France, forcing them out of the conflict while adding their resources and economics to the greater Reich. This early advantage in the vertical continuum, based on war preparations and superior military competence, offset initially the growing Allied efforts set in train by Churchill in the horizontal plane of grand strategy.

But the British leader's success in that plane, most notably in the formation of the Grand Alliance, began in 1942 to affect the amount of resources that could be applied to the vertical continuum in one theater after another. Added to this was the growing military quality of Allied forces after that year which deprived the Axis powers of their former advantage at the tactical and operational levels. Moreover, as the victorious forces moved closer to the Axis homelands, Allied military power on the vertical continuum increasingly was used to diminish enemy strength in the horizontal dimension by the strategic bombing of industry, marshalling yards and capital infrastructure. In the end, the final victory in the Second World War was due to this mutually reinforcing confluence of the vertical and horizontal dimensions of strategy, as the Allies increasingly outmatched the Axis forces at the tactical and operational levels of war in all theaters of operations, the very multiplicity of which attested to the Axis failure in the realm of grand strategy.[258]

NOTES

This essay was first published in 1990 by the Strategic Studies Institute, US Army War College.

1. Charles Eade, ed., *Churchill By His Contemporaries* (New York: Simon and Schuster, 1954), p. 173.
2. Carl von Clausewitz, *On War*, ed. and trans. by Michael Howard and Peter Paret (Princeton: Princeton University Press, 1976), p. 182.
3. James J. Schneider, "The Theory of Operational Art," *Theoretical Paper No. 3* (Ft. Leavenworth, KS: School of Advanced Military Studies, 3 March 1988), p. 9.
4. David G. Chandler, *The Campaigns of Napoleon* (New York: Macmillan, 1966), p. 161.
5. *On War*, p. 177.
6. David M. Glantz, "The Nature of Soviet Operational Art," *Parameters*, Spring 1985, p. 65.
7. Department of the Army, *Field Manual (FM) 100-5, Operations* (Washington, DC: 5 May 1986), p. 9.
8. *JCS Pub 1-02, Department of Defense Dictionary of Military and Associated Terms* (Washington, DC: USGPO, 1 December 1989), p. 264.
9. Winston S. Churchill, *A Roving Commission* (New York: Charles Scribner's Sons, 1951), p. 112. Churchill retained this early interest in Macaulay despite what he considered to be that historian's unfair treatment of John Churchill, the first Duke of Marlborough. "I am deep in Macaulay," he wrote to his mother on 21 January 1897. "... It is easier reading than Gibbon and in quite a different style .... Both are fascinating and show what a fine language English is." Randolph S. Churchill, *Winston S. Churchill, Companion Volume I, Part II, 1896–1900* (hereafter C.V. I, II) (Boston: Houghton Mifflin, 1967), p. 726.
10. Janet and Peter Phillips, *Victorians at Home and Away* (London: Croom Helm, 1978), p. 166.
11. Randolph S. Churchill, *Winston S. Churchill. Companion Volume I, Part I. 1874–1896* (hereafter C.V. I, I) (Boston: Houghton Mifflin, 1967), p. 181.
12. Ibid., p. 522.
13. *Roving Commission*, p. 44.
14. Ibid., p. 43. One title, Sir Edward Bruce Hamley's *The Operations of War Explained and Illustrated*. C.V. I, II, p. 480.
15. *Roving Commission*, p. 44.
16. C.V.I, II, p. 548. In an interview with General Eisenhower after Churchill's death, Alistair Cooke referred to the fact that Churchill was at the bottom of his class at Harrow, but excelled at tactics and fortifications at Sandhurst. Eisenhower replied that he had been a mediocre student at West Point, but graduated at the top of his Leavenworth class at the Command and General Staff School. "So you can't tell when

people begin to mature," he concluded. James Nelson, ed., *General Eisenhower on the Military Churchill* (New York: W. W. Norton, 1970), p. 22.

17. Ibid., p. 26.
18. William Manchester, *The Last Lion: Winston Spencer Churchill: Visions of Glory, 1874–1932* (Boston: Little, Brown, 1983), p. 317.
19. Winston S. Churchill, *London to Ladysmith via Pretoria* (London and Bombay: Longmans, Green, 1900), pp. 416 and 429. C.V. I, I, p. 1147. *Roving Commission*, p. 309.
20. *London to Ladysmith*, p. 420.
21. Winston S. Churchill, *Thoughts and Adventures* (London: Odhams Press, 1949), pp. 51–2. In the 1909 maneuvers, which Churchill also attended, he pointed out that the absurdities of the Silesian maneuvers were not repeated, ibid., p. 55. Nevertheless, Churchill wrote to Lord Elgin from Silesia that he was impressed with the army whose "members, quality, discipline and organization are four good roads to victory." And upon returning to England, he told his Aunt Leonie: "I am very thankful there is a sea between that army and England." Randolph S. Churchill, *Winston S. Churchill. Volume II. 1901–1914. Young Statesman* (Boston: Houghton Mifflin, 1967), p. 191.
22. Lord Moran, *Churchill. Taken from the Diaries of Lord Moran: The Struggle for Survival, 1940–1965* (Boston: Houghton Mifflin, 1966), p. 254. On 20 September 1940, after a particularly grueling day that included a visit to one of the worst bombed areas of London, Churchill sat up late at night, reminiscing about the Boer conflict, "the last enjoyable war." Martin Gilbert, *Winston Churchill. Volume VI. Finest Hour. 1939–1941* (Boston: Houghton Mifflin, 1983), p. 800.
23. Winston S. Churchill, *Ian Hamilton's March* (London: Longmans, Green, 1900), p. 386.
24. Ibid., p. 244.
25. Frederick Woods, *Young Winston's Wars. The Original Dispatches of Winston S. Churchill, War Correspondent 1897–1900* (New York: Viking Press, 1972), p. 192.
26. Ibid., p. 194.
27. Winston S. Churchill, *The World Crisis 1911–1914* (New York: Charles Scribner's Sons, 1928), p. 247.
28. Ibid., pp. 211 and 225. B. H. Liddell Hart, "The Military Strategist," *Churchill Revised. A Critical Assessment* (New York: Dial Press, 1969), pp. 182–3.
29. *World Crisis 1911–1914*, p. 248. Churchill's 13 August 1911 memorandum entitled "Military Aspects of the Continental Problem" anticipated the use of "four or six British divisions in these great initial operations," particularly since France "will not be strong enough to invade Germany. Her only chance is to conquer Germany in France." Ibid., p. 59.
30. Robert M. Epstein, "The Practice and Evolution of Operational Art," *Program Course No. 4* (Ft. Leavenworth, KS: School of Advanced Military Studies, n.d.), pp. 4–7/8–2.
31. Winston S. Churchill, *The World Crisis 1916–1918*, Parts I and II (London: Thornton Butterworth, 1927), pp. 18–19.
32. Winston S. Churchill, *Great Contemporaries* (London: Thornton Butterworth, 1937), p. 189.
33. Winston S. Churchill, *The Unknown War* (New York: Charles Scribner's Sons, 1931), p. 274.
34. Winston S. Churchill, *Marlborough. His Life and Times. Volume V. 1705–1708* (New York: Charles Scribner's Sons, 1937), p. 404.
35. Emphasis added. Winston S. Churchill, *Marlborough. His Life and Times. Volume IV. 1704–1705* (New York: Charles Scribner's Sons, 1935), pp. 31–2.
36. *Unknown War*, p. 176.
37. Ibid.
38. *FM 100–5*, pp. 2–11.
39. Original emphasis. Ibid.
40. *Unknown War*, p. 193.

41. Maurice Ashley, *Churchill as Historian* (New York: Charles Scribner's Sons, 1968), p. 131. See also Churchill's analysis of the 23 August–12 September 1914 Lemberg campaign in the East in which three Austrian armies were pitted against four Russian armies. "This prodigious event," he wrote, "comprised seven separate hard-fought battles between individual armies each lasting several days, reacting upon each other." *Unknown War*, p. 144.

42. *World Crisis, 1916–1918*, Part I, p. 40.

43. Ibid. Gordon A. Craig, "Delbrueck: The Military Historian," *Makers of Modern Strategy From Machiavelli to the Nuclear Age*, ed. Peter Paret (Princeton, NJ: Princeton University Press, 1986), p. 351. Ludendorff stated that "tactics were to be valued more than pure strategy." Ibid. "Ludendorff started from the assumption that tactics were more important than strategy; it was a question above all of launching an offensive at a point where a tactical breakthrough was possible, not where a strategic one was desirable." Martin van Creveld, *Command in War* (Cambridge, MA: Harvard University Press, 1985), p. 172. Ludendorff concluded that "tactics have to be considered before purely strategical objects." B. H. Liddell Hart, *Strategy*, 2d ed. (New York: Praeger Publishers, 1967), p. 205.

44. *World Crisis 1916–1918*, Part II, p. 458. See also Churchill's analysis at ibid., p. 448:

> He had thus resigned all the decisive strategic objects for which the German armies had been fighting since March 21. He had first abandoned the great roll-up of the British line from Arras northwards and the general destruction of the British armies, in favour of the more definite but still vital aim of taking Amiens and dividing the British from the French armies. Arrested in this, he had struck in the north to draw British reserves from the Amiens battlefield. But the Battle of the Lys, begun as a diversion, had offered the lesser yet still enormous prize of the northern Channel ports. Now he must abandon that; and his strategic ambition, already thrice contracted, must henceforward sink to an altogether lower plane. The fourth German offensive battle of 1918 was to a large extent a mere bid for a local victory, and apart from its usefulness in diverting Allied troops from the fateful fronts, offered no direct deadly strategic possibilities.

Although often critical of Haig, Churchill approved of the General's "backs to the wall" message which, he noted, the British commander had written himself without any aid. *Great Contemporaries*, p. 228.

45. Ibid., p. 420. Craig, p. 351, and Creveld, pp. 180 and 183. For the reasons behind the tactical successes, see Timothy T. Lupfer, "The Dynamics of Doctrine: the Changes in German Tactical Doctrine During the First World War," *Leavenworth Papers*, No. 4 (Fort Leavenworth, KS: U.S. Army Command and General Staff College, July 1981), *passim*, Chapter 2.

46. *Unknown War*, p. 76.

47. Winston S. Churchill, *The World Crisis 1915* (New York: Charles Scribner's Sons, 1929), pp. 421–2. G. D. Sheffield, "Blitzkrieg and Attrition: Land Operations in Europe 1914–1945," *Warfare in the Twentieth Century. Theory and Practice*, ed. Colin McInnes and G. D. Sheffield (London: Unwin Hyman, 1988), p. 61.

48. Ibid., p. 422.

49. Churchill took the quote from Hindenburg's post-war memoirs, *Out of My Life*. *Unknown War*, p. 299.

50. *World Crisis 1916–1918*, Part II, p. 509.

51. All quotes from Churchill's 1 June 1915 "Note on the General Situation," *World Crisis 1915*, pp. 404–5.

52. *World Crisis 1916–1918*, Part I, pp. 40–2.

53. Winston S. Churchill, *Marlborough. His Life and Times, Volume III. 1702–1704* (New York: Charles Scribner's Sons, 1935), p. 99.

54. *Unknown War*, p. 131.

55. Ibid., p. 190. "It is not to their neglect to enter Paris or seize Calais that their fatal defeat was due, but rather to the withdrawal of two German army corps to repel the

Russian invasion of East Prussia." *World Crisis 1911–1914*, p. 357. And later, referring to the initial Russian offensive, Churchill noted: "We know that the effects of this offensive upon the nerves of the German Headquarters Staff had led to the withdrawal of two army corps from the German right in Belgium during the crisis before the Marne. It may very well be argued that this event was decisive upon the fate of the battle." *World Crisis 1915*, p. 9. The Brusilov offensive in 1916, of course, played a similar fortunate role for the Entente at Verdun – achieving, as Churchill pointed out, "results far beyond anything dreamt of." *Unknown War*, p. 362.

56. *Great Contemporaries*, p. 114. *Unknown War*, p. 239.
57. *Unknown War*, p. 301.
58. *Marlborough*, III, p. 258.
59. *World Crisis 1915*, p. 14.
60. *World Crisis 1915*, p. 36, and *1916–1918*, Part II, p. 461.
61. Winston S. Churchill, *Marlborough. His Life and Times. Volume II. 1688–1702* (New York: Charles Scribner's Sons, 1933), p. 208.
62. *World Crisis 1916–1918*, Part II, p. 336. When the main theater became the decisive theater again in 1918, resources were pulled from the indecisive theaters, no matter the extent of operational successes. "In the actual event . . . Ludendorff's offensive of 1918 dissipated in a day all of Allenby's careful plans for the spring campaign. Not less than sixty battalions with many batteries were incontinently snatched from Palestine to plug the shot hole of the Twenty-first of March." Ibid.
63. Ibid., Part I, p. 74.
64. *World Crisis 1915*, p. 446.
65. Ibid., p. 413.
66. Ibid., p. 544.
67. *Unknown War*, p. 287.
68. *World Crisis 1916–1918*, Part I, p. 195.
69. Ibid., p. 179.
70. *Great Contemporaries*, p. 227. "He was rarely capable of rising to great heights," he wrote of Haig; "he was always incapable of falling below his standards." Ibid., p. 230.
71. *Thoughts and Adventures*, p. 108.
72. *World Crisis 1916–1918*, Part I, p. 464.
73. All quotes from Ibid., Part II, p. 401.
74. Ibid., Part I, p. 458.
75. Winston S. Churchill, *Marlborough. His Life and Times. Volume VI. 1708–1722* (New York: Charles Scribner's Sons, 1938), p. 172.
76. Ibid., p. 175. The Allies lost 24,000; the French about 15,000 casualties. Ibid., p. 172. All the years of war had not been able to create in Marlborough, as Churchill pointed out, "that detachment from human suffering which has often frozen the hearts of great captains." Ibid., p. 173.
77. *World Crisis 1916–1918*, Part I, p. 195.
78. *World Crisis 1915*, p. 22.
79. Ibid., p. 423.
80. *World Crisis 1916–1918*, Part I, p. 59.
81. Ibid., p. 57.
82. Ibid., p. 59.
83. Ibid., p. 60.
84. Ibid.
85. Ibid., p. 85.
86. Ibid., Part II, p. 423.
87. *Marlborough*, III, p. 337.
88. Ibid., p. 107.
89. *World Crisis 1915*, p. 5.
90. Patrick Beesly, *Room 40* (London: Hamish Hamilton, 1982), p. 16. Beesly pointed out that "'Excessively Secret' would have been a more accurate description." Ibid.
91. *World Crisis 1915*, p. 51.

92. *World Crisis 1916–1918*, Part I, p. 118. But see Beesly, p. 69, who also points out:

> The problem for the British was that Room 40's information was now so good, that signs of any German movement could be detected in advance and could not be ignored if the Grand Fleet was to be sailed in time to intercept any offensive operation. That the movements, throughout 1915, always turned out in the end to be purely local ones, to cover minesweeping in the Helgoland Bight, or to guard against possible British attacks, could rarely be ascertained until the *Hochseeflotte* was already on its way back to port. It was all most frustrating for those who credited the Germans with more aggressive intentions than they in fact possessed, and equally for those longing to bring on the great naval battle which would ensure Britannia's supremacy once and for all.

93. *Unknown War*, pp. 201–202.
94. *Marlborough*, III, p. 338. See also ibid., p. 337, ibid., II, p. 183 and ibid., IV, p. 128.
95. Ibid., III, p. 61.
96. *World Crisis 1911–1914*, p. 334–5.
97. Ibid., p. 335. Gilbert, III, p. 56.
98. Liddell Hart, "Military Strategist," p. 185. Ashley, p. 81. *World Crisis 1911–1914*, p. 336.
99. *Marlborough*, II, p. 209.
100. Ibid., VI, p. 111.
101. Ibid., pp. 421–8. Marlborough was involved personally in the deception plan that resulted in the rupture of the Ne Plus Ultra line. In the opening gambit of that operation, the British commander, protected by a large troop of cavalry, conducted an ostentatious reconnaissance of the French right flank at close quarters, even while his pioneers were preparing the approaches to the enemy's left. Ibid., pp. 426–7 and 438.
102. Martin Gilbert, *Winston S. Churchill. Volume V. The Prophet of Truth 1922–1939* (Boston: Houghton Mifflin, 1977), p. 438.
103. *Marlborough*, IV, p. 65.
104. Ibid.
105. Ibid., p. 67.
106. Ibid., p. 87.
107. Martin Gilbert, *Winston S. Churchill. Volume IV. The Stricken World 1916–1922* (Boston: Houghton Mifflin, 1975), p. 8, and Ashley, p. 101.
108. Gilbert, IV, p. 62.
109. *World Crisis 1915*, p. 64. Liddell Hart has pointed out that "without Churchill's impulsion, and the naval experiments he initiated, however misguided, the new idea might never have survived the chill of that first winter in official Army quarters." Liddell Hart, "Military Strategist," p. 196. Nevertheless, Churchill never claimed primary advocacy. He acknowledged that a 1903 H. G. Wells article "had practically exhausted the possibilities in this sphere." Moreover, there was also the small band of armor pioneers in the British Army. "There never was a moment which it was possible to say that a tank had been 'invented,'" Churchill wrote later. "There never was a person about whom it would be said 'this man invented the tank.'" *World Crisis 1915*, pp. 69–70.
110. Ibid., p. 425.
111. Ibid., p. 71.
112. Ibid., *1916–1918*, Part II, p. 564.
113. Ibid., *1915*, p. 496. See also ibid., *1916–1918*, Part II, p. 555.
114. Ibid., Part I, p. 185.
115. Ibid., *1915*, p. 82.
116. Ibid., *1916–1918*, Part II, p. 345.
117. Ibid., Part I, p. 194.
118. Ibid., Part II, p. 347.
119. Ibid., pp. 517–518.
120. Emphasis added. *JCS Pub 1*, p. 240.

121. Liddell Hart, *Strategy*, p. 322.
122. For the basic concept of this horizontal-vertical paradigm, see Edward N. Luttwak, *Strategy. The Logic of War and Peace* (Cambridge, MA: Harvard University Press, 1987).
123. C.V. I, I, p. 587. *Roving Commission*, p. 131.
124. Ashley, p. 49.
125. Herbert Tingsten, *Victoria and the Victorians* (London: Allen & Unwin, 1972), p. 473.
126. R. Churchill, II, p. 19.
127. Winston S. Churchill, *The Aftermath* (New York: Charles Scribner's Sons, 1929), p. 483 and *World Crisis 1915*, p. 6.
128. *Marlborough*, III, p. 353.
129. Michael Howard, *Studies in War and Peace* (New York: The Viking Press, Inc., 1971), p. 189.
130. *Marlborough*, III, p. 13.
131. Ibid., V, p. 90.
132. Ibid., III, p. 262.
133. Ibid., IV, pp. 31 and 40.
134. *World Crisis 1915*, p. 6.
135. Ibid., *1916–1918*, Part II, p. 454.
136. Ibid., *1911–1914*, p. 362.
137. *Unknown War*, pp. 231–2. The Austrian attack had been a conscious effort to prod the neutrals. "The defeat of Serbia and her collapse," Churchill pointed out, "might be the signal which several at least of the neutral States seemed to await before answering the long-drawn and several times repeated trumpet call." Ibid., p. 117. See also ibid., p. 273.
138. Ibid., p. 285.
139. *Marlborough*, VI, p. 124. "Neither in his headquarters at the front nor behind him at home did he have that sense of plenary authority which gave to Frederick the Great and to Napoleon their marvelous freedom of action." Ibid., V, p. 6.
140. Ibid., p. 246.
141. Ibid., III, p. 290. See also ibid., p. 355: "Surveying the general war," Churchill wrote of the situation just before Blenheim, "we can see that matters had now come to such a pitch that without a great victory in two or three months, the Grand Alliance was doomed."
142. Ibid., IV, p. 84.
143. *World Crisis 1916–1918*, Part I, pp. 237 and 240.
144. Gilbert, IV, p. 50.
145. Winston S. Churchill, *Savrola. A Tale of the Revolution in Laurania* (New York: Random House, 1956), p. 82.
146. *Marlborough*, VI, p. 76. Chapter 24 in the fifth volume of *Marlborough* is entitled "The Home Front."
147. *Great Contemporaries*, p. 113.
148. Howard, *Studies*, p. 108.
149. *World Crisis 1916–1918*, Part I, p. 58.
150. Ibid., pp. 79–80.
151. *Aftermath*, p. 476.
152. *World Crisis 1916–1918*, Part I, p. 88.
153. Ibid., p. 84, "In a military sense, Verdun had no exceptional military importance....It was two hundred and twenty kilometers from Paris, and its capture would not have made any material difference to the safety either of the capital or of the general line." Ibid.
154. Ibid., p. 86.
155. *Unknown War*, pp. 352–3.
156. Ibid., p. 363.
157. *World Crisis 1916–1918*, Part I, p. 112.
158. Ibid., p. 110.

159. Ibid., p. 109.
160. Geoffrey Till, "Naval Power," *Warfare in the Twentieth Century*, p.91.
161. *World Crisis 1916–1918*, Part II, p. 396.
162. Ibid., Part I, pp. 226–7.
163. Ibid., pp. 238–9.
164. *Great Contemporaries*, p. 254.
165. *World Crisis 1916–1918*, Part I, p. 239.
166. R. Churchill, Vol. II, pp. 26–7.
167. *Thoughts and Adventures*, p. 100.
168. Ibid., p. 92.
169. *World Crisis 1915*, p. 6.
170. Ibid., p. 526. "There was no supreme authority."
171. Martin Gilbert, *Winston Churchill. Volume III. The Challenge of War 1914–1916* (Boston: Houghton Mifflin, 1971), p. 543.
172. Ibid., IV, p. 111. A similar position to that of Marlborough in the winter of 1703 when "responsibility with odium, but without power, was all that was offered." *Marlborough*, III, p. 266.
173. Ibid., p. 113.
174. *World Crisis 1916–1918*, Part I, p. 244.
175. Ibid.
176. Ibid., p. 243.
177. *Unknown War*, p. 287.
178. *World Crisis 1916–1918*, Part I, pp. 277 and 279.
179. *Great Contemporaries*, p. 117.
180. *Thoughts and Adventures*, p. 109.
181. *World Crisis 1916–1918*, Part I, p. 199.
182. *Thoughts and Adventures*, p. 109.
183. *World Crisis 1915*, p. 349.
184. *Aftermath*, p. 467.
185. *Great Contemporaries*, p. 118.
186. *World Crisis 1916–1918*, Part I, pp. 213–214.
187. Ibid., Part II, p. 511.
188. Ibid., p. 405, and *Thoughts and Adventures*, p. 113.
189. *World Crisis 1915*, p. 527.
190. *Marlborough*, III, p. 141. Michael Howard, *War in European Society* (New York: Oxford University Press, 1976), p. 113.
191. *World Crisis 1916–1918*, Part II, p. 305.
192. Ibid., p. 338.
193. This led the King to comment: "What good words for a recent subaltern of Hussars!" R. Churchill, Vol. II, p. 93.
194. Sir Julian Corbett, *Some Principles of Maritime Strategy* (London: Longmans, Green, 1911), p. 236.
195. Sir Julian Corbett, *The Successors of Drake* (London: Longmans, Green, 1900), p. vii. Churchill pointed out in the First World War that "the Admirals who thought only of the Grand Fleet and the Generals who thought only of the Main Army may learn how cruel are the revenges which Fortune wreaks upon those who disdain her first and golden offerings." *World Crisis 1915*, p. 539.
196. Ibid., *1911–1914*, p. 93
197. Ibid., p. 58–62. But see D. M. Schurman, *The Education of a Navy. The Development of British Naval Strategic Thought, 1867–1914* (Chicago: University of Chicago Press, 1965), p. 185, who points out that the pre–1914 naval strategists may have had some indirect influence on the conduct of both World Wars, but adds that "their obvious influence was not overwhelming."
198. Michael Howard, *The British Way in Warfare, A Reappraisal* (London: Jonathan Cape, 1975), p. 5.
199. Schurman, p. 190.

200. *Aftermath*, p. 474.
201. *Unknown War*, p. 304. See also *World Crisis 1916–1918*, Part I, p. 66.
202. *Aftermath*, p. 475.
203. Emphasis added. *Marlborough*, III, p. 95.
204. Winston S. Churchill, *Marlborough. His Life and Times. Volume I. 1650–1688* (New York: Charles Scribner's Sons, 1933), p. 77.
205. Ibid., III, p. 299.
206. Ibid., V, p. 242 and IV, p. 135.
207. Ibid., V, pp. 242–3. "Nevertheless," Churchill concluded about the failure of the 1707 Toulon amphibious operation, "good strategy even in failure often produces compensations." Ibid., p. 290.
208. Ibid., p. 553 and ibid., III, p. 95.
209. Ibid., IV, p. 149.
210. Ibid., V, pp. 453 and 456.
211. Ibid., pp. 456–7.
212. *Great Contemporaries*, p. 326.
213. *Thoughts and Adventures*, p. 113.
214. Gilbert, IV, p. 915.
215. Ibid., p. 914.
216. *Thoughts and Adventures*, p. 197.
217. *Great Contemporaries*, p. 23.
218. Ibid., p. 103.
219. Gilbert, V, p. 319.
220. *Marlborough*, IV, p. 129.
221. Ibid., III, p. 112.
222. *Aftermath*, p. 1.
223. *Marlborough*, IV, p. 50.
224. Ibid., p. 68.
225. Ibid., III, p. 16.
226. Ibid., V, p. 242. Marlborough also believed "that any peace offer from France would only be an attempt to. . .cheat the Allies." Ibid., VI, p. 42.
227. Ibid., IV, p. 128.
228. Ibid., VI, pp. 177 and 185.
229. Ibid., V, p. 9.
230. Winston S. Churchill, *The Story of the Malakand Field Force. An Episode of Frontier War* (New York and Bombay: Longmans, Green, 1901), p. 208.
231. *London to Ladysmith*, p. 98.
232. John Wheeler-Bennett, ed., *Action This Day. Working With Churchill* (New York: St. Martin's Press, 1969), p. 106. Churchill was certainly more magnanimous than his ancestor. When King George restored Marlborough to office after the death of Queen Anne, the earlier Churchill immediately attacked the two men most responsible for driving him from power and staining his reputation. Henry St. John, Viscount Bolingbroke, the former Tory Secretary of State, fled England to escape from Marlborough, who simultaneously pressured the government to impeach the second of his enemies, Robert Harley, Earl of Oxford and former Lord Treasurer. Ashley, p. 156.
233. Wheeler-Bennett, p. 82.
234. Winston S. Churchill, *The Second World War. Volume I. The Gathering Storm* (Boston: Houghton Mifflin, 1948), p. 4.
235. *World Crisis 1915*, p. 1.
236. Ibid., *1911–1914*, p. 174.
237. *Unknown War*, p. 288.
238. A. J. P. Taylor, "The Statesman," *Churchill Revised*, p. 43.
239. *Marlborough*, III, pp. 13–14.
240. Eade, p. 285.
241. Ronald Lewin, *Churchill as Warlord* (New York: Stein and Day, 1973), p. 35.

242. Ibid., p. 50.
243. Gilbert, VII, p. 75.
244. Churchill's emphasis. *Unknown War*, p. 132. "But it was no use sitting down and waiting for a year while these preparations were completing," Churchill commented in the First World War concerning the efforts to meet the Zeppelin threat. "Only offensive action could help us." *World Crisis 1911–1914*, p. 339.
245. Martin Gilbert, *Winston Churchill. Volume VIII. Road to Victory 1941–1945* (Boston: Houghton Mifflin, 1986), pp. 133 and 151–2.
246. Ibid., p. 139.
247. Ibid., p. 140.
248. See Michael Howard, *Grand Strategy. Volume IV. August 1942–September 1943* (London: HMSO, 1972), *passim*, pp. 191–221.
249. Michael Howard, *The Mediterranean Strategy in the Second World War* (New York: Frederick A. Praeger, 1968), p. 69. "In after months," Eisenhower commented many years later in reference to Torch, "I became absolutely certain that Mr. Churchill had been right in his argument for that particular operation; and so did General Marshall." Nelson, p. 33.
250. Lewin, p. 139.
251. Winston S. Churchill, *The Second World War. Volume V. Closing the Ring* (Boston: Houghton Mifflin, 1951), p. 38.
252. Howard, *Mediterranean Strategy*, pp. 70–1.
253. Wheeler-Bennett, p. 96.
254. Maurice Matloff, "Allied Strategy in Europe," *Makers of Modern Strategy*, p. 695.
255. Winston S. Churchill, *The Second World War. Volume VI. Triumph and Tragedy* (Boston: Houghton Mifflin, 1953), p. 455.
256. Matloff, p. 702.
257. *Marlborough*, VI, p. 90.
258. Luttwak, Part III, *passim*.

3

The Paradox of Duality: Adolf Hitler and the Concept of Military Surprise

The First World War was a major event in Adolf Hitler's life. "To me," he wrote in a famous passage six years after the conflict, "those hours seem like a release from the painful feelings of my youth. Even today I am not ashamed to say that, overpowered by stormy enthusiasm, I fell down on my knees and thanked Heaven from an overflowing heart for granting me the good fortune of being permitted to live at this time."[1] Hitler may have been influenced by the romantic nihilism of Friedrich Nietzsche, whose *Also sprach Zarathustra* was sold to thousands of German soldiers going off to war in 1914.[2] What is certain is that for Hitler at the end of that conflict the "romance of battle had been replaced by horror ... and the exuberant joy was stifled by mortal fear."[3]

Hitler was a *Meldegänger*, a runner, in the 16th Bavarian Infantry "List" Regiment for the entire war.[4] It was a particularly dangerous and thankless job in the attritive, mass technology warfare of the stalemated Western Front. Runners would hunch forward in trenches and shell-holes and then spring up, as one officer in the List Regiment recalled, "like rabbits" between artillery salvoes and run in semi-crouched positions to other shell-holes, which if they had calculated correctly they would reach before the next torrent of steel descended.[5] By most accounts, Hitler was a brave and dependable soldier who refused promotion and remained in his dangerous job in order to stay with his regiment.[6] He seldom took leave, only rarely received mail and increasingly withdrew into the narrow world of his unit. "For Corporal Hitler," the regimental adjutant later recalled, "the list Regiment was homeland."[7]

By the end of the war, Hitler had been wounded and gassed as well as awarded the Iron Cross First Class, the former common enough occurrences for the average German infantryman, the latter an extremely rare honor for *Untersoldaten*.[8] Equally important, Hitler had begun to form opinions on warfare that were to remain with him for the rest of his life. "Even a man of thirty," he wrote at that age while serving a prison term, "will have much to learn in the course of his life, but this will only be a *supplement*."[9] This is particularly true of the two different approaches to war that crystalized in Hitler's mind as a result of his First World War experiences. One focused on the decisive head-on battles of annihilation; the other on what Liddell Hart would later refer to as "the indirect approach."[10] The purpose of this chapter is to demonstrate why Hitler was drawn to these two approaches and how his experiences and personality led him to the concept of military surprise as a means to bridge the gap between these warfighting dichotomies. It was not always an easy bridge to build, despite Hitler's well publicized surprise attacks in the early years of the Second World War. And in fact, as this study also demonstrates, the early experiences and personality traits that drew him to the concept of military surprise also resulted in a leadership style and an organization for war that ultimately destroyed the bridge, leaving the Nazi leader vulnerable to surprise at every level of war.

The Direct Approach

The stalemated Western Front of the First World War was anathema to Hitler's nature. To begin with, there was his inner compulsion to change entire situations suddenly, not gradually. As a boy in Linz, for instance, he was furious that improvements to the town's concert hall did not begin with demolitions. "It seems," he told his boyhood friend, August Kubizek, "they intend to patch up once more the old junk-heap."[11] A few years later in Vienna, the young Hitler developed a grandiose project for completely rebuilding the entire city district by district.[12] Not surprisingly, Hitler considered Georges Haussmann the "greatest city planner in history." Haussmann had torn down most of central Paris at the behest of Napoleon III between 1853 and 1870 and had subsequently built elegant façades along gleaming, extra-ordinarily wide boulevards.[13]

Equally important, there was the all-or-nothing aspect of Hitler's character that could only chafe in the trenches of northern France and Belgium. "*Another symptom of decay*," he wrote after the war, "*is a half-heartedness in all things.*"[14] Added to this was a nervous, aggressive temperament that was only comfortable with quick decisive actions. "Patience," even the normally docile Kubizek noted, "did not seem to be one of Adolf's outstanding characteristics."[15] It is in this context that Hitler could look back nostalgically after the war on the Ludendorff offensive of March 1918, "when out of the cool nights the allied soldiers already seemed to hear the dull rumble of the advancing storm units."[16] That offensive, Hitler noted almost lyrically, created:

the most tremendous impressions of my life; tremendous because now for the last time, as in 1914, the fight lost the character of defense and assumed that of attack. A sigh of relief passed through the trenches and the dugouts of the German Army when at length, after more than three years' endurance in the enemy hell, the day of retribution came. Once again the victorious battalions cheered and hung the last wreaths of immortal laurel on their banners rent by the storm of victory. Once again the songs of the fatherhood roared to the heavens along the endless marching columns.[17]

Hitler's preference for the offensive was strengthened over the years by his admiration for Clausewitz and Frederick the Great. It is now generally concluded that the Nazi leader did not study Clausewitz in either a systematic or critical manner; but instead, in the way of semi-educated people, used the Prussian philospher as a source for quotations.[18] Nevertheless, there is abundant evidence that Hitler also used Clausewitz as much as any other purported spiritual predecessors as a source and justification of his thoughts, particularly those dealing with military affairs.[19] This was certainly true in his view of Clausewitz as the primary exponent of the offensive, a misperception he shared with the majority of the German military.[20] In a similar manner, Hitler's works and conferences were replete with references to Frederick the Great, whose major departure from the other military leaders of the baroque age was, as Gerhard Ritter pointed out, "the unflagging offensive spirit, pressing for rapid decisions."[21]

Hitler remained remarkably true to this spirit, even after losing the strategic offensive in the Second World War. In August 1943, he stated that "defensive operations alone are not enough. We must resume the offensive."[22] And at a conference just before the 1944 German counterthrust in the Ardennes, he pointed out to his division commanders that "in the long run the principle that defence is stronger than attack does not hold."[23] At the end of that conference, Hitler stressed the need for offensive operations if decisive results were to be obtained. Once again his personal battlefield experiences dominated his thinking. "From the outset of the war therefore," he concluded, "I have striven to act offensively whenever possible, to conduct a war of movement and not to allow myself to be maneuvered into a position comparable to that of the First World War."[24]

Clearly linked to this spirit of the offensive in Hitler's mind was the decisive battle of annihilation, the so-called *Vernichtungsschlacht* of Clausewitz. And although a scholastic conflict had been waged for years as to whether Frederick the Great belonged to this school of warfare, Hitler obviously had no doubt. Had not Thomas Carlyle, the Nazi leader's favorite chronicler of the Prussian King's life, quoted Frederick's call for "a great battle to decide our fate?"[25] Such battles could come about by great mass movement and envelopment. The key was to avoid stalemate. "Interlocked frontal struggles lasting for years on petrified fronts," Hitler stated in 1932, "will not return. I guarantee that. They were a degenerate form of war."[26]

Not only degenerate but for Hitler an aberration. As a boy he had devoured popular histories of the battles in the wars of German unification. "It was not long," he wrote concerning one such history of the Franco-Prussian War, "before the great heroic struggle had become my greatest inner experience."[27] It was the battles of annihilation that were decisive. Without Königgrätz and Sedan, there was always the specter of a two-front war.[28] Without these decisive battles Germany's "union policy would no more have led to anything than did the chattering of the men of 48 in the Cathedral of Frankfurt."[29] There had been no such battles of annihilation on the Western Front in the First World War. For Hitler, the infantryman with the "worm's-eye view" of that struggle, this void still rankled decades later as his troops engaged in such battles throughout Russia. "Apart from the great victories ... of Tannenberg and the ... Masurian Marshes," Hitler reminisced in the fall of 1941, "the Imperial High Command proved itself inadequate."[30]

The Indirect Approach

The carnage of the First World War made an enormous impression on Hitler. As early as 1915 he referred to "the sacrifices and agonies which now so many hundreds of thousands of us endure every day" and to "the river of blood which flows here daily."[31] The feelings of loss and pain were genuine enough for a man who felt for the first time a sense of mission and belonging in the comradeship of his regiment.[32] Years after the war, Hitler returned again and again in his memoirs to "how those boys of seventeen sank into the earth of Flanders."[33] By then, his own brand of Social Darwinism had evolved into the conviction that the shedding of precious German blood was a biological crime. In this context, the war "drained the extreme of the best humanity almost entirely of its blood. For the amount of irreplaceable German heroes' blood that was shed in those four years was really enormous."[34] This theme dominated Hitler's so-called "Secret Book" of 1928, which in myriad blood-filled images hammered home the point that the "most unprecedented blood sheddings in history" were "sacrileges committed against a nation, a sin against a people's future."[35] By the early 1930s, all this had taken on the characteristics of dogma for Hitler in addition to its obvious propaganda use. "Whoever has experienced war at the front," he confided to Hermann Rausch-ning, "will want to refrain from all avoidable bloodshed. Anything that helps preserve precious German blood is good."[36]

Seen in this context, any type of warfare that avoided direct assaults against weapons of mass technology would appeal to Hitler. In 1924, the heavy human toll of that technology still weighed heavily on him as he reminisced about the Battle of the Somme, which his unit had entered in September 1916. "For us it was the first of the tremendous battles of material which now followed, and ... it was more like hell than war."[37] Battles of attrition had replaced those of annihilation. Offensive action was no longer decisive. As late as 1941, the subject was an emotional one for Hitler.

> The offensive at Verdun ... was an act of lunacy. From begin-ning to end, all the commanders responsible for that operation should have been put in straitjackets. We've not yet completely got over those mistaken notions.[38]

The answer to these "mistaken notions" was an indirect approach based on cunning, mobility and maneuver, much of which was designed to dislocate the enemy's moral, mental and material balance even before the actual decisive engagement.[39] There was a natural appeal for Hitler in such a concept. As a young boy at the time of the Boer War, he and his schoolmates became rabid partisans of the hard-riding Boers who effectively out-maneuvered the British by guerrilla tactics.[40] Added to this was his fascination with the frontiersmen and especially the Indians of North America. In 1942, Hitler recalled how he would read the works of James Fenimore Cooper and Karl May late at night by candle or by moonlight with the aid of a huge magnifying glass. Karl May was particularly influential. "I owe him my first notions of geography," Hitler stated, "and the fact that he opened my eyes on the world."[41] He remained enthralled all his life with the cunning and bravery of such May heroes as Winnetou and Old Shatterhand, the latter of whom possessed a 48 shot rifle with which he could break a hangman's rope at 50 yards, and who used such expletives as "Hang it all, fellows" (*Zum Henker, Kerle!*). During the war, Hilter told his valet, "I have ordered every officer to carry with him ... Karl May's books about fighting Indians. That's the way the

Russians fight – hidden like Indians behind trees and bridges, they jump out for the kill."[42]

There were, of course, other intellectual influences that drew Hitler to the indirect approach. T.E. Lawrence's description of his raid-and-run tactics against the Turks in *Seven Pillars of Wisdom* made a "lasting impression" on him, as did Machiavelli's *The Prince*, which he kept beside his bed and found "simply indispensable."[43] All of these reinforced a natural conspirator's love of the daring and cunning in Hitler. These were the traits of Cola di Rienzo, the medieval rebel and hero of his favourite Wagner opera. And there was more than one passing reference by the Nazi leader during the Second World War that revealed a grudging admiration for the craftiness and cunning of Stalin.[44] It is perhaps only natural in this regard that wolves were a subject of fascination to a man who possessed, as Rauschning pointed out, "a peasant cunning that one is tempted to describe as sublime."[45]

Hitler's preference for the indirect approach was reinforced by his political experience during the *Kampfzeit*. The failure of his 1923 *putsch* and his subsequent 14-month imprisonment acted as a catharsis in this regard. "When I was younger," he ruefully recalled in 1941, "I thought it was necessary to set about matters with dynamite. I've since realized that there's room for a little subtlety."[46] That subtlety included a legal policy which, as Theodor Heuss pointed out, was only a "moratorium on illegality."[47] Hitler never denied this. "It will take longer, to be sure," he stated in 1924, "to outvote our opponents than to outshoot them, but in the end their own Constitution will give us success."[48] And of course that came to pass by dint of persistence, patience and a large measure of luck. In 1936, Hitler could look back with satisfaction on his accomplishments in a speech that captured the essence of the indirect approach.

> We have conquered our State without, I believe, the breaking of a window pane ... and the greatest miracle of all: it is ... due solely to the experience of 1923 that we were able to sail around the rock which faces any revolution.[49]

After assuming power in 1933, Hitler continued his political indirect approach, particularly the duality that was typical of his psychological method of overcoming opposition. Unconstitutional developments in the form of the *"coup d'état* by installment" process of *Gleichschaltung* as well as diplomatic surprises were always accompanied by assurances of respect for law and love for peace. In terms of military matters, Hitler was increasingly preoccupied with the idea of achieving decisive results, previously associated with direct assaults, by means of the indirect approach. His conversations with Hermann Rauschning between 1932 and 1934 revealed an acute appreciation of psychological warfare complemented by unexpected, offensive *coups de main* from within as well as from without. Throughout these talks, the concept of military surprise emerged repeatedly as the key to decisive battle as well as the most humane concept. "The most horrible warfare is the kindest," Hitler confided. "I shall spread terror by the *surprise* employment of all my measures. The important thing is the sudden shock of an overwhelming fear of death."[50] This was the general concept, then, that for Hitler would avoid any repetition of First World War attrition despite technological advances. "Who says I am going to start a war like those fools in 1914?" Hitler asked in some agitation when reminded by Rudolf Hess of those advances. "Are not all our efforts bent toward preventing this? Most people have no imagination .... They are blind to the new, the *surprising* things."[51]

Surprise and Risk

Hitler had, as Liddell Hart has pointed out, "a deeply subtle sense of surprise."[52] This sense was inherent in his love of conspiracy and secrecy, and was a major factor in his only overt attempt to seize power. Even his co-conspirators were not prepared for the sudden timing of his November 1923 *putsch*. "Gentlemen," General Ludendorff stated as he arrived at the *Bürgerbräukeller* on the fateful night of 8 November, "I am as surprised as you are."[53] Surprise continued to play a major role for Hitler after he assumed power, whether in the domestic realm in his suppression of the Röhm faction or in the diplomatic sphere with his phenomenal run of March successes. For him, military as well as diplomatic surprise was nothing more than the extension of the professional revolutionary's credo. "Do you know the doctrine of the coup d'état?" he asked Rauschning in 1934. "Study it. Then you will know our task."[54] That task, once war began, was to achieve decisive victory quickly at the least cost to German manhood.

> I shall determine the correct movement for attack. There is only one most favorable movement. I shall await it – with iron determination. I shall not miss it .... If I succeed in that, then, I ... shall have saved as many lives then as could be saved.[55]

Military surprise was for Hitler indelibly linked with his concept of risk. He liked to think of himself as a cold-blooded, ruthless gambler, yet once confessed to a compulsion to go against the odds and court disaster. "You know, I am like a wanderer who must cross an abyss on the edge of a knife," he said. "But I must, I just must cross."[56] In planning a surprise attack, the more risky a project the higher his interest. The Polish and Balkan campaigns, for instance, involved the Nazi leader very little since he had assured that the opposition could easily be overcome. In the Norwegian, Western and Russian campaigns, on the other hand, Hitler was considerably more active, as one historian has pointed out, "not only because he had gained in experience and self-confidence but because they were dangerous, risky affairs which would be almost certain to lead to disaster unless he could fully apply his own qualities of daring and surprise to deal swift, decisive blows."[57]

At the same time, Hitler firmly rejected unnecessary risk, particularly when prestige was involved. More than once he recalled bitterly how he had risked his life in the First World War to deliver a postcard.[58] And after learning of Reinhard Heydrich's assassination, he termed the public rides of that SS official through the streets of Prague "stupid and idiotic."[59] For the same reason he once grounded Rudolf Hess whose daredevil flying was good enough to win Central European competitions as well as to bring him ultimately and unerringly to Scotland.[60] In a similar manner, Hitler detested mountain-climbing and Alpine skiing. And when he heard that a detachment of German mountain troops had planted a German flag on Mount Elbrus, at 19,000 feet the highest mountain in the Caucasus, he was furious, raging for hours about "crazy mountain-climbers" who "belong before a court martial."[61]

Risk for a legitimate purpose, however, was another matter for the Nazi leader. "First off," he pointed out at a conference in August 1939, "is recognition of the fact that any political or military success involves taking risks."[62] This was particularly true if potential adversaries were unwilling to make such recognition. "*The Men of Munich*," Hitler concluded at the conference, "*will not take the risk*."[63] But the will to act, to take risks, had to exist if success were to be achieved. This lack of will, Hitler noted in 1924,

> prevents any decision with which a risk is connected, as through the greatness of a deed did not consist precisely in the risk. Without suspecting it, a German general succeeded in finding the classic formula for his miserable spinelessness: "I act only if I can count on 51 per cent likelihood of success." In these "51 per cent" lies the tragedy of the German collapse; any one who

demands of Fate a guarantee of success, automatically renounces all ideas of a heroic deed.[64]

By this criteria, "heroic" deeds were the norm at the strategic level in the early war years. For Hitler during this period, the more risky the indirect strategic approach, the greater the role of military surprise and thus the less risky the venture. Years before, he had summed up this paradox by pointing out that "the impossible is always successful. The unlikely thing is the surest."[65] The April 1940 operation against Norway was no exception to the paradox, involving as it did less than 9,000 initial assault forces landing from ships and airplanes to seize Oslo and five other port cities as far north as Narvik. "Weakness in numbers," Hitler's war directive emphasized, "will be made good by skillful action and surprise."[66] To achieve this surprise, all attacks required simultaneity, and there was no disguising the risks involved in the daring plan. The loss of half of the naval forces in the operation was an accepted part of the calculus of surprise as was the need for reckless commitment at all levels if such hazardous operations as the occupation of distant Narvik and the forcing of the heavily defended Oslo Fjord were to be successful.[67]

The key to decisive results was the unexpected. "It is of supreme importance," Hitler concluded his Norwegian campaign directive, "that our operations should come as a *surprise* to the Northern countries as well as to our enemies in the West."[68] And Admiral Raeder's implementing decree for the execution of operations in the north gave top priority to the qualities of "boldness, tenacity and skill. The pre-requisites for the success of the operation are surprise and rapid action."[69] How well this all succeeded was summed up by General Ismay, Churchill's Chief of Staff. "I realized for the first time," he recalled, "the devastating and demoralizing effect of surprise."[70]

Surprise continued to be a prime consideration as preparations for the western campaign progressed from the autumn of 1939 through the spring of the following year. During a conference with army commanders on 20 January 1940, for instance, Hitler insisted that the German forces launch the offensive on the shortest notice as a "leap from wherever they stand" rather than spoil surprise by mass movements of troops before the attack. As a result, the first wave on D-Day would consist primarily of the comparatively weak forces on the frontier who would only be capable of carrying out special *Aktionen* against a number of particularly important preliminary objectives ranging from bridges to road blocks. When the OKH leaders predictably insisted on an alert and mobilization period before hostilities, Hitler's reply captured the essence of achieving decisive results by indirect means through the "leap from stand" concept. "More may be expected from attack with small forces, using maximum surprise," he stated, "than with larger forces, but against an enemy poised for defense."[71]

How important the concept of military surprise was in Hilter's strategic calculations was demonstrated in the crisis concerning the news on 10 January 1940 that a forced landing of a German plane in Belgium had placed the current OKH plans for the western campaign in Belgian hands. When the news arrived General Jodl commented, "No surprise (neither operational nor tactical)" and in describing Hitler's initial reaction pointed out that in "future the Führer will divulge his plans only to a very few."[72] The compromise was just one of the factors that led Hitler to the so-called "Manstein Plan" later that winter. Enemy intelligence efforts and the consequent possible loss of the element of surprise continued to be major concerns as he rearranged his force in accordance with the new plan. "Surprise," he remarked with some relief on 18 February 1940, "may now be regarded as assured. It took the enemy 10–14 days to learn about some of our regrouping movements."[73]

Surprise, of course, was not assured for Operation Sea Lion, the plan for the invasion of England, despite the fact that Hitler's preparatory directive on 16 July 1940 emphasized that the landing would be "in the form of a surprise crossing."[74] Five days earlier, Admiral Raeder had already pointed out the extreme difficulties Sea Lion presented and had recommended "that an invasion should be used only as a last resort to force Britain to sue for peace."[75] This advice, added to his uneasiness with naval operations, his own ambiguous *Hassliebe* feelings towards Britain, and his false appraisal of the British mind, which led him to believe that the island nation would come to terms, caused Hitler, in Field Marshal von Manstein's judgement, *"to recoil from taking the risk."*[76] The interplay of this risk with his concept of surprise was summed up by Hitler at a conference on 21 July 1940.

> The invasion of Britain is an exceptionally daring undertaking, because even if the way is short, this is not just a river crossing, but the crossing of a sea which is dominated by the enemy. This is not a case of a single crossing operation as in Norway; operational surprise cannot be expected; a defensively prepared and utterly determined enemy faces us and dominates the sea area which we must use.[77]

The result was a series of desultory conferences during the remainder of the summer marked by increased army – navy friction as well as continued warnings by Admiral Raeder as to the difficulties and risks associated with the entire operation.[78] A deception plan was created at the end of August; but on 17 September Hitler ordered the postponement of the operation.[79] And although he renewed the order for Sea Lion preparations on 27 September, that was also the date on which he ordered reinforcements to the east. If the element of strategic and even operational surprise was missing in the proposed Sea Lion invasion as a bridge between the direct and indirect approach, at least the prospect of that invasion could serve the Nazi leader as the mortar to build such a bridge in the east. By 12 October, Sea Lion preparations were firmly established as the basis for the principal Barbarossa deception plan.[80] But there was no disguising the failure of the primary operation against Britain. The following January, Hitler attempted to rationalize it in terms that ignored the synergistic relationship between surprise and risk that had formed the basis of his military victories thus far. "The success of an invasion must be absolutely assured." Otherwise, he argued, "it is a crime to attempt it."[81]

The Unitary Actor – Military Surprise and Decision-Making

Hitler began in the early *Kampfzeit* as a *Trommler*, a drummer for the nascent Nazi party. During the period of his imprisonment from November 1923 to February 1925, the squabbling factions in his party as well as in the entire *völkisch* movement increasingly oriented on the Nazi leader and sought his approval. Hitler remained above these internecine struggles, however, and emerged from prison well on his way to becoming the "myth person," the leader to whom all factions turned.[82] Through this *Führerprinzip* in the following years, Hitler was able to use his charismatic leadership to manipulate various factions with relatively different pragmatic orientations, thus allowing the movement to appeal to diverse classes and interests. The key to all this was a basic "divide and rule" *modus operandi* that he would use until his death. In this way, he remained the sole source of legitimacy for the NSDAP and later for the Nazi state in peace and war.[83]

Hitler, the Surpriser: The Advantages of the Unitary Actor

Before the move against Poland, General Halder noted that Hitler "reckons with the possibility that the French and British might adopt a passive attitude in the face of our invasion. This belief, he feels, is justified by the difficulties of prompt communication between the political and military authorities."[84] The Führer was, of course, well aware even in 1939 of the advantages that accrued to him as the possessor of both these authorities. It was their combination that had allowed him to move from one triumph of diplomatic surprise to another.[85] And it was the unitary interaction of his political–military command system that continued to provide him the ability to surprise once war began. Political and military intentions, for instance, were easy for Hitler to conceal, and could be changed at the last minute. In this regard, the speed of the unitary decision-making process allowed him to react in time to British moves toward Norway and launch his surprise assault on that country. In a similar manner, within hours of a coup that had dethroned his puppet in Belgrade on 27 March 1941 and had thus exposed a German flank in the impending attack on Russia, Hitler issued his directive for the invasion of Yugoslavia.[86]

Structure and Character

To this decision-making process, Hitler brought a great many traits that contributed to the decisiveness of his indirect approach, at least in the opening years of the war. As early as the 1923 *putsch*, he had demonstrated an ability to improvise that, combined with a flexible and opportunistic pragmatism on the part of the Nazi leader, invariably caught opponents off guard. "*So oder so*" (in this way or in that) was a favorite saying, as was "there are two possibilities."[87] He was a political animal who knew how to delay decisions and play for time while he looked for weaknesses in his opponents. This tendency was reinforced by chronic procrastination, a reflection of his bohemian and artistic temperament that had been formed during his early Vienna days. "Problems are not solved by getting fidgety," Hitler told his staff when approached on this subject. "If the time is ripe, the matter will be settled one way or another."[88]

All of this contributed in the fall and winter of 1939–40 to Hitler's series of stop and go decisions concerning the western campaign. Certainly, as has been demonstrated, the Nazi leader's position allowed him to react quickly to such unforeseen events as the January 1940 compromise of the German operational plan. And there was more than a passing resemblance between the three French mobilizations during this period and Hitler's method of working up a crowd before the war. This was particularly true in terms of the artificial delay, while the audience was whipped into a frenzy by music, banners and shouts, before his sudden entrance. The French were no exception to this appeal, and fell victim to what Barton Whaley has termed the "cry wolf" syndrome in which, as Herman Kahn has pointed out, "if one reacts whenever there is ambiguous evidence, the number of false alarms increases the likelihood of being taken by surprise if the event proves genuine."[89]

It was, however, in Hitler's traits of imagination and daring that the advantages of the unitary actor over pluralistic decision-making were most evident. In this regard, as Klaus Knorr has suggested, "a truly daring attempt at surprise runs the risk of emasculation in any decision-making system that is collegial even if elitist, and accords veto power to cautious minds."[90] There was nothing cautious about Hitler's mind. Because he had never learned a profession, there was an amateurism about him that served him well in the early war years. Without any idea of what actual specialized knowledge meant, he normally did not sense the complexities of any great task. The key was to get to the heart of the matter, he wrote in *Mein Kampf*, "*to retain the essential, to forget the non-essential.*"[91] And a decade later, the Nazi leader admonished Rauschning not to "falter over trifles" before summing up his approach:

> I have the gift of reducing all problems to their simplest foundations. War has been erected into a secret science. . . . What is war but cunning, deception, delusion, attack and surprise?[92]

This approach had certain advantages, particularly in the early years of the war when, unburdened by standard ideas and approaches, Hitler often conceived unusual methods and measures far outside the conceptual framework of a specialist. For instance, he was deeply involved in the spring 1940 preparations for the seizure of bridges across the Albert Canal and the capture of Fort Eben Emael, which imposing structure lay a few miles south of Maastricht near the confluence of the Meuse and that canal. Less than an hour after the offensive in the west began, German glider forces reinforced by paratroopers had seized the bridges, while one glider group, landing on the roof of Eben Emael, had used shaped charges to put large sections of the fort out of action and to render the entire structure operationally blind. There was much speculation at the time about secret weapons and nerve gas. But as Telford Taylor has pointed out, "the crucial weapons were simply gliders, special explosives, and imaginative planning coupled with bold and speedy execution."[93]

This pattern continued throughout the war. The Nazi leader enthusiastically plunged into the planning minutiae for Operation Felix, the assault on Gibraltar, in the fall of 1940. The pattern of vertical and

horizontal air bombing, the clearing of minefields, and particularly
the scaling of the apparently impregnable eastern or seaward side of
the rock by mountain troops – no detail was too insignificant for
Hitler. "It was his kind of job," one historian has noted, "... the only
kind he was capable of: tricky, small commitment but full of surprises."[94] In a similar manner in 1943, he was completely caught up in the
planning details for the rescue of Mussolini; and during the 1944
Ardennes counter-offensive he helped select the commandos who
would operate far behind Allied lines as part of Operation Greif. Not
surprisingly, Hitler's personal choice to lead both operations was Otto
Skorzeny, a man who, like Hitler, did not play by the rules.[95]

It was this ignorance of the "rules of the game" in the Nazi leader's
self-taught, autocratic mind that played a major role between 1939
and 1941 in surprising adversaries who knew only how to play by
them. Not surprisingly, the preponderance of the German military
was also used to these rules. Despite their vaunted organizational
abilities, however, the military leaders could not establish a workable
high command structure, much less an organization to govern overall
strategy.[96] The field was thus left to Hitler who quickly placed his
surprise-oriented, offensive stamp on initial operations. It was he who
insisted over his generals' objections in the fall of 1939 that the *drôle
de guerre* must be broken by offensive action, a solution Field Marshal
von Manstein later assessed as "fundamentally correct."[97] And it was
Hitler's well-known adoption of the so-called Manstein Plan in the
winter of 1940 that shaped the nature of that offensive against the
wishes of most military leaders who wished to wait until the enemy
reaction could be gauged. The problem with that approach, as Hitler
well realized, was that it would divorce surprise from decisiveness in
the indirect approach – a problem later articulated by the plan's
author. "To wait and see" which way the cat jumped "before deciding
where to place one's main effort," von Manstein wrote, "was tanta-
mount to abandoning the chance of annihilating the enemy forces in
northern Belgium by an outflanking movement from the south."[98]

The imagination and daring that Hitler demonstrated to his
generals were also supplemented by a large admixture of amoral
pragmatism. Simply put, there was no principle, no idea, no article of
faith the Nazi leader would not have been willing to accept or abandon
to reach his goals. "Hitler's ability to repudiate his own conscience in
arriving at ... decisions," Walter Langer pointed out in his wartime
psychoanalysis of Hitler, "has eliminated the force that usually checks
and complicates the forward-going thoughts and resolutions of most
socially responsible statesmen."[99] Added to this was a lifelong con-

spirator's love of plots, intrigue and surprise. It was only natural, in this regard, that he would be intimately involved in the *coup de main*-like operation of 1943 that established the puppet fascist government of Italy and that he would turn to the Arab Freedom Movement in the Middle East as a "natural ally" against England.[100] In a similar manner the Nazi leader would rail at the inability of the Wilhelmstrasse to predict what Britain would do next, simply because the Foreign Office played by the rules. "The best way of accomplishing it," he said, "would be by means of a little flirtation with Churchill's daughter. But our ... gentlemanly diplomats consider such methods beneath their dignity and they are not prepared to make this agreeable sacrifice."[101]

The Rational Actor

Hitler's thought-processes normally began with the emotional and then proceeded to the factual – the reverse of the rational decision-making process. Instead of focusing on a particular problem, he would avoid it and occupy himself elsewhere until "unconscious processes" provided a solution. At that point he would begin to search for facts that proved his solution correct. By the time he was ready to present these facts, the entire product had the appearance of rational judgement. It was an orientation, as Langer pointed out, of an artist. "It is this characteristic of his thinking process," the psychiatrist concluded, "that makes it difficult for ordinary people to understand Hitler or to predict his future actions."[102]

Unpredictability and surprise, as has been demonstrated, however, can be as much the product of imagination and daring as apparent deviation from the rational norm. Certainly, the discounting by the French military leaders of a possible thrust through the Ardennes in April 1940 was not an unsound professional view considering the opposition to that route by the majority of the German High Command. It is also in this context and compared with the behavior of the Western Allies in the early stages of the war, that Hitler's "irrational" decision not to surrender after 1943 can better be gauged. One example is Britain's decision not to yield or ever seek peace in 1940 after the fall of France and before the entry into the war of both the United States and Russia. And of course there was Stalin's decision to continue fighting when the German Army stood ten miles from Moscow in 1941.[103]

Finally there was the fact that adversaries often projected their concept of a rational pattern on to Hitler's actions without understanding that often there was another equally rational pattern behind the decisions leading to these actions. For example, most of the Nazi leader's peacetime coups were preceded by demands before he struck, as in the cases of Austria, Czechoslovakia and Poland. In 1940, the Norwegian Foreign Minister believed that Hitler would issue similar demands before any attack. What he overlooked was that Hitler's behavior in the last years of the preceding decade was formed in a peacetime context, in which he believed he had to justify the use of coercion. After the war began, however, the incentive shifted sharply for Hitler toward assaults without warning. The Norwegian minister, in short, had detected the previous pattern without understanding that the element of military surprise had become dominant in the wartime context as a means of bringing decisiveness to the indirect approach.[104]

The Deceiver

Hitler was by nature a theatrical character. Deception and dissimulation came naturally to him and stood him in good stead in planning military and diplomatic operations requiring surprise, indirect action, force and fraud. It was not an idle boast when he referred to himself as "Europe's greatest actor" (*den grössten Schauspieler Europas*).[105] Certainly, from his *Kampfzeit* period throughout his six pre-war years in power, Hitler's belief that people and nations could be manipulated was reinforced by his diplomatic triumphs as well as the countless party gatherings, whether in small villages or at gigantic Nazi rallies. At the great annual Nuremberg *Parteitag*, for instance, there was always this element of deceptive manipulation with the psychotechnical virtuosity that pulled the masses through forests of banners and cones of searchlights, against a backdrop of trumpet fanfares, night fires, and endless marching columns, to a National Socialist nirvana that could only be *erlebt* without critical self-searching.[106]

It was thus a natural consequence that Hitler continued his involvement in deception operations during the war. The Nazi leader normally initiated all major deception plans by devising the broad outline and even specifying the main theme of each strategic operation, leaving it to his personal military staff to effect the detailed coordination with the appropriate military, propaganda and foreign affairs bureaucracies. In this manner, as early as 1940, deception became a standard procedure in any OKW co-ordination and by the next year, "was a practiced, institutionalized, and routine part of Hitler's strategic planning."[107]

Already in 1939, Hitler had been the driving force behind the construction of the west wall "partly for reasons of military utility and partly for purposes of deception."[108] The deception concerned not so much the construction, but the crude exaggeration of the wall's technical skill and impregnability. Shortly after the initial construction phase, Hitler personally launched a media blitz to leave the impression that similar mammoth works existed along the entire defensive front with the west. Tours of reporters were organized in sectors that would make a strong impression. High-ranking government and party leaders also took part in these tours and uniformly expressed their amazement at the west wall's strength and size to the press. During one such tour, an escort officer learned the quantity of sand used in one sector for making concrete. This quantity seemed considerable to the official, who as a consequence provided the information to visiting German reporters. The problem was that a specialist could have learned from the apparently "large" quantity of sand described that at the most only one-tenth of the alleged bunker could have been built in the sector in question. Publication of the figure was only stopped at the last minute by the personal and vigorous intervention of Hitler himself.[109]

The lavish means and extensive time he devoted to the west wall deception were the principal reasons, Hitler believed, why the German High Command was able to conclude the Polish campaign without being disturbed by its Western opponents. In subsequent discussions of his ideas concerning strategic deception, Hitler returned to the west wall example and the advantages his position as the unitary decision-maker offered in deception operations. It was "particularly important and effective to maintain the deception even during the beginning and initial phase of the operation," he emphasized, in order

> to get through the inevitable moment of weakness which occurs almost three or four days after the beginning of an operation and before the enemy has completely realized the deception and can redeploy his forces against the critical points. It is only seldom that a command has the moral strength to accept great risks and concentrate his [sic] forces at one point in the face of threats at several critical points – whether these threats are real or simulated. Therefore, the forces employed in, and the timing of, the deception must be synchronized with those of the real operation in such a way that this far-reaching effect can be achieved.[110]

It was this personal involvement and direction as a unitary decision-maker that made Hitler so effective in deception operations. As has been demonstrated, it was the Nazi leader's conception in the summer and fall of 1940 that caused preparations to continue for the aborted invasion of England in order to disguise the German build-up against Russia. And it was Hitler who carried off the rare double bluff beginning on 15 February 1941 by presenting the movement of German troops to the east as a deception to cover plans for the invasion of England.[111] In all this, his centralized control allowed him to maintain secrecy effectively by keeping his strategic plans from many top officials and their middle-rank subordinates – a rare example, as Barton Whaley has pointed out, of "a gigantic military operation being effectively concealed from the very civil–military bureaucracy conducting it."[112]

This concealment, in turn, facilitated an incredible "business-as-usual" approach before the invasion of Russia that could only be sustained under the strict rigor of the *Führerprinzip*. On 30 March 1941, for instance, Hitler ordered top priority of war materials to Russia over that of the *Wehrmacht*.[113] Just how far this approach had permeated to the lower echelons was illustrated by a Reich Labor Service Camp for girls near the Russian border that was not evacuated despite the fact that the German attack was scheduled to be launched on both sides of the camp. The division that was to attack suggested that the camp be transferred to the rear a few days before the invasion under the pretext that a maneuver was to occur. This was categorically rejected for reasons of deception. As a consequence, the girls took bicycle trips to the Russian customs barrier on 21 June and carried on normal camp routine until retiring to their tents after a strenuous evening of singing camp songs. During the night the girls were awakened and led to the rear.[114]

The centrality of Hitler to German deception was underscored in the last year of the war by two operations that were successful because on one occasion the Allies forgot to focus on the Nazi leader, and on the other occasion because they focused excessively on him. In December 1944, the Allies were looking for rational military reasons for the Germans to attack in the west; and given the situation in the east, they could find none. But they were lulled by the fact that the very sensible and professional Field Marshal von Rundstedt was the Commander-in-Chief West, ignoring the dominant nature of Hitler's centralized rule.[115] And if they did consider Hitler, it was still in the military sense. No one considered the fact that he might be thinking of a political gambit as a means of obtaining a separate peace in the west.[116] As a consequence, German deception in the Ardennes was achieved for the second time in the war.

The rumors of a possible retreat by German forces into a National Redoubt in the Bavarian Alps were given credence by the Allies in

1944 despite the lack of any evidence, because such action seemed in keeping with Hitler's *modus operandi*. On learning of these rumors, Goebbels began to encourage the speculation on his own and organized within his propaganda ministry a special section that, beginning in January 1945, produced a constant stream of reports concerning an impregnable Alpine position with supplies hidden in bomb-proof caves, underground factories and a force of elite troops to man the entire bastion. When Hitler, for whom a defensive fortress concept was unthinkable, became aware early in the new year that the *Festung Alpenland* was becoming an *idée fixe* with the Americans, he also encouraged the continuation and expansion of the deception. The fictitious operation had momentous consequences for the post-war world, as United States and French troops poured into southern Germany the following spring. That spectacle, as Russell Weigley has pointed out, made "the weight assigned the campaign into the Redoubt appear all the more incongruous in comparison with the weakness of the spearheads that Eisenhower and Bradley extended so tentatively toward an objective the Germans still considered worth defending, Berlin."[117]

Hitler, the Surprised: The Disadvantages of the Unitary Actor

There was an irresolute side to Hitler that counter-balanced the normal picture of the decisive decision-maker moving from one surprise attack to another. This tendency was due, in part, to a fundamental difficulty that Hitler had all his life in maintaining a disciplined focus on any project. In his Vienna days, he would sit up half the night dreaming up grandiose projects that would never come to fruition. There was the rebuilding of the Hofburg, there were designs for concert halls, theaters and museums, social reform projects, and even a stab at writing an opera, "Wieland the Smith," once planned by Richard Wagner.[118] But Hitler carried no project through to completion. Feverish euphoria alternating abruptly with depression was matched by a swing back and forth between hyperactivity and indolence.

This pattern continued throughout his life. During the critical first hours of the 1923 *putsch*, for example, Hitler turned lethargic after his successful speech, issuing no orders from the beer cellar, content to let events take their course.[119] And in the Second World War, there was the postponement of a decision for months in the late spring of 1943 concerning the counter-attack at the Kursk salient, which eventually doomed any possible potential for surprise. Even the operation that began that war was subject to this tendency. Surprise was critical to the Polish campaign, initially scheduled to begin in late August 1939. But Hitler's last-minute cancellation of the operation on 26 August was too late to stop the movement of troops into their final assembly areas near the frontier. "There could be no question now," von Manstein observed, "of catching the enemy unawares .... The element of surprise was lost."[120]

Hitler's intuition, what he termed his "inner voice," contributed to his unpredictability throughout his career. On the other hand, as Walter Langer's wartime psychoanalysis makes clear, this inner voice also caused a rigidity that made it difficult for Hitler to modify his course when faced by unexpected developments or, as in the Russian campaign, by firm opposition. "When he was then confronted by contradictory facts," Otto Strasser noted, "he was left floundering."[121] Added to this rigidity was the fact that Hitler had a typical Austrian *Schlamperei*, an "all-embracing disorderliness," as one early colleague termed it, and in the bohemian manner of the artist would not force himself to work regularly.[122] He never used a desk which he apparently viewed only as a decoration. And normal office hours were foreign concepts. "Hitler literally turned day into night," his press chief wrote, describing the endless night-time conferences of trivial emptiness, in which Hitler indulged his constant need for companionship and from which visitors would emerge in the early hours of the morning, as Speer described it, "dead tired, exhausted from doing nothing."[123]

This pattern meant that the machinery of the authoritarian Nazi state ground to a halt in the critical morning hours in which Hitler slept. More importantly, Hitler's lack of daily routine meant that last-minute decision-making would become the norm as the war progressed, just as he always waited until the eleventh hour to finish preparations for his speeches in order to generate the pressure he needed. Such last-minute improvisation, in other words, did not indicate so much Hitler's flexibility as his inability to exercise his unitary decision-making powers in a timely manner.

The Decline of Intelligence

Under the surface, chaos reigned in the Third Reich, reflecting the basic totalitarian maxim that all authority must be unreliable. Like other dictators, Hitler's internal imprecision in delimiting authority served not only to place the individual German citizen in a state of anguished uncertainty, but also created power centers at every level that expended their energy, dangerous in other circumstances to the Führer's authority, on conflicting rivalries. "Once again I was right,"

Hitler stated, summing up the essence of this concept after the abortive 20 July 1944 attempt on his life. "Who wanted to believe me when I objected to any unification of the *Wehrmacht* leadership? Under one man, the *Wehrmacht* is a menace."[124] This chaotic state of affairs also appealed to Hitler because he believed compartmentalization aided secrecy and because of his concept of what can only be termed bureaucratic Darwinism. At one time, for example, Albert Speer noted that in discussing an armaments organization, the Nazi leader had no established clear lines of jurisdiction. "That way," Hitler explained, "the stronger one does the job."[125]

In these competitive circumstances, co-operation in the field of intelligence among Nazi decision-makers was virtually non-existent, since control of intelligence assessments meant access to Hitler, and such access meant power. The military service, for instance, developed their distinct approaches to intelligence and its use, primarily oriented on providing simple answers to high-ranking officials for immediate use in dealing with the Führer. These tendencies, particularly on the part of the army, also damaged the ability of the OKW to co-ordinate and evaluate intelligence at the joint level.[126] In a similar manner, the OKW had to compete with four outside propaganda offices in attempting to co-ordinate military deception operations, a problem only compounded by the fact that "Hitler himself frequently ... and indiscriminately assigned problems of military deception to some other agencies."[127] The consequences of all this, as Michael Geyer has pointed out, was a perversion of the meaning of intelligence.

> By fragmenting his intelligence network and making the services compete for his favors, Hitler reshaped and transformed the role of intelligence much more radically than by simply demanding one or another ideological bias. When top officials ventured forth, it was not their expertise *per se* that counted, but the "market value" of their current information. This information lost its intrinsic character. It was used to establish credibility in competition, not to discern "truth."[128]

Hitler's feeling about the intelligence structure that he spawned appeared to fit in with his anti-bourgeois distaste for traditional processes, whether diplomatic or bureaucratic. Once in the early 1930s, Speer took him on a tour of the new chancellery in Berlin, remarking to the Nazi leader that he was reluctant to cover the polished marble floor with a runner. "That's exactly right," Hitler replied, "Diplomats should have practice in moving on a slippery surface."[129] Bureaucracies inspired similar contempt. "You must keep free of red tape," he advised Rauschning. "... You must keep your vision clear."[130] And during the war, in one of his endless night-time sessions, Hitler reiterated, "I am often urged to say something in praise of bureaucracy – I can't do it."[131] For the intuitive Nazi leader, such organizations would always be staffed with "experts" who could never possess the true instinct, but were simply "caught in their routine like the spider in its web, incapable of spinning anything but eternally the same web."[132]

Given this attitude, intelligence was bound to suffer in the Third Reich, even had the structure evolved differently. In addition, the actual gathering of intelligence was not perceived as vital for at least the first half of the war, in which Germany was on the offensive and could thus choose the time and place of attack. In such circumstances, other factors dominated intelligence considerations. In August 1939, for example, once the non-aggression pact had been signed with the Soviet Union, Hitler issued an order that effectively forbade any espionage efforts against the new ally.[133] Most importantly, the Nazi leader was not generally inclined to relate his own calculations to the probable intentions and capabilities of the enemy, since he was convinced that his will would always be triumphant in the end. In the face of this will, von Manstein concluded, "the essential elements of the 'appreciation' of a situation, on which every military commander's decision must be based were virtually eliminated. And with that Hitler turned his back on reality."[134]

"What we need is something like the British secret service," Hitler once complained, "an order doing its works with passion."[135] The irony, of course, was that this was impossible in the Nazi state, and that the very intelligence apparatus that did develop in the Third Reich was so mistrusted by the supreme commander as the war progressed, that surprise was virtually assured in many cases against German forces. During the weeks before the Allied landings at Normandy, for example, Hitler received contradictory predictions on the time and location of the invasion from the rival intelligence agencies within the SS, the *Wehrmacht* and the Foreign Office. "How many of these fine agents are paid by the Allies, eh?" he demanded in one instance when confronted with information concerning the possibility of landings in locations other than Calais. "Then they deliberately plant confusing reports. I won't even pass this one on to Paris."[136] As a consequence, Hitler remained fixated on Calais, convinced that the invasion was merely a feint, the purpose of which was to trick him into a wrong deployment of his forces.

Hubris and Surprise

Hitler received abundant, if not necessarily correct, information from the cumulative effect of the routine information flow as well as from digests of the foreign press and excerpts from wire services. "There has probably never been a head of government," his press chief wrote later, "who was so swiftly informed on public opinion throughout the world as Hitler."[137] Moreover, he could be surprisingly open to new information at times. During the war, for instance, the naval liaison officer to OKW headquarters was given the assignment of briefing Hitler on the previous day's naval events, which had been disastrous from the Mediterranean up to Norway. Jodl had already warned him that he should be prepared for an "eruption of the volcano." Instead Hitler took the briefing calmly and gave a reply that was revealing not only concerning his openness to information, but his conception of his own decision-making process as well.

> Gentlemen, you have just heard the report of the naval situation. It is a deplorably unfavorable picture. I wish that all branches of the *Wehrmacht* would give me such an unvarnished picture of the situation. Bear in mind that my brain works in about the same way as a calculating machine. Each of the officers who makes a presentation here introduces into this calculating machine a small wheel of information. There is formed a certain picture; or a number, on each wheel. At the conclusion of a conference, I press a button and there flashes into my mind the sum of all this information. This is the estimate of that particular day. When one of these wheels gives me a faulty picture as the result of an incorrect presentation, then I must necessarily make an incorrect estimate of the situation, thus drawing false deductions, making a false decision and issuing completely faulty orders. You yourselves can appreciate what that means for the High Command, yes for the entire conduct of war. Consequently, I would like to take this opportunity to request most earnestly that all persons giving presentations here should give a really clear and unvarnished picture of the situation.[138]

The reality, of course, was quite different. Hitler's growing belief in his own omnipotence combined ironically with a fundamental ego weakness meant that criticism and openness normally gave way to servility and sycophancy. This was particularly true with the ever-narrowing inner staff circle around the Nazi leader who protected him

from unwelcome information and made personal networks and con-
tacts even more critical for reaching the Führer's ear. Martin Bormann was a prime example, in this regard, as the war progressed; but
it was Field Marshal Keitel who consistently and over the longest
period gave new meaning to the term "toady."[139] In January 1938,
when Field Marshal von Blomberg advised Hitler that Keitel was
"suitable only for the position of administrative assistant," Hitler
immediately replied: "That's exactly the man I'm looking for."[140] In
subsequent years, as he advanced within Hitler's entourage, Keitel set
new standards for obsequious servility. He soon earned the nickname,
"*Lakaitel*" (lackey) and was also known as the "nodding ass."[141] In the
winter of 1943-44, Hitler's chief adjutant led an attempt to replace the
Field Marshal that was firmly rejected by the Nazi leader who stated
that Keitel was irreplaceable because he was "loyal as a dog."[142] After
the bomb attempt on Hitler's life on 20 July the following summer,
there were no more attempts to replace Keitel. In the immediate
aftermath of that explosion, the Field Marshal picked himself up out
of the dust and saw Hitler standing relatively uninjured in the
wreckage. He immediately rushed to the Nazi leader and, ignoring all
convention, wildly embraced him, exclaiming: "Mein Führer, you're
alive, you're alive!"[143]

The development of this inner circle staffed by subordinates such as
Keitel caused a further gap to develop between the administrative
level of the intelligence community and the decision-making tier.
This, in turn, exacerbated rivalries between individuals and groups
which were already, as has been demonstrated, fixtures in the structure of the National Socialist regime. The inner staffs only further
personalized a system for processing information that enhanced the
likelihood of unrealistic intelligence and threat assessments. This
tendency became more pronounced after 1941 as Hitler and his
entourage moved to a series of isolated field headquarters that
confined the decision-making process to even smaller circles.[144]

"Yes-men" only further reinforced an all-embracing hubris in
Hitler that had begun years before. Already in 1924, there was a
feeling of mission and vision that outstripped mere mortals, for "the
greater a man's work for the future, the less the present can comprehend them."[145] And ten years later, after being greeted by thousands of people while traveling by car in the countryside, Hitler
commented to Speer: "Until now only one German has been hailed
like this: Luther."[146] These feelings, reinforced by his triumphs in the
remainder of that decade and by a romantic streak that caused him to
view himself as the embodiment of the *Furor Teutonicus*, a reborn

Attila, spilled over into Hitler's role as war-lord after 1939. This was particularly evident after the 1940 western campaign when he began to comment *ex cathedra* on military affairs as he always had done on politics. It was after that campaign that Keitel bestowed on the Nazi leader the title, "*Grösster Feldherr aller Zeiten*," a phase that always elicited a pleased grin from Hitler. By the end of the war, his generals were referring to the Führer as "*Gröfaz*," an ugly acronym for Hitler's title that could have described, as Gordon Craig has pointed out, some monster of mythology.[147]

All this fueled a growing sense of omniscience in Hitler, which made him increasingly resistant to new or contrary information as the war progressed, and thus more susceptible to surprise. "He was unteachable," his press chief recalled, and General Halder described Hitler's "almost wild-animal perception for anything which ran counter to himself."[148] Once during the war, the Nazi leader was whistling a tune when a secretary had the courage to suggest that he had made a mistake in the melody. "I don't have it wrong," Hitler raged in reply. "It is the composer who made a mistake in this passage."[149]

Overwork and isolation played a part in all this as did an increasing sense of his indispensability. "It's easy to advise me to take a vacation," Hitler frequently replied to the urging of his staff, "but it's impossible. I cannot leave current military decisions to others even for 24 hours."[150] An air of unreality began to pervade the decision-making process. In one instance, warned that the Soviets could easily render a key road impassable on the southern front, Hitler immediately went far beyond the immediate goal. After securing the road, he concluded, German forces would "start an offensive into the underbelly of the British Empire. With a minimum of effort we can liberate Persia and Iraq."[151] In such an atmosphere, even military surprise perpetrated on German forces could not be accepted immediately by the Nazi leader and his subordinates. Hitler's reaction to the Normandy invasion is one example. Another occurred when Reich Marshal Göring was told earlier in the war that an Allied fighter was shot down over Aachen, thus demonstrating that long-range fighter escorts for bombers over Germany did exist. "I'm an experienced fighter pilot," Göring replied. "... I officially assert that American fighter planes did not reach Aachen .... I herewith give you an official order that they weren't there."[152]

Ethnocentrism contributed heavily to this air of unreality and resulted in a string of military surprises against the Nazi leader throughout the war by both his enemies and his allies. In this regard,

Hitler remained at heart an Austrian *Landsman*, never fully comprehending the sea and the vast spaces to the east, who could recall in
lyrical tones his initial train trip to the war front in 1914 when "for the
first time I saw the Rhine as we rode westward along its great waters to
defend it, the German stream of streams."[153] This tendency was
summed up by his military attaché in the United States who pointed
out that Hitler "never had any experience outside of Central Europe
and eagerly clutched at everything which fitted into his own picture of
the world."[154]

Part of that picture was formed by the Nazi leader's temperament.
A putschist at heart, a master of the *coup de main*, he never could
understand the cautious Western strategy and always expected the
Allies to act more boldly than they did throughout the war. The Allied
landings in North Africa were one example. There were many intelligence reports on the evening of 7 November 1942 of naval units on
the move in and around the Mediterranean, enough to surmise that a
large operation was under way. That night, Hitler proposed several
different explanations for the movements, all of which revealed not
only the fundamental deficiencies of German military intelligence,
but how the Nazi leader's perception of the enemy was colored by his
own earlier experiences and his concept of risk. He considered the
most likely enemy option to be a huge supply operation to reinforce
the offensive against Rommel's African forces. His alternative
reflected his penchant for risky operations, considering the German
dominance of Europe. "The enemy will land in central Italy tonight,"
he stated, warming to the subject. Once more the revolutionary with
his love of duplicity and surprise emerged. "I would occupy Rome at
once and form a new Italian government. Or, and this would be the
third possibility, I would use this great fleet to land in southern
France."[155]

Hitler's guesses were wrong, of course, because he could not
comprehend a strategy so alien to his background. Placing troops on
land in safe positions in order to begin a methodical extension without
unnecessary risk was so foreign to this survivor of the *Kampfzeit* as to
be almost incomprehensible. The element of surprise in such a
situation could only be writ small, even in a strategic sense, for the
man who had conceived the sudden onslaught of Barbarossa.

The Nazi leader's ethnocentrism was also rooted in his ideological
and racist views, which in turn contributed to his vulnerability to
surprise. He believed, for instance, that the Jews "were the source of
all evil . . . whom he had been given a divine mission to destroy."[156] The
persecution of this group eliminated many skilled personnel both

from the various intelligence agencies and from the vast array of scientists on whom Hitler depended for achieving technological surprise. To the Jews, Hitler added Slavs and other "mongrelized" groups that he considered to be racially, not to mention politically, inferior, as a principal basis for the prism through which he viewed potential enemies. The result was consistent underestimation of his most powerful adversaries, the United States and the Soviet Union.

Hitler viewed America as the primary example of the negative results of a mongrelized society. To him, this admixture of ethnic groups accounted for the general panic in 1938 that resulted from the Orson Welles Mercury Theater radio broadcast of H.G. Wells' *War of the Worlds*. This impression was reinforced by John Ford's 1939 movie of Steinbeck's *The Grapes of Wrath*, a grim picture of the depression, which Hitler viewed several times and which he believed represented the entire United States for all time, a nation "permanently on the brink of revolution."[157] The apotheosis of this degenerate state was Franklin Roosevelt who, Hitler often asserted, suffered from syphilitic, not infantile, paralysis, and whose 1933 inauguration was described by the Nazi leader as "the last disgusting death-rattle of a corrupt and outworn system."[158] By 1941, Hitler could confidently assert that if the United States were to work feverishly for four years, it could not replace the material that the Russian Army had lost thus far.[159] This underestimation was not limited to production and supply capability. "I'll never believe," Hitler stated shortly after Pearl Harbor, "that an American soldier can fight like a hero."[160]

In a similar manner, the Nazi leader viewed the Slavs of Soviet Russia as *Untermenschen*, sub-humans, against whom war would be "child's play."[161] This perception was buttressed by Hitler's memories of the Russian effort in the First World War and compounded by the incredibly vast 1936–38 purges of the Soviet military and the poor Soviet performance in the 1939–40 winter war against the Finns.[162] For him, the internal rottenness of the Soviet state would crumble of itself without any positive political and propaganda appeal to the people, a belief reinforced by the OKH leadership confident of its own ability and anxious not to repeat the pessimistic enemy overestimations of previous years that had brought Hitler's wrath down on the General Staff.

Added to this were the various intelligence agencies all vying to please their leader and his preconceived notions of the Slavic enemy. The results were appallingly bad intelligence estimates. "At the beginning of the war we reckoned with about 200 divisions," Halder wrote in August 1941. "We have already counted 360."[163] In addition, there was the vast underestimation of Soviet armor, not only in terms of quantity, but quality as well. "If someone had told me that the Russians had ten thousand tanks," Hitler said in early 1942, "I'd have answered: 'You're completely mad'."[164] In fact, the Russians possessed many more tanks than this, many of which were the new T-34 and KV models. And although German intelligence estimated Russian tank strength to be 15,000 after the unpleasant surprises in the summer of 1941, the total was probably closer to 24,000.[165]

There is some evidence that despite his racial and ideological predispositions, Hitler, along with the world, held his breath when Operation Barbarossa began. Several months before, he had cautioned that "the Russians should not be underestimated, even now."[166] Nevertheless, there was a feeling of ideological release when the invasion began. The night before, the Nazi leader wrote to Mussolini that he felt more "spiritually free" since he was terminating "the hypocritical performance" which was "a break with my whole origin, my concepts and my former obligations."[167] This outlook continued to dominate his thinking even after the Nazi tide began to turn in the east. The result was continued and heightened vulnerability to surprise.

In late November 1942, for instance, Hitler was at the Berghof in Obersalzburg when he received news of the great Russian winter offensive that was to culminate nine weeks later in the capitulation at Stalingrad. As details were recounted of Soviet troops beginning to overwhelm German divisions, Hitler paced the great hall of the Berghof, making tactical and operational intelligence reports fit his ethnocentric strategic appraisals.

> Our generals are making their old mistakes again. They always overestimate the strength of the Russians. According to all the front-line reports, the enemy's human material is no longer sufficient. They are weakened; they have lost far too much blood. But of course nobody wants to accept such reports. Besides, how badly Russian officers are trained! No offensive can be organized with such officers. We know what it takes! In the short or long run the Russians will simply come to a halt. They'll run down. Meanwhile we shall throw in a few fresh divisions; that will put things right.[168]

Ethnocentrism and egocentrism also played a part in Hitler's failure to use his allies effectively, a failure which created situations in which those allies often repaid the Nazi leader in his own currency of surprise. This was particularly true in the relationship between Germany and Japan, both of which, as one historian has noted, "practiced such secrecy and deception concerning their own objectives that even on those few occasions when their interests genuinely converged they were unable to coordinate their policies."[169] In such circumstances, a common conspiracy against peace was merely the fortuitous criss-crossing of two lines of force, each in the pursuit of different objectives. These two moments of confluence occurred in September 1940 when the Tripartite Pact was concluded and in December 1941 when both powers declared war on the United States.

Despite the Tripartite Pact, Hitler failed to inform the Japanese in the spring of 1941 of the impending invasion of Russia and in fact deliberately misinformed his erstwhile ally. At a conference on 20 April, for instance, Hitler stated that a Japanese envoy had been "informed that Russia will not be attacked as long as she maintains a friendly attitude in accordance with the treaty."[170] This gratuitous insult had momentous consequences. Had Hitler requested assistance from, or even consulted with, Japan before the German invasion, that nation might have been persuaded to invade Russia simultaneously from the east.[171] As it was, the Japanese High Command decided in September 1941 to postpone any military action against the USSR until at least the spring of 1942 and plunged immediately into southeast Asia, a move that would place Japan increasingly on a collision path with the United States.[172]

Hitler was both surprised and initially elated when the Japanese paid him back in kind with the attack on Pearl Harbor. When he heard of it, he asked his assembled staff, comprising all the military services, for the location of the American naval base. Not one member could find Pearl Harbor, a fact that, as Williamson Murray has pointed out, "speaks volumes on the subject of the strategic Weltanschauung of people who aimed at the conquest of the world."[173] As a result of Pearl Harbor, and for a variety of reasons that certainly included his skewed perception of the United States, Hitler went to war against that country on 11 December. Had the Nazi leader received advance warning of the Japanese attack, he might have arrived at a different, less viscerally reactive course of action – one more in keeping with an earlier declaration that he would "never be so foolish as the old Kaiser and declare war on anybody."[174] What is certain is that the surprise of Pearl Harbor brought no closer strategic co-ordination between the two allies. "What is happening in the Far East," Hitler commented less than two weeks after the surprise attack, "is happening by no will of mine."[175]

Similar situations occurred with the Italians. Nevertheless, Hitler retained a fierce kind of Nibelungen loyalty to Mussolini, despite a series of unpleasant surprises throughout the war, the most egregious of which was the Italian attack on Greece. Hitler was worried about the possibility of this attack and traveled to Florence in order to dissuade the Duce from any such venture. When he alighted from his train on 28 October 1940, however, he was met by an elated Mussolini who immediately announced that Italian forces had crossed into Greece. Thus began Germany's deepening involvement in the Balkans, which was to cost valuable, possibly decisive, time for Barbarossa as Hitler, in Jodl's words, "found himself forced to hatch the cuckoo eggs that Italy had laid in the nest of the joint war effort."[176]

Technological Surprise

Hitler was aware of the value of technological surprise. His own First World War experiences with gas as well as the front-line stories exchanged about confrontations with British tanks were indelible memories. His subsequent experiences only reinforced his tendency to consider the entire matter in Darwinistic terms. "Necessity teaches men not merely to pray, but ceaselessly to invent," he stated. "Every new invention so much reduces the value of the previous material that it is a ceaselessly renewed struggle to introduce a novelty."[177] It was this novelty, the ability to surprise, that was crucial to the Nazi leader in terms of developing weapons and equipment and the techniques to use them. At a conference in 1943, for example, Hitler reminded Admiral Raeder that both the navy and the air force in 1938 had convinced him that it would not be advisable to develop an aerial torpedo force. Together, probably the only time the two services were in full agreement, he pointed out, they worked up "wretched demonstrations" that made him abandon his plans "for a strong torpedo force which could have been used as a surprise measure at the outbreak of war."[178]

Hitler and the Success of Technical Surprise

Hitler had always been interested in new technology. The works of Jules Verne had fired his imagination as a boy. Later, before becoming chancellor, he grew animated in conversations concerning inventions and technological breakthroughs, pointing out bitterly that "we do not develop them. We allow the possibilities to moulder." Ad hoc approaches to such breakthroughs were not the answer, he concluded; "what was once an accident, must become planned."[179] The Nazi leader's interest in technology continued throughout the Second World War as did a general openness to new discoveries since he was "firmly convinced that technical changes will continue alternately to favour or impede offensive or defensive warfare. One must therefore not become discouraged but ever be receptive to new ideas. A man whose mind is closed thereby admits his defeat."[180]

Hitler combined an incredible grasp of detail with a gift for creative fantasy in technological issues and matters of armament and equipment. He was particularly knowledgeable about the traditional weapons of the army and the navy that had evolved since the First World War, and frequently proposed, as Speer admitted, "convincing and usable innovations."[181] It was Hitler's suggestion, for instance, to fit the Stuka dive bombers with sirens for psychological effect, and it was at his insistence that the highly successful 77mm anti-tank gun was selected to replace the 37mm and 50mm guns. Moreover, it was the Führer's initiative that modernized German armor in the forms of the Panther, Tiger and King Tiger tanks. General Jodl summed up his leader's involvement: "Hitler's astounding technical and tactical vision caused him to become the creator of modern weaponry for the Army."[182]

Technological surprise could also result from using existing weapons and equipment in unorthodox ways. At the strategic level, for example, Hitler set his sights early in the war on the Azores as the only facility for launching surprise strategic air attacks on the United States if that country should enter the war. "Thereby," he concluded, "America would be forced to build up her own anti-aircraft defence, which is still completely lacking, instead of assisting Britain."[183] In another example, at the tactical and operational level, Hitler was attuned from the outset of the war to the potential for surprise inherent in such unorthodox means of attack as airborne and glider assaults. In this regard, the Nazi leader personally suggested the use of gliders against the Aegean Islands, particularly on Lemnos off the Dardanelles, during Operation Marita. And he was intimately involved in the planning for the use of both glider and airborne forces for the rescue of Mussolini in 1943.[184]

Only Hitler's central position in German decision-making allowed him to play such roles. When his interest was aroused in any piece of technology, whether weapons or equipment, his energy and enthusiasm focused through his dominant unitary position could quickly impel a developmental project to the top of any priority list. At the strategic level, for instance, there was his support for such massive projects as the V-l/V-2 rockets and the jet plane. In contrast, there was his critical role at a lower level in the development and production of the first tactical assault rifle for the basic infantryman. The proponents of such projects could expect to be inundated with requests for progress reports, technical data and a myriad of associated details. When, for example, Admiral Raeder reported to Hitler on the new "electro" submarine, he and his staff were bombarded with intricate questions on such diverse technical matters as radius of action, recharging requirements and depth of submergence.[185] "Again and again," the original keeper of the OKW Diaries noted of such meetings, "he astonished his audience by his comprehensive and thorough knowledge of technological matters, in which he was excellently supported by his memory."[186]

Hitler and the Failure of Technological Surprise

The centrality of Hitler in decision-making during the Second World War was the principal reason for Germany's failure to achieve technological surprise in a consistent and meaningful way. Hitler, the Bohemian, the unsystematic artist dashing enthusiastically from one pet project to another, could no more set priorities in the technological field than he could in the strategic field. True to form, the Nazi leader was always looking for the simple, quick, technological solution, a miracle secret weapon that could turn the tide of war. In the early summer of 1942, for instance, he personally ordered the first six Tiger tanks to be used in battle. He then regaled his entourage with vivid descriptions of how the Soviet 77mm anti-tank gun, which could penetrate the German Panzer IV's front armor, would be useless against the Tiger.[187] Later in the war, Hitler demonstrated a similar yearning for a technological panacea when he met Field Marshal Rommel, whom he had appointed as technical inspector of the coastal defenses in the west. During the discussions, Rommel pointed out that unless the Allied bombers could be stopped, the invasion would succeed once the Allies had a lodgement on the continent. In reply, the Führer showed Rommel an experimental armored vehicle on which an 88mm anti-aircraft gun was mounted. "You see," Hitler concluded, "with this armored flak weapon we can take care of the concentration of bombers over our divisions."[188]

In other more major technological areas, Hitler's unsystematic fascination with creating surprise weapons was equally disastrous. The idea of the V rockets, for instance, appealed to his romantic streak. But his interference and mercurial moods concerning the weapons, as early as 1941, meant the late arrival of flawed rockets in

1944. On 22 June, the over-anxious Führer fired the first ten V-ls prematurely. Only five reached London. Hitler was on the point of halting production of the weapon, which he had supposed would reshape the course of the war, when his press chief handed him some exaggerated and sensationalized reports from the London press on the surprise effect of the V-l. Hitler immediately changed his mind.[189] On the other hand, the Nazi leader's commitment to the huge V-2 rocket from July 1943 was, as David Irving has noted, "an extravagant irrelevance" that siphoned off a large amount of industrial capacity which could have been used in manufacturing more mundane but useful weapons such as the ground-to-air defensive rocket that had been developed in 1942. That weapon, with its ability to hit enemy bombers at 50,000 feet, as Albert Speer pointed out, would have achieved complete technological surprise and, combined with the jet fighter, might have beaten back the 1944 Allied spring air offensive.[190]

The jet fighter, of course, also fell victim to Hitler's interference. The Me-262 fighter plane, with two jet engines and a speed of over 500 miles per hour, was the most valuable of the German "secret weapons." In September 1943, however, without explanation, Hitler ordered preparations for large-scale production of the jet to stop. In January 1944, he ordered resumption of production, but directed that the plane, which was built to be a fighter, was to be used as a fast bomber. Despite objections, Hitler, not surprisingly, prevailed. But the element of useful surprise had been lost through this change in role and the concomitant production delays.[191] By that time, Hitler's enthusiasm for new gadgets and his lack of discipline meant that Germany was suffering from an excess of "secret weapons" ranging from a rocket that hunted enemy planes by tracking heat waves from their engines to a torpedo that reacted to sound and was thus capable of pursuing a ship seeking escape in a zigzag course.[192] Priorities for all these projects depended on the whim of the technically illiterate Führer.

Despite this "illiteracy," Hitler possessed a remarkable technological aptitude combined with an equally phenomenal memory and mastery of details. But since he lacked serious training and inclination for the hard sciences, his comprehension stopped where simple technology could not proceed without such aids as physics and chemistry. Years before in Vienna, Kubizek had commented on a similar tendency in the young Hitler. Kubizek was a musician and noted how rapidly Hitler acquired all the technical terms of music and was able to speak, in this regard, "about everything without having studied it systematically .... How often was I surprised at his opinions, seeing that he did not know a thing about the question at hand."[193] Added to this lack of training was Hitler's experience as a First World War infantryman, which made him slow to appreciate the significance of the new scientific developments. It was impossible for Hitler ever to deal rationally as a unitary actor with all the emerging technological possibilities out of his own mental resources, a condition summarized by one of his personal physicians:

> He had little interest in other technical questions, particularly in the area of physics, such as high-frequency technology or atomic physics and the like, even though they were brought to his attention. Consequently, he came to recognize their significance in warfare only when enemy progress in these areas had already won a decisive influence in the sea and air war.[194]

Hitler's hubris and personal prejudices also spilled over into this arena. His belief in his own superiority in technological questions had an inhibiting effect on the development of many weapons, the early surprise introduction of which might have altered the war. As early as 23 June 1942, for instance, Hitler discussed the possibility of an atomic bomb. But other projects, particularly the development of the V rockets at Peenemünde, were drawing off resources. More importantly, Hitler's anti-Semitic policies had depleted the ranks of atomic scientists; and the entire atomic bomb effort still had Jewish associations in the mind of the Nazi leader who occasionally referred to nuclear physics as "Jewish physics."[195] Without funds and scientists, Speer scuttled the project to develop an atom bomb by the fall of 1942. At that time he was told that it would be at least three or four years before the project could come to fruition.

Hitler's hubris also played a major part in one of the great Allied technological surprises perpetrated on Germany during the war. In August 1942, the Nazi leader established his guidelines for defense in the west based on the fundamental assumption that the Allies could not succeed in an invasion of Europe unless they were able to seize a sizeable port. Without such a port, he emphasized, an Allied force landing along French, Belgian or Dutch coasts could not receive enough supplies for long enough to withstand German counter-attacks. As a consequence, Hitler began to plan in detail the defensive installations of the larger ports himself, thus assuring a lack of attention to the widely scattered observation bunkers on the intervening coastal areas. Often working late into the night, he personally designed all of the various types of bunkers and pillboxes that ringed the ports. His designs, though only sketches, were naturally adopted by the engineer commander almost without revision. As construction proceeded, the Nazi leader, working from a complete set of 1:25,000 maps that showed all German coastal installations in the occupied countries, demanded a constant stream of detailed reports on the fortifications. He studied these maps frequently, and in one subordinate's judgment, "probably knew the location of defenses in detail better than any single army officer."[196]

All this effort was wiped out within two weeks of the initial Allied landings by the use of artificial harbors. German intelligence had spotted the harbors in England, but believed, as General Jodl reported, that their "purpose was to form new quays in place of those destroyed in the port .... It came as a surprise to us."[197] By bringing their own ports with them, the Allies rendered Hitler's plan of defense irrelevant. Technological surprise was thus a key ingredient in effecting the Allied lodgement in Europe.

The use of poison gas by Germany in the Second World War would certainly have added another element of technological surprise, if not at the strategic, at least at the operational and tactical levels of war. Why Hitler did not use it can only be a matter of conjecture. Certainly there was his own experience of a gas attack before Ypres on 13 October 1918, from which he "stumbled and tottered back with burning eyes."[198] But years later he could sit on the veranda of Wachenfeld House in the Obersalzberg humming the motif, "by pity enlightened (*durch Mitleid wissend*)" from *Parsifal* as he discussed gas warfare. "The new poison gases are horrible," he admitted at that time. "But there is no difference between a slow death in barbed-wire entanglements and the agonized death of a gassed man."[199]

What is certain is that Hitler generally opposed the offensive use of gas to the extent that he prohibited the transfer of any gas munitions outside the pre-war Reich borders for fear of accidental use. These munitions were extensive, primarily consisting of mustard gas and Tabun, a new secret nerve-gas that penetrated the filters of all known gas masks and produced fatalities after even limited contact.[200] Hitler continued to keep absolute control over these supplies throughout the war. How concerned he was on this matter was demonstrated in its closing months, as the Germans retreated on both fronts, by the issuance of an order over his signature that no toxic chemical dumps were to be blown up in case such an action might be construed as initiation of gas warfare. There is evidence that Hitler did give some consideration to gas warfare. He refused, for example, to let Speer stop poison gas production in 1944. And at one meeting, he brought up the subject of using gas against Russian troops, but when no one responded, never raised the possibility again.[201] For the rest of the war, he remained opposed to initiating gas warfare.[202]

If Germany had used the Tabun nerve-gas in V weapons it could have been a serious threat to the Allied war effort. But the retaliation would have been worse, a fact that must have been obvious to Hitler. By June 1944, the Allies had both sufficient gas and proven delivery systems to depopulate Germany's major cities.[203] And the will to use these weapons, not just in retaliation but offensively as well, was certainly not lacking. In July 1944, Churchill suggested that the British Chiefs of Staff consider attacks against rocket sites and German cities, concluding that "we could drench the cities of the Ruhr and many other cities in Germany in such a way that most of the population would be requiring medical attention."[204]

There was a more straightforward reason for Hitler's failure to use airborne troops consistently throughout the war, particularly after the opening successes in Belgium and Holland. Once more the Führer confronted the same situation that had stymied him during Operation Sea Lion. Once again in another operation against the British, this time in the Greek campaign, it was precisely when the indirect approach became costly at the tactical and operational levels, because of the lack of military surprise, that Hitler turned away from this approach at the strategic level. The operation against Crete (Mercury) on 15 May 1941 had all the attributes of daring and imagination that ensured Hitler's approval, if not undivided attention because of the impending invasion of Russia.[205] Nevertheless, German paratroopers ran into unexpectedly stiff resistance at their four drop zones. The result was 6,580 dead, wounded and missing out of a 22,000-man assault force, a relatively low price for the seizure of the critical island, yet still involving more fatal casualties for the *Wehrmacht* than the entire Balkan campaign.[206] It was these casualties that made Crete, as General Student, the airborne commander, summed it up, "the grave of Germany's parachutist."[207] But it was Hitler who supplied the missing pieces. "Crete proved that the days of parachute troops are over," he advised Student the following July. "The parachute arm is one which relies entirely on surprise. In the meantime, the surprise factor has exhausted itself."[208]

On 20 June 1941, Goebbels published in the Nazi party organ, the *Völkischer Beobachter*, an article entitled "Crete as a Model," from which it was easy to conclude that Crete was only a final rehearsal for the imminent invasion of England. Goebbels had written the article at the behest of the OKW, presumably with Hitler's clearance. The OKW then pretended to attempt to stop publication and confiscate the article.[209] There was a certain irony in using the Crete operation to bolster Germany's successful double deception against Russia. For the failure to achieve surprise in that operation also dissuaded Hitler from ever again attempting major parachute operations, thus depriving the eastern *Blitzkrieg* of a significant force that would have greatly contributed to psychological as well as operational surprise.[210] In this regard, the tactical British defeat could be considered a strategic success.

Doctrinal Surprise

Doctrine can be a source of military surprise in warfare. In the 1973 Yom Kippur War, for instance, Israel was unaware of important alterations by Egypt and Syria in their military doctrines concerned with control of the air above troop dispositions. The changes, combined with lack of information on new anti-aircraft weapons, adversely affected evaluation by the Israelis of both enemy capabilities and intentions.[211] No such secrecy surrounded the tactical and operational military doctrine of *Blitzkrieg*, which thanks to propaganda became popularized even before it was fully accepted by the *Wehrmacht*. "The expression '*Blitzkrieg*' is an Italian invention. We picked it up from the newspapers," Hitler commented sarcastically during the war. "I've just learnt that I owe all my successes to an attentive study of Italian military theories."[212]

Blitzkrieg had its source in earlier German doctrine. After the First World War, the German Army returned to its pre-war offensive doctrine with only minor changes to reflect not only the experience of that conflict, but Weimar Germany's military weakness as well. In this sense, it was a continuation of the style of warfare particular to the Germans-cum-Prussians since the age of Frederick II, with its emphasis on mobility as a means of conducting flank attacks, envelopments, and encirclements – all forms of the battles of annihilation. Political and geographical realities had contributed to this doctrine over the centuries. The one constant during this time had been the existence of German enemies on multiple fronts with a totality of superior military force that could only be met with swift, sequential attacks before those adversaries could co-ordinate and conduct coalition warfare. The evolvement of the offensive German military doctrine, then, lay primarily with the problem of a two-front war. Only by means of speed could this problem be solved. Only by exploiting speed could Germany dictate the time and place of battle to the potential adversary.[213]

Thus, many of the key ingredients in *Blitzkrieg* were not new. Combined arms operations, for instance, had been used in the infiltration tactics devised for the Ludendorff offensive of 1918 and had been subsequently enlarged upon in the Weimar era by General von Seeckt, the chief of the *Reichswehr*. The 1936 manual, *Die Truppenführung*, continued to emphasize such operations as it did other doctrinal elements that were later to be associated with *Blitzkrieg*. There was the focus on lower-level initiative in mission-type orders that were "to avoid going into detail when changes in the situation cannot be excluded by the time they are carried out."[214] And the manual changed the purely support role of armor by emphasizing that "if tanks are too closely tied to the infantry, they lose the advantage of their speed and are liable to be knocked out by the defense."[215] In fact, the *Blitzkrieg* method of attack could also be characterized as nothing more than the First World War tactics of infiltration elevated to the operational level of the campaign and supplemented by technological advances in air and land weaponry. Seen in this light and the fact that *Blitzkrieg* also stressed the importance of decisive offensive operations, there is certainly a strong case to be made that the new doctrine was a logical evolutionary development of traditional German military strategic thought.

But there was an important difference. The focus of the *Blitzkrieg* as it moved through distinct phases turned as much on the disorientation and dislocation of the enemy command, control, communications and intelligence (C^3I) as it did on the annihilation of enemy forces. In the breakthrough phases, tactical battles fought along the front were merely the preconditions for penetration and exploitation. In this context, so long as the mobile units could gain entry into the defender's rear, it hardly mattered how the battles along the front away from the *Schwerpunkt* proceeded. As a consequence, attacking commanders could choose the breakthrough points opportunistically, which meant eventually that a successful defense against any one *Schwerpunkt* would result in encirclement and capture once penetration was executed.

Speed and momentum were the keys to penetration and exploitation. At the tactical level, the long thin columns of vehicles presented almost no front and seemingly vulnerable flanks. At the operational level, however, the tempo of these columns, moving opportunistically along whatever routes offered the least resistance, maintained a swift and ceaseless forward momentum that made it hard for defenders to react in a timely manner. Moreover, those columns were normally moving vertically across a horizontally organized defensive front, and their advance often overran enemy command centers and cut lines of communications. To all this was added the fact that attacking units were guided by the traditional mission-type orders that decentralized decision-making to commanders at the lowest possible tactical and operational levels. This so-called *Auftragstaktik* allowed these commanders to use initiative and flexibility in reacting swiftly to any opportunities that presented themselves, thus creating a continuous succession of tactical and operational surprises for the enemy.[216]

Typically, all this allowed the *Blitzkrieg* attacker to get within the intelligence-decision-action cycle of the defenders who, overwhelmed by a surfeit of intelligence "noise" from the myriad sightings across the front, normally vacillated between the paralysis of doing nothing while attempting to make sense from the chaos, and the dispatch of mobile forces in pursuit of sightings that appeared most credible. In this environment, as long as the invasion columns maintained a high tempo of forward movement, "their apparent tactical vulnerability was dominated by their operational advantage since the defender's intercepting and blocking actions would always be one step behind."[217] Moreover, the bulk of the defending forces still holding on to those parts of the front between the attacker's axes of penetration would begin to suffer moral disintegration as news from the rear was received. A not uncommon result was abandonment of frontal defenses still intact by demoralized troops racing to the rear. The final stage was left to the infantry advancing across this abandoned frontage as it fell upon the fragmented and disoriented defensive forces trapped within the encirclements of the mobile units that had penetrated earlier.

Hitler and the Use of Doctrinal Surprise

After the Second World War, J.F.C. Fuller pronounced judgement on Hitler's contribution to the concept of *Blitzkrieg*:

> As a tactical theorist, Hitler was as clairvoyant as he was astute as a politician. He had watched the last war closely and had absorbed its tactical lessons – a remarkable thing for a corporal to do. But what was more remarkable, he projected them into the future and built his military power on them.[218]

This is certainly a vast overstatement. Nevertheless, the fact remains that the concept of *Blitzkrieg* required some major changes in the German Army in terms of armor and mechanized warfare which met with institutional resistance. The army preferred to superimpose new technology on its offensive doctrine, rather than to experiment and innovate doctrinally in order to exploit to a greater degree all its potential.[219]

This resistance, most historians agree, was overcome through Hitler's major commitment to the *Blitzkrieg* cause. When that conversion actually took place is open to debate. But the fact remains that by the mid to late 1930s, Hitler was extremely interested in the machinery of *Blitzkrieg*. "That's what I need," he is reported to have stated in February 1935 after his first glimpse of tank maneuvers. "That's what I want to have."[220] Hitler was also interested in the interaction of the new technology. "Practice has shown that any one arm acquires its maximum efficiency only when used in collaboration with other arms," he stated. "The various weapons are indeed so interdependent that success in war is achieved by the skillful and combined use of all of them."[221] The result was outstanding co-operation between the air force and the army as the concept of *Blitzkrieg* developed. And while this was also due to the extreme youth of the Luftwaffe as well as the large number of army officer transfers, it was primarily the overwhelming political and budgetary support provided by Hitler and his political hierarchy that prevented inter-service rivalries. "While one must be careful not to overstate how fully or correctly Hitler conceptualized the *Blitzkrieg*," Barry Posen has concluded, "his support for its various elements must be credited as the key pressure that brought the doctrine to operational fruition."[222]

The attraction of a concept like *Blitzkrieg* for Hitler, however, went far beyond the technology and machinery of war. It was ideally suited to his personality and temperament. To begin with, there was the re-emphasis on speed, a quality that had always fascinated Hitler. In *Mein Kampf*, he listed "swift as greyhounds" as a key military virtue.[223] And before the Second World War, he loved to be driven for miles at high speed along the German autobahns, the building of which he had initiated. Later, he looked back longingly on those rides, once remarking "it hurts me no longer to be able to bowl along those lovely white tracks."[224] And that relationship was intimately involved with the concept of time, which was always on the enemy's side. This was why the *Sitzkrieg* of 1939–40 was unacceptable to Hitler, and it explains why, even in 1934, discussing an imaginary invasion of Holland, he would comment that in "less than eight hours we shall break through to the coast."[225]

Another attraction of *Blitzkrieg* for Hitler was that it was not a systematic approach to military decisions, normally the essence of doctrine. It was, instead, a kind of operational opportunism without prearranged rules and standardized methods oriented on surprise and complete exploitation – "an avalanche of actions that were sorted out less by design than by success."[226] And if that was not enough to satisfy the unsystematic and opportunistic Nazi leader, the doctrine appeared to be an ideal means for using the indirect approach to conduct decisive wars of annihilation successfully, an answer to all the problems of attrition and indecisive static warfare from his experiences with the *Materialschlacht* on the elongated fronts of the First World War as well as to the more traditional strategic problem of a two-front war. Certainly the doctrinal emphasis on avoiding the enemy as much as possible appealed to Hitler, with attrition being limited almost entirely to the breakthrough phase, and even then, at a very low level. German forces would only be required to attack a small part of the enemy holding the full frontage in order to create breakthrough gaps for the penetration columns. Moreover, these gaps would not be at strategically important locations for the enemy, but rather those points which happened to be both least well-defended and conducive to subsequent deep penetrations.[227]

The initial thrust, as Hitler envisaged it early on, would be to violate an enemy politically, using propaganda and sabotage to undermine morale before launching swift, overwhelming surprise attacks by forces headed by "highly qualified special formations."[228] Warfare, in this context, was a series of *coups de main*, in which the Nazi leader could use the revolutionary principles and tactics to which he was suited. In particular, there was the absolute centrality of deception in every phase of *Blitzkrieg* in order to keep the enemy command in a state of uncertainty beyond the outbreak of hostilities. During the breakthrough attacks on selected narrow sections of the front, deception operations ranging from feints to demonstrations were necessary to prevent the majority of the enemy forces along the unattacked frontage from moving towards the intended axes of these attacks. During penetration, it was the speed and directional unpredictability that created the "noise" of multiple sightings which in turn prevented the tactical vulnerability of the racing mobile columns from becoming an operational one as well. Finally, as exploitation proceeded and disoriented masses of the enemy were encircled by German forces often inferior in numbers, as in the 1941 Ukraine campaign, it was only the deception caused by the psychological disruption of *Blitzkrieg* that could prevent breakouts by the numerically superior encircled forces.[229]

Surprise, for Hitler, was the major factor in winning this psychological battle. This in turn meant that *Blitzkrieg* operations had to be limited in both time and space for the psychological (and material) effects to be large enough to force another country to capitulate – to ensure, in short, that surprise would not wear off. In theory, this pure form of the new doctrine with its emphasis on paralysis of the enemy could be seen as supplanting battles of annihilation. But in fact during the early years of the war, swift encirclement maneuvers in which great hordes of the enemy were captured served also as a means of paralysis. In this regard, *Blitzkrieg* "provided a highly levered, low-cost, high-speed way to achieve the encirclement battle of annihilation."[230]

Blitzkrieg, however, was not for Hitler just an operational and tactical military doctrine that sought to avoid the static conflict of the First World War. It was also by its speed and surprise designed to avoid a repetition for Germany of the political, economic and psychological strains of that war. In the fall of 1916, Hitler had taken convalescent leave in Berlin. "Clearly there was dire misery everywhere," he recalled. "The big city was suffering from hunger. Discontent was great."[231] By the end of that war, the interrelationship of foreign and domestic policies was fixed in his mind. "What was the army fighting for," he asked in *Mein Kampf*, "if the homeland itself no longer wanted victory?"[232] And a decade later, that concern was pointed to the future. "How can I expect to wage war," Hitler asked Rauschning, "if I drive the masses into the same state of apathy that they were in during 1917–1918?"[233]

Blitzkrieg provided an answer to Hitler's concern with the German home front, since it was economically a means around the seeming impasse of "total warfare," a method of conducting war that did not bear too heavily on civilians. Moreover, as Hitler well realized, a war of attrition by its very prolongation would pass the economic advantage to an enemy coalition whose economic potential far surpassed Germany's.[234] As a consequence of all this, *Blitzkrieg* economics were ideally suited to Hitler's unitary but ad hoc administrative methods. Moreover, the Führer command system could cut swiftly through red tape. Hitler could thus make violent shifts of priority, a necessity in pure *Blitzkrieg*, which was geared to short but intensive bursts of economic effort. The attack on France, as an example, was preceded by high production of vehicles and armor and the planning for Sea Lion by increased output of naval vessels and airplanes. Each war in this economics of priorities required a different type of *Wehrmacht*. Most important, *Blitzkrieg* provided a type of war in the opening years of conflict that did not entail restrictions on economic goods. "Psychologically, it had all the sound and fury of total war," Barry Leach has commented, "but only the material cost and duration of limited war."[235]

Although it was popularized and well known, the doctrine of *Blitzkrieg* surprised adversaries during the early part of the Second World War. The key was Hitler's ability early on to take the offensive and exercise his initiative. In this way, he was able to dictate the style of the conflict and structure the individual battles. The advantages inherent in surprise were thus more than psychological, since a military organization fighting the war that it created and planned is likely to do better than one that is simply in the reactive mode. Certainly this was true in the Polish and French campaigns, particularly the latter, where despite the adversary's knowledge of just how effective German military doctrine could be, Hitler's choice of where and when to apply that doctrine was decisive. Even the code name "Sichelschnitt," applied to the particular operation in that campaign by the southern army group, reflected the decisive aspects of achieving surprise by taking the initiative in the indirect approach. For it was, in the final analysis of that operation, the difference between pushing against the maximum opposition in the north and cutting off that opposition decisively at the roots in the south, as a sickle does with wheat.

Hitler and the Failure of Doctrinal Surprise

Hitler had a good feel for tactical questions from his First World War "worm's-eye view" reinforced by a phenomenal memory that allowed him to absorb data on a large number of weapons and equipment. And at the strategic level, he embodied both the political and military authority as a unitary decision-maker to an extraordinary degree, as we have seen. The field where he was most lacking, as Martin van Creveld has pointed out, was everything in between, to include supplies, logistics, organization and use of operational reserves.[236] For the most part, as the Second World War progressed, the Nazi leader was unable to view separate battles and engagements as part of the operational whole. As a consequence, he was unable to bring these tactical pieces together consistently into a coherent mosaic of major operations and campaigns at the operational level in order to achieve his own military strategic objectives. "He had a good grasp of how a single division moved and fought," General von Manteuffel later recalled, "but he did not understand how armies operated."[237] And at one time, the Nazi leader dismissed all the army's group commanders on the Eastern Front as being interested in *"immer nur operieren."*[238]

Hitler attempted to bridge this gap by demonstrating his mastery of technical detail, just as he had discussed music in detail as a young man without being able to play an instrument. At that time, however, when he had attempted to play the violin, his friend had dissuaded him. But when the Nazi leader began to take a hand in "this little affair of operational command" in December 1941, there was nobody to discourage him.[239]

Certainly, by that time the military was in no position to accomplish this. Hitler had always considered his First World War experience as an infantryman as the *sine qua non* for military expertise. As early as 1928, he had been unable to hide his contempt for those who had not had this experience:

> I did not learn about the war at a restaurant table reserved for regular customers. Nor was I in this war one of those who had to give orders or to command. I was an ordinary soldier who was given orders for four and a half years, and who nevertheless honorably and truly fulfilled his duty. But I thereby had the good fortune to know war as it is and not as one would like to see it.[240]

This contempt grew stronger as the next war progressed. "And just what has your front-line experience been?" he screamed at Halder. "Where were you in the First World War? To think of you trying to tell me I don't know about the front!" These feelings extended to the entire General Staff, which Hitler asserted, "is the only Masonic Order that I haven't yet dissolved; those gentlemen with the purple stripes down their trousers sometimes seem to me more revolting than Jews."[241]

Without an implicit trust in his subordinates, Hitler began to interfere at the operational and tactical levels early in the war. "The Führer has left for consultation with Mussolini," Halder wrote in June 1940 with an undisguised air of relief. "For a day and a half we will be our own masters."[242] But it was the extremely sophisticated German communications equipment of the period that made this interference possible, not the Nazi leader's active involvement. Normally, von Manstein recalled, Hitler remained with "his far too detailed situation maps" in the rear, never venturing near a front line. "It was hard enough to persuade him to visit an Army Group headquarters."[243] Nevertheless, the interference became worse as the war progressed. At one point toward the end of that conflict, Field Marshal von Kluge was directed to consult with Hitler before any operation involving forces of a battalion or higher could be initiated.[244] "His mistrust of his subordinate commander," von Manstein concluded in this regard, "prevented him from giving them, in the form of long-term directives, freedom of action, which they might put to a use that was not to his liking. The effect of this, however, was to do away with the very essence of leadership."[245]

Exploitation and Risk in Military Surprise

Exploitation is an integral part of military surprise. Without it, as Michael Handel has pointed out, "the whole *raison d'être* of launching a strategic surprise will collapse."[246] But it was precisely at the exploitation stage after surprise had been initially effected that Hitler often was at his worst, displaying a lack of nerve and self-confidence, totally at variance with the decisiveness and aplomb with which he planned the daring operations. "He could never sit back with the confident calm of the great general," Halder pointed out in this regard, "and allow a developing situation to come to fruition."[247]

This tendency was apparent early in the Norwegian campaign, when the events at Narvik "brought Hitler to the point of a nervous breakdown."[248] As early as 14 April 1940, Hitler became, in Jodl's words, "terribly agitated" over the situation of General Dietl and his force of one mountain infantry regiment that had invested that port far to the north and were without outside support.[249] For hours on end, the Nazi leader bent over the situation map contemplating how he could help the *Gebirgsregiment*. By 17 April, Hitler was prepared to evacuate the Narvik force by air or to order it to start moving to the south through the mountains. Jodl, whose field experience had been primarily as a mountain artillery officer, did not consider the situation hopeless and talked Hitler out of abandoning the key strategic position. On 18 April, Dietl was ordered to hold out at Narvik for as long as possible,[250] but Hitler remained extremely nervous until the situation stabilized at the end of the month. Ironically, the official British history of the campaign noted "the success with which the capture of Narvik had been exploited" and pointed out that General Dietl had been ordered to hold out "since considerations of prestige had given Hitler a passionate interest in the event."[251] That "passionate interest," of course, had almost caused the Nazi leader to evacuate needlessly one of the most important strategic objectives in the entire Norwegian operation.

Although not so evident in Norway, part of Hitler's problem with the exploitation stage was his lack of full commitment to the pure indirect approach of *Blitzkrieg*, one which he shared with the majority of the German High Command. In the undiluted doctrinal concept, the primary aim was paralysis of the enemy's C^3I. Hitler was well attuned to this objective and had discussed it extensively before the

war as the preferred goal instead of the outright physical destruction
of annihilation, which could lead to the attritive and static shedding of
German blood. It was in the method, however, that Hitler vacillated
between the traditional flanking and encirclement movements and
Blitzkrieg's unsupported thrusts deep into the enemy's rear. For the
Nazi leader, both methods could complement each other, with en-
veloping movements providing a quick and decisive way of capturing
the enemy – annihilation, in other words, without the cost to German
manpower. And if it were a double envelopment, so much the better
in terms of low cost decisiveness. "Has anything changed since the
battle of Cannae?" he once asked Rauschning.[252]

Hitler's approach to the doctrinal method also meant a compromise
with the basic rules of operations. On the one hand, there were the
traditional guarded flanks and unbroken, if strained, supply lines; on
the other hand, there was the commitment to velocity and unpredictability of the new doctrine, which was the key to penetration and
exploitation. "The secret of *Blitzkrieg*," Liddell Hart wrote to
Guderian after the war, "lay ... above all in the 'follow-through' – the
way that a breakthrough ... was exploited by a deep strategic penetration carried out by an armoured force racing ahead of the main
army, and operating independently."[253] But this pure form of the
doctrine involved substantial risk to the spearhead mobile units
without infantry support which could only be accepted by a man with a
full understanding and a deep commitment to the "armored idea."
And this was lacking in Hitler despite his early support of tank
development. "In a campaign," he commented, "it's the infantryman
who, when all's said, sets the tempo of operations within reasonable
limits."[254]

Hitler's vacillation on doctrinal matters can be seen in the months
leading up to the invasion of the West. In the fall of 1939, the Nazi
leader was prepared for a vast Schlieffen-type encirclement to the
North and later to dissipate his limited armored strength into three
widely spaced army groups. This, of course, changed the following
winter with the acceptance of the so-called Manstein plan, which in its
initial *Blitzkrieg*-oriented, surprise execution appealed to Hitler's
sense of audacity and daring.[255] But by 16 May, the Nazi leader had
become obsessed with the flank protection of the armored force that
he had thrust so boldly through the Ardennes six days earlier. Already
on that day, he toyed with the idea of holding up the armor spearhead
until a "pearl necklace" of infantry could be marched and horse
transported into flank protective positions.[256] On 17 May, Hitler
insisted that the main threat to the German forces was from the south.

"The Führer is terribly nervous," Halder wrote in his diary that day. "Frightened by his own success, he is afraid to take any chance and so would rather pull the reins on us." The following day Halder noted:

> The Führer unaccountably keeps worrying about the south flank. He rages and screams that we are on the best way to ruin the whole campaign and that we are leading up to a defeat. He won't have any part of continuing the operation in a westward direction.[257]

Hitler's halting of von Kleist's armored group illustrated the interplay of risk and military surprise at all levels of war. For by interfering in the penetration and exploitation phase achieved by strategic surprise, the Nazi leader limited the ability of his subordinate commanders to achieve successive operational and tactical surprise, which is the essence of decentralized command operating under the mission-type orders of *Auftragstaktik*, a core concept not only of *Blitzkrieg* but of the traditional German offensive doctrine as well. The initiative, responsibility and flexibility necessary to maintain surprise, as von Manstein summed it up, could not exist at the lower levels under such interference.

> Worst of all, in its preoccupation with security it waives the opportunity that may occur through the independent action of a subordinate commander in boldly exploiting some favourable situation at a decisive moment. The German method is really rooted in the German character, which – contrary to all the nonsense talked about "blind obedience" – has a strong streak of individuality and – possibly as part of its Germanic heritage – finds certain pleasure in taking risks.[258]

Hitler's interference also lessened the decisiveness of action at all levels. By assigning the 12th Army to the defensive flank role in the south, for example, the Nazi leader allowed the enemy time to build up a new front which had to be surmounted all over again in the second phase of the French campaign. "The choice of finally putting an end by offensive action to any coherent French defense," von Manstein later observed in this regard, "... had been needlessly sacrificed."[259] This pattern continued at the operational and tactical level through the remainder of the entire campaign. "Here we have the same old story again," Halder noted on 6 June as German forces moved south into France during Operation Red. "On top there isn't a spark of the spirit that would dare to put high stakes on a single throw."[260]

The problems with exploitation in the 1940 western campaign also demonstrated how Hitler's lack of understanding of *Blitzkrieg* led to a serious disconnect between the operational and strategic level of war. The Nazi leader was joined in this misunderstanding of that operational doctrine by the majority of the German High Command which, as a consequence, lavished most of the attention in the plan for the invasion of the west on the actual breakthrough, and very little on the immediate operational aftermath. The possibility that the plan would lead to total victory over France, as Alistair Horne has indicated, "seemed so remote that beyond the operation itself no thought whatsoever had been given to how a knockout blow might be administered to Britain."[261] The successful evacuation of that country's forces from Dunkirk was the immediate consequence. And what appeared to be a spectacular operational success in the French campaign actually meant that Hitler failed in his principal strategic aim of coercing Britain into accepting German hegemony on the continent.

Blitzkrieg and Russia

Hitler's directive for Barbarossa contained vintage *Blitzkrieg*, emphasizing that the "bulk of the Russian Army stationed in Western Russia will be destroyed by daring operations led by deeply penetrating armoured spearheads."[262] To succeed, that doctrine required a sustained, uninterrupted offensive in Russia. But that had not even been possible in the French campaign, where, at the tactical and operational levels, military success had resulted primarily from the skillful use of economy and concentration of force as well as mobility under the protective umbrella of strategic surprise. The psychological and physical effects of *Blitzkrieg* surprise in the Soviet Union were mitigated by time and space factors that dwarfed the German experience in the west, a fact recognized by Stalin a few months after the German invasion, even as his forces continued to retreat into the huge Russian hinterland.

> The element of surprise and suddenness, as a reserve of the German Fascist troops, is completely spent. This removes the inequality in fighting conditions created by the suddenness of the German Fascist attack. Now the outcome of the war will be decided not by such a fortuitous element as surprise, but by permanently operating factors.[263]

The vastness of time and space is one of the most striking aspects of the German–Soviet conflict in the Second World War. The fighting lasted three years, 10 months and 16 days with hardly a break. The length of the front from fall 1941 to fall 1943 was never less than 2,400 miles and for a period in late 1942 stretched to 3,060 miles. Finally, the German invasion thrust 1,200 miles into the Soviet Union, and the Soviet counter-offensive extended 1,500 miles to Berlin.[264]

For Hitler, time and space were "only vague ideas that should not be allowed to affect the determination of a man who knew where he was going."[265] In fact, he had only been dissuaded from an attack on the Soviet Union in late summer 1940 by Jodl and Keitel, both of whom were able to convince him that these factors combined with weather conditions rendered the plan totally impractical. Nevertheless, the Nazi leader was still unprepared for what he found. "How small Germany looks from here," he mused over a year after the invasion had begun.[266]

In the Russian context, the French campaign seemed in retrospect to be of infinitely manageable proportions to Hitler. "The breakthrough at Abbeville was an advance of a mere 350 kilometres, which is nothing in comparison with distances in the East."[267] Nor, despite warnings, was the full impact of the Russian winter anticipated. "The staggering blow for us," Hitler stated in January 1942, "was that the situation was entirely unexpected, and the fact that our men were not equipped for the temperatures they had to face."[268] This lack of winter gear also pointed out the problems inherent in the single-minded emphasis under the *Blitzkrieg* doctrine on combat operations to the detriment, if not neglect, of logistics and other supportive functions. Typically, the Führer was satisfied with a temporary and unsystematic answer to these problems. "The supplying of the front creates enormous problems," he acknowledged during the first winter in Russia. "In this matter, we've given proof of the most magnificent gifts of improvisation."[269]

In these circumstances, the basic operational premises of *Blitzkrieg* did not obtain. To begin with, the accumulation of successes did not ensure victory. The initial 1941 campaign was wildly successful, resulting in the destruction or capture of the main Soviet forces in the opening weeks. And yet the Russians continued to fight, trading space for time. Hitler summed up the problem on the thirty-fifth day of the campaign. "You cannot beat the Russians with operational successes," he argued, "because they simply do not know when they are defeated. On that account it will be necessary to destroy them bit by bit, in small encircling actions of a purely tactical nature."[270]

This approach, in turn, ended any possibility, if it existed at all, of achieving the fundamental *Blitzkrieg* goal of paralysing the Soviet C^3I. Once again Hitler's lack of understanding of the doctrinal relationship between risk and exploitation in achieving surprise at the operational level was key. Already, on the twelfth day of the campaign, the specter of a flank attack on his armor spearhead arose again. "Führer is afraid that the wedge of AGp. South now advancing eastward might be threatened by flank attacks from north and south," Halder noted on 3 July. "Tactically speaking, of course, this fear is not at all unwarranted, but that's what we have Army Corps CGs for."[271] Halder attempted throughout the initial summer in Russia to replace this tactical view with an operational perspective. "We must not allow the enemy's strategy," he argued, "to dictate our operational conceptions .... We must aim at complete victory by keeping our forces together for distant, decisive objectives, and crippling blows, and must not fritter ourselves away on trivial objectives."[272] How dismally the Army Chief of Staff failed in this endeavor was illustrated by Hitler's directive for the spring campaign the following year.

> Experience has sufficiently shown that the Russians are not very vulnerable to operational encircling movements. It is therefore of decisive importance that ... individual breaches of the front should take the form of close pincer movements. *We must avoid* closing the pincers too late .... *It must not happen* that by advancing too quickly and too far, armoured and motorised formations lose connection with the infantry following them; or that they lose the opportunity of supporting the hard-pressed, forward-fighting infantry by direct attacks on the rear of the encircled Russian armies.[273]

The result of all this was tactical surprise and decisiveness without operational coherence, a situation not designed to induce Soviet surrender. And without capitulation, the only alternative for the German forces was occupation, a far cry from the basic concept of *Blitzkrieg*, as every square mile had to be extracted from a fanatical enemy and held against the resistance of the occupied. In this context, time and space began to dominate and ultimately crushed German military doctrine.

Operational Art and Strategic Defense

Hitler's lack of expertise at the operational level of war was particularly evident in his inability to view defensive operations holistically after Germany lost the strategic initiative in the winter of 1942–43. That same winter, as an example, the Nazi leader directed the SS "Reich" Division to launch a surprise attack from Kharkov into the rear of the enemy forces advancing against the German Donetz front. Even with this tactical surprise, however, it would have been impossible for that solitary division to have accomplished its far-ranging operational level objective; and it was only saved by a more urgent commitment elsewhere.[274] This type of tactical interference without an overall operational framework for guidance increased in direct proportion to the strategic and operational pressure exerted by the Russians, astonishing Speer on one of his trips to the Eastern Headquarters with the aimless way "Hitler in the course of hearing ... reports made deployments, pushed divisions back and forth, or dealt with petty details."[275] Halder attempted to put the conduct of defense in operational level perspective for his leader by pointing out that:

> penetration in a single Div. sector, tactically a setback, need not necessarily have an adverse effect on the overall operations; as long as it could be contained, it might even react to one's advantage as it drew away enemy forces. The commander must not allow himself to become nervous and view the situation with the eyes of e.g., the Bn. Commander.[276]

But Hitler did not see it that way. For him any retreat was a catastrophe, no matter what level of war. There was, of course, his fear of losing prestige. Added to this was an intense dislike of giving up anything, particularly in regard to Russia, concerning which, as General Warlimont pointed out, "Hitler always took the stand on the principle of 'everything or nothing'."[277] This was illustrated by a Russian army commander who had been taken prisoner in the Ukraine and offered to conduct propaganda against the Soviet forces. Hitler was furious when he heard of this offer and refused to exploit it. To take such a man into the confidence of the Germans would diminish the hatred that he believed was necessary to maintain toward Russia.[278]

Hitler's First World War experiences also played a major role in his attitude toward any pullback of German forces. He was aware of incidents in that conflict such as occurred in the intentional German retreat from the Arres–Soissons line in spring 1917, in which the infantry avoided costly rearguard actions because they knew prepared positions awaited them on the Siegfried Line. The lesson for the Nazi leader was that any defensive line to the rear exerted a "magnetic"

effect on troops to the front and that, moreover, it was extremely
difficult to stop an infantryman once he was free of the trenches and in
open country.[279] In the summer of 1944, for instance, Hitler ordered
that no preparatory directives were to be issued regarding the withdrawal of German troops from the western coastlines to rear
positions. In this way, he believed he could prevent OB West and his
troops from "squinting" to the rear.[280] More and more as the war
proceeded, the Führer viewed any tendency not to hold in place at any
level in terms of his own previous experience. "It's becoming a mania,
a real disease," Hitler told General Zeitzler at the end of 1943.

> After all, this isn't a trifle. It's 220 miles, half the width of the
> entire Western Front. We talk about the battle of the Marne, but
> we're doing it all the time. There's a break-through and we give
> up a colossal front of 220 miles. 220 miles was the entire right
> wing of the army in 1914.[281]

It is generally agreed that when Hitler gave his initial order to hold
before Moscow in December 1941, he probably saved his troops.[282]
But this attitude was disastrous at the operational level after Germany
went on the strategic defensive. For the only means to effect
operational surprise in that situation, as von Manstein repeatedly
emphasized, was to conduct a strategic withdrawal that, by shortening
the front, would allow the creation of an operational reserve. This
type of retrograde movement would also allow for the development of
the full freedom of maneuver at the operational level so necessary for
that reserve to deliver a decisive counter-attack against an over-
extended enemy advance. And this Hitler would not allow. Moreover, he was constantly ordering into existence new units that had to
be filled by existing forces. The result was ill-trained reserves and
weakened front-line defensive forces – hardly the prescription for
unexpected counter-attacks at the operational, much less the tactical,
level. In any event, without a flexible defense and a string of
operational reserves sufficient to influence major operations, Hitler
lost the ability to shape the battlefield and reverted to the positional
defense of the First World War in which reserves remained close to
the front to be poured, as needed, piecemeal into the line. "This
mistake of his was to cost us dear when the Russians broke through in
January 1945 ...," Guderian wrote after the war. "Main line of
defence, major defensive positions, reserves, all were buried beneath
the tidal wave of the initial Russian breakthrough."[283] The mistake had
already long since cost Germany operational initiative and surprise.

There were times, of course, in the closing years of the war when enemy pressure did force Hitler to make substantial pullbacks of his forces. But these were unpalatable decisions, which brought out the Nazi leader's tendency to procrastinate just when boldness and decisiveness were necessary if there were to be timely commitment of forces either to forestall an enemy operational success or to prevent enemy exploitation. Protracted arguments, often lasting weeks, occurred over the abandonment of such untenable positions as the Donetz area in 1943 and Dnieper Bend in 1944, or over the evacuation of unimportant salients on quiet stretches of the front in order to acquire substantial enough reserves. Finally, Hitler would issue his orders, but only, as Warlimont reported, "after interminable periods of weighing pros and cons, during which he held long monologues . . . instead of carefully examining the operational and tactical elements of the situation such as space and timing."[284]

The most fateful consequence of Hitler's procrastination and lack of operational vision, however, concerned a decision to attack. In early 1943, von Manstein offered Hitler two alternatives. The first, the "backhand" approach which Manstein favored, was to wait for the expected Soviet attack in the South Ukraine, give ground before that attack and then counter-attack on the exposed Russian north flank. The second approach, the "forehand" method, involved concentric attacks to cut off the immense Soviet salient at Kursk and, after annihilating Soviet armor reserves, move south to destroy the Soviet front in the South Ukraine. Predictably, Hitler rejected the risk and the loss of territory in the "backhand" alternative, opting for the limited objective in the "forehand" stroke of destroying the Russian forces within the salient with only the vaguest references to exploitation and subsequent operations.[285]

Even then Hitler was nervous about the attack which, he admitted, gave him "a queasy stomach feeling" whenever he thought of it.[286] On 2 May, he postponed the attack from mid-May until June, by which time he hoped the German armor divisions would be reinforced with new Tiger and Panther tanks. But surprise was the price that had to be paid for making the Nazi leader more comfortable with the operational risks. "Though every deception and camouflage measures had been taken," von Manstein commented concerning the final attack date of 5 July, "we could no longer expect to catch the enemy unawares after a delay of that length."[287] As a consequence, by the time the attack started on the Kursk positions, the German forces were up against an enemy ensconced in a salient 100 miles wide and 159 miles deep with six defense belts consisting of anti-tank posts, dense minefields and 3,500 miles of trenches backed by 3,306 tanks and 20,000 artillery pieces. The inevitable result was a battle of attrition – just what Hitler had hoped to avoid at any level of war since his experiences of 1914–18.[288]

Kursk marked the passage of the strategic initiative to the Allies. Nevertheless, as the German position grew weaker in the fall of 1944, Hitler resolved to make one last, high-risk thrust in the Ardennes to regain that initiative in terms of ending the war in the west. But surprise as a means of conflict termination by psychological shock is only effective when the stronger adversary does not have absolute interests and open-ended commitments. When these conditions do not obtain, as Richard Betts has pointed out, "stunning surprises by the weaker side are only a dent in the curve toward victory by the stronger."[289]

Such was the case with Hitler's 1944 Ardennes offensive. The Allies did not perceive the conflict with Nazi Germany as limited. Despite the very great shock of the German surprise attack in mid-December, there never was a move to reassess war aims. The offensive thus remained stunted at both the tactical and strategic levels of war. For without a truly overwhelming force, Hitler was left with tactical surprise that did not allow him to coalesce the scattered engagements into military conditions at the operational level that would achieve strategic goals. And even if it had been possible to seize military strategic objectives such as Antwerp, the political preconditions, as we have seen, were lacking at that level. In fact, the very lack of these preconditions, which might have made the German attack a reasonable gamble at the strategic level, contributed to the success of Hitler's surprise.[290]

Hitler personally supervised the Ardennes deception measure which, instead of simulating something that did not exist, actually screened something that did exist up to the point of attack. But, as we have seen, *Blitzkrieg* is dependent on deception at more than the point of attack, a fact lost on the Nazi leader who neglected to use effective deception methods to pin down enemy forces in the neighboring sectors of the front. As a consequence, the Allies were able to shift troops swiftly as operational reserves from those parts of the line not under attack, without risking the abandonment of valuable territory which they had only just captured and which they needed as a springboard for further operations against Germany. The result of all this was the expenditure of German forces that were, in Halder's words, "the last penny in the pauper's purse."[291] Even when news of German defeats poured in toward the end of December, Hitler remained true to form and refused to cut his losses by trading space for time. "I get an attack of the horrors," he stated, "whenever I hear that there must be a withdrawal somewhere or other in order to have room to manoeuvre. I've heard that sort of thing for the past two years and the results have always been appalling."[292]

Conclusion

In the end, *Blitzkrieg* failed because Hitler failed as a strategist. Strategy is the calculated relationship of means to ends, but this connection hardly existed for Germany in the Second World War. In terms of means, the German military machinery, which emphasized rearmament in "width" rather than "depth," was adequate in the pre-war period to further Hitler's coercive diplomacy.[293] But Hitler's obsession with this type of showy rearmament was to have negative results once war began. As late as 1939, when informed that ammunition stocks were extremely low, Hitler replied, "Nobody inquires whether I have any bombs or ammo, it is the number of aircraft and guns that counts."[294]

In addition, as we have seen, there was almost a total absence of any systematic economic planning for war; and like so many other aspects of the Third Reich operating under the *Führerprinzip*, economic preparations were a welter of competing interests and makeshift administrative methods. This chaos was rendered even more unmanageable and inefficient because Hitler gave pride of place to such party hacks and sycophants as Goebbels and Himmler, who thus had considerable advantage over the service chiefs in gaining the Nazi leader's support for their own projects. Added to this were Hitler's insistence on devoting large funds to maximizing civilian prosperity and his concomitant refusal to mobilize the population until the fall of 1944. Even then, luxury goods were still being produced and rolling stock continued to transport Jews to concentration camps and plundered art to hiding places. This "guns and butter" policy meant that, as Matthew Cooper has pointed out, "while the German economy was in part geared *to* war, it was not geared *for* war."[295]

The means in Germany's strategic equation were, in any case, ultimately irrelevant because of the transformation of the ends. Although *Blitzkrieg* was initially closely meshed with Hitler's political goals, his very success led him beyond strategic aims that could be met by the limited opportunistic aggression of that doctrine. Instead, those aims became limitless in terms of achieving domination and entirely lacked a conception of an "ultimate status quo." "The essential thing for the moment, is to conquer," Hitler stated after the successful summer campaign of 1941 against Russia. "After that everything will be simply a question of organization."[296] Because Hitler's strategic ends were infinitely expansive, no military doctrine, much less any type of force-programming, could keep up with his policy in the end.

That lack of moderation, of a sense of limitation, was instinctive in Hitler. "For wherever our success ends," he stated as early as 1928, "it will always be only the point of departure for a new struggle."[297] As long as he kept these tendencies in check, he was able with his own Hobbesian amalgam of force and fraud to achieve military surprise at all levels of war. But that surprise became sterile and indecisively isolated at the lower levels as the war progressed and the Führer became dazzled by the plethora of strategic alternatives. Like a child, he could not bring himself to disregard any attractive objectives, nor to give up ground once gained. In this context, no amount of tactical surprises and successes could be molded operationally to fit strategic goals that never ceased to expand. "He who cannot reject, cannot select," Telford Taylor has pointed out, "and the downfall of the Third Reich was due in no small measure to Adolf Hitler's inability to realize that in strategic terms, the road to everywhere is the road to nowhere."[298]

NOTES

This paper was first published in *Intelligence and National Security*, Vol. 3, No. 3, July 1988, pp. 55–117.

1. Adolf Hitler, *Mein Kampf*, trans. Ralph Manheim (Boston: Houghton Mifflin, 1943), p. 161. In 1942, Hitler could still describe sending young men into the army, "whence they will return refreshed and cleansed of eight years of scholastic slime." H.R. Trevor Roper, ed., *Hitler's Secret Conversations, 1941–1944* (New York: Farrar, Straus and Young, 1953). See also Hans Frank, *Im Angesicht des Galgens. Deutung Hitlers und seiner Zeit auf Grund eigener Erlebnisse und Erkenntnisse* (Neuhaus bei Schliersee, 1955), p. 230.
2. Robert G.L. Waite, *The Psychopathic God Adolf Hitler* (New York: Basic Books, 1977), p. 322.
3. *Mein Kampf*, p. 165.
4. For Hitler's First World War experiences, see Fritz Wiedemann, *Der Mann, der Feldherr werden wollte* (Dortmund: Blick & Bild, 1964); B. Branmayer, *Mit Hitler Meldgänger, 1914–1918* (Überlingen, 1940); Hans Mend, *Adolf Hitler im Felde 1914–1918* (Munich: Jos. C. Hubers, 1934, 2nd ed.).

5. Wiedemann, p. 28.
6. Ibid., p. 26. Hitler was a *Gefreiter*, roughly equivalent to a corporal. According to Mend, the regimental staff was also reluctant to promote Hitler for fear of losing somebody with his skill and sense of responsibility. Mend, p. 180. Mend, who served with Hitler until 25 July 1916 and who described Hitler's exploits as *übermenschliches* (superhuman), was not an altogether unbiased observer despite his disclaimer: "I am completely independent from Hitler and I have no interest in celebrating him as a hero," p. 185.
7. Wiedemann, p. 29; Branmayer, p. 52. At one point, Hitler was among six runners who were wounded. As Wiedemann knelt over him, Hitler said: "It's not so bad, Lieutenant, really. I'll stay with you, stay with the regiment": Wiedemann, p. 29.
8. Hitler was awarded the Iron Cross First Class on 4 August 1918 for capturing 12 Frenchmen. He was carrying a message when he came upon the French soldiers unexpectedly. By arm motions and fractured French, Hitler persuaded the soldiers that he was backed up by an entire company – not the last time that he would successfully resort to bluff: Mend, p. 190. Ironically Hitler received the Iron Cross through the efforts of Wiedemann's successor as regimental adjutant, First Lieutenant Gutmann, a Jew: Wiedemann, p. 25.
9. Emphasis added: *Mein Kampf*, p. 67. See also Walter C. Langer, *The Mind of Adolf Hitler. The Secret Wartime Report* (New York: Basic Books, 1972, 2nd printing), p. 126, who after describing Hitler's life through the First World War and his return to Munich, stated: "This is the foundation of Hitler's character. Whatever he tried to do afterwards is only superstructure, and the superstructure can be no firmer than the foundations on which it rests."
10. Liddell Hart, *Strategy* (New York: Frederick A. Praeger, 2nd printing, 1968). For Hitler's distinction between the two approaches, see Martin van Creveld, "War Lord Hitler: Some Points Reconsidered," *European Studies Review* (Jan. 1974), pp. 57–79.
11. August Kubizek, *Young Hitler. The Story of our Friendship*, trans. E.V. Anderson (London: Allan Wingate, 1954), p. 54. Hitler's boldest project at this time was a "grandiose bridge which would span the Danube at a great height": ibid, p. 56.
12. Bradley F. Smith, *Adolf Hitler, His Family, Childhood and Youth* (Stanford: Stanford University Press, 1967), p. 118.
13. Albert Speer, *Inside the Third Reich*, trans. Richard and Clara Winston (New York: Macmillan, 1970), p. 75. Louis Napoleon, of course, was anxious to be rid of the narrow streets that had been so effectively blockaded in the 1848 revolution.
14. Hitler's emphasis: *Mein Kampf*, p. 240. "*One of the worst symptoms of decay in Germany of the pre-War era was the steadily increasing habit of doing things by half*" (Hitler's emphasis): ibid., p. 236.
15. Kubizek, p. 7. Aggression can also be a manifestation of exaggerated masculinity. As Robert G. L. Waite points out in his afterword to the Langer study, this might be explained by the fact that Hitler suffered from monorchism (lack of testicle) – in Hitler's case the left testicle, as revealed in the Russian autopsy of May 1945. Langer, p. 227. British soldiers anticipated this during the war with a song to the tune of "Colonel Bogey" that began: "Hitler has only got one ball."
16. *Mein Kampf*, p. 197.
17. Ibid., pp. 198–9. But Hitler also thought the offensive was premature. "A month later the ground would have been dry and the meteorological conditions favorable": *Secret Conversations*, 10–11 October 1941, p. 43.
18. Jehuda L. Wallach, "Misperceptions of Clausewitz' *On War* by the German Military," in Michael I. Handel, ed., *Clausewitz and Modern Strategy* (London: Frank Cass, 1986), p. 219. "His other politico-military master was Clausewitz, whom he could quote by the yard." Ernst Hanfstaengel, *Unheard Witness* (New York: J.B. Lippincott, 1957), p. 41. But see also Waite, p. 71, who says the edition of *On War* in Hitler's library "was especially well thumbed" and Trevor-Roper in his introduction to *Secret Conversations*, p. xxv, who states that Hitler first studied Clausewitz before 1923, probably in the First World War, and later in the next war studied the works of

Clausewitz constantly in his headquarters.
19. P.M. Baldwin, "Clausewitz in Nazi Germany," *Journal of Contemporary History*, 16 (Jan. 1981), p. 11. "In his political testament to Admiral Dönitz just before he committed suicide, Hitler exhorted his countrymen to continue the fight in accordance with "the ideals of the great Clausewitz": ibid., p. 10. For examples of Hitler's references to Clausewitz, see *Mein Kampf*, p. 668, and Adolf Hitler, *Hitler's Secret Book*, trans. Salvator Attanasio (New York: Grove Press, 1961), p. 120.
20. Wallach, p. 225.
21. Gerhard Ritter, *Frederick the Great. A Historical Profile*, trans. Peter Paret (Berkeley: University of California Press, 1974), p. 142. Frederick cultivated the offensive battle as the basis for Prussian tactics in stark contrast to the Austrian emphasis on the defensive battle. Although Frederick's strategic offensives remained limited in scope, he brought the tactical offensive, in Ritter's judgement, "to a state of perfection": ibid., p. 143. For examples of Hitler's references to Frederick II, see *Mein Kampf*, p. 238, and *Secret Conversations*, 20 August 1942, p. 525.
22. H.G. Thursfield, ed., *Brassey's Naval Annual 1948* (New York: Macmillan, 1948), p. 352. At the strategic level, of course, Hitler often had to overcome the defensive orientation of his commanders. See his speech on 23 November 1939 concerning the forthcoming campaign in the West, in which he pointed out the necessity for taking the offensive to his reluctant major commanders: Franz Halder, *The Halder Diaries. The Private War Journals of Colonel General Franz Halder*, ed. Arnold Lissance (Boulder, Colorado and Dunn Loring, Virginia: Westview Press and T.N. Dupuy Associates, 1976), I, p. 132.
23. Walter Warlimont, *Inside Hitler's Headquarters 1939–1945*, trans. R.H. Barry (New York: Frederick A. Praeger, 1964), p. 486.
24. Ibid. It was, of course, another matter when applied to the enemy. On 25 September 1941, Hitler commented that "the offensive spirit that inspires the Russian when he is advancing, does not surprise us. It was the same during the First World War, and the explanation for it is their bottomless stupidity": *Secret Conversations*, 25 September 1941, p. 34.
25. Jay Luvaas, ed., *Frederick the Great on the Art of War* (New York: Free Press, 1966), p. 23. One school led by Theodor von Bernhardi maintained that "Frederick, alone among his contemporaries looked to the destruction of the enemy's forces as the one decisive thing in war." The other school, led by Hans Delbrück, associated Frederick with a strategy of exhaustion, the so-called *Ermattungsstrategie*. For the debate, see ibid., pp. 22–6. "Each side," Professor Luvaas concludes, "could point to enough examples from Frederick's campaigns to present a convincing case": ibid., p. 23.
26. Hermann Rauschning, *Hitler Speaks. A Series of Political Conversations with Adolf Hitler on his Real Aims* (London: Thornton Butterworth, 1939), p. 13.
27. *Mein Kampf*, p. 6. The book was *Eine Volksausgabe des deutsch-französischen Krieges von 1870–71*. See also ibid., p. 94, and *Secret Book*, p. 60, in which he lists the great battles of the war from Gravelotte to Sedan, and p. 47, in which he discusses how the Franco-Prussian War gave Germany "a position of infinite esteem in Europe." For Hitler's early interest in history, see Smith, p. 80, and *Mein Kampf*, pp. 15 and 118.
28. In his 1928 *Secret Book* (p. 81), Hitler wrote that Germany should have launched a preventive war against France in 1904 while Russia was in conflict with Japan. Without an Eastern Front, decisive battles of annihilation were possible against France. See also *Mein Kampf*, p. 141.
29. Rauschning, p. 127.
30. *Secret Conversations*, 10–11 October 1941, p. 43. In a similar vein, Hitler commented the previous summer, "World history knows three battles of annihilation: Cannae, Sedan and Tannenberg. We can be proud that two of them were fought by German armies": ibid., 8–11 Aug. 1941, p. 21.
31. Eberhard Jäckel, *Hitler's Weltanschauung. A Blueprint for Power*, trans. Herbert Arnold (Middletown, CT.: Wesleyan University Press, 1972), p. 118.
32. In a 4 June 1939 speech, Hitler greeted the German troops "as an old soldier with the

feeling of comradeship which in its deepest sense is disclosed only to one who experienced this conception in the War. For the glorious sense of a brave comradeship reveals itself in its most compelling force only to him who has seen that companionship hold under this severest test of man's courage": Norman H. Baynes, ed., *The Speeches of Adolf Hitler, April 1922–August 1939*, II (New York: Howard Fertig, 1969), p. 1666. See also *Secret Conversations*, 4 April 1942, p. 322: "My attachment and sympathy belong in the first place to the frontline German soldier."

33. *Mein Kampf*, p. 205.
34. Ibid., p. 520. See also Hitler's discussion of the period after the First World War when the "absence of the young German intelligentsia which found its death on the fields of Flanders in the fall of 1914 was sorely felt later on. It was the highest treasure that the German nation possessed": ibid., p. 335. For his thoughts on Social Darwinism, see ibid., p. 134. These beliefs, like so many others of Hitler's, endured throughout his life. See, for instance, his 1941 reminiscences of the war. "Then I saw men falling around me in thousands. Thus I learnt that life is a cruel struggle, and has no other object but the preservation of the species": *Secret Conversations*, 25–26 September 1941, p. 37.
35. *Secret Book*, pp. 77 and 10. For instance, the "World War burst forth blood-red": ibid., p. 75. See also ibid., pp. 170 and 184.
36. Rauschning, p. 19.
37. *Mein Kampf*, p. 191.
38. *Secret Conversations*, 13 October 1941, pp. 46–7.
39. Liddell Hart, pp. 223–6. For Liddell Hart's reaction to the First World War, see Jay Luvaas, "Clausewitz, Fuller and Liddell Hart" in Michael I. Handel, ed., *Clausewitz and Modern Strategy* (London: Frank Cass, 1986), p. 208.
40. Smith, p. 66. "The Boer War was like summer lightning to me": *Mein Kampf*, p. 158.
41. *Secret Conversations*, 17 February 1942, p. 257. Hitler's favorite work of Cooper's was *The Last of the Mohicans*. The first May book that Hitler read was *The Ride through the Desert*. "I was carried away by it. And I went on to devour at once the other books by the same author": ibid. See also pp. 547 and 257.
42. Waite, p. 11. After he became chancellor, Hitler reread the entire May series: Smith, p. 67. In describing the Hitler *putsch* of 1923, one police observer commented that the Nazi leader behaved "like a Red Indian": William Carr, *Hitler. A Study in Personality and Politics* (New York: St. Martin's Press, 1979), p. 170.
43. Percy Ernest Schramm, *Hitler: The Man and the Military Leader*, trans. Donald S. Detwiler (Chicago: Quadrangle Books, 1971), p. 53, and Rauschning, p. 267.
44. The overture from Rienzi was always used as a musical backdrop for the annual opening of the Nuremberg rally. "Stalin, too, must command our unconditional respect. In his own way he is a hell of a fellow!" *Secret Conversations*, 22 July 1942, p. 476. "He is a beast, but he's a beast on the grand scale": ibid., 22 August 1942, p. 534. See also pp. xx, xxviii, 504 and 507.
45. Rauschning, p. 31. In his youth, Hitler noted that his first name came from the old German *Athawolf*, a compound of *Athal* (noble) and *Wolfa* (wolf). "Herr Wolf" was his cover name in the early days of the *Kampfzeit*; and his favorite dogs – the only ones he would allow in Führer photographs – were *Wolfshunde* (alsatians). The names of his field headquarters during the Second World War were derived from wolf (e.g. *Wolfsschlucht* at Bruly-Pêche in Eastern France; *Werwolf* at Winnitza in the Ukraine; *Wolfschanze* at Rastenburg in East Prussia. Waite, p. 28, Langer, pp. 217–18, and *Secret Conversations*, pp. x–xi). Although normally formal with his secretaries, he often addressed one of them, Johanna Wolf, as *Wölfin*. Finally, one of his favorite tunes after he became chancellor was "Who's Afraid of the Big Bad Wolf?" from the Disney cartoon, which he would often absentmindedly whistle. Waite, p. 28. For Hitler, of course, it was not just the cunning of the wolves that attracted him. In *Mein Kampf*, he described the work of his storm troopers in a 1921 dance hall brawl against Socialists and Communists: "Like wolves they flung themselves in packs of eight or ten again and again on their enemies" (p. 505).

46. *Secret Conversations*, 13 December 1941, p. 117.
47. Theodor Heuss, *Hitler's Weg: Eine historisch-politische Studie über den National-sozialismus* (Stuttgart, 1932), p. 142.
48. Joachim Fest. "On Remembering Adolf Hitler," *Encounter* 4 (October 1973), p. 30, and Kurt Lüdecke, *I Knew Hitler* (London, 1938), pp. 217ff.
49. Baynes, I, p. 157. See also ibid., p. 155. The key, of course, was patience and persistence, qualities not often associated with Hitler, but ones which he began to develop after the aborted *putsch*. A person, he wrote in *Mein Kampf*, will be "better able to conquer the endless highway if he divides it into sections and attacks each one as though it represented the desired goal itself" (p. 250). Without persistence, he wrote four years later, "one will never learn how to have patience in particulars, and also to renounce them if necessary, in order finally to be able to achieve the vitally necessary aim on a large scale": *Secret Book*, p. 112.
50. Emphasis added. Rauschning, p. 90. See also ibid., p. 14, in which Hitler discussed wearing down the enemy "*before* the war," and p. 18, in which he stated that his strategy was "to destroy the enemy from within, to conquer him through himself" (original emphasis).
51. Emphasis added. Ibid., p. 16.
52. B.H. Liddell Hart, *The German Generals Talk* (New York: William Morrow, 1948), p. 3.
53. *The Hitler Trial Before the People's Court in Munich*, Vol. I, trans. H. Francis Freniere, Lucie Karcic, Philip Fandek (Arlington, VA, 1976), p. 5.
54. Rauschning, p. 19.
55. Ibid., p. 20.
56. Waite, p. 456. Albert Speer, commenting on Hitler's decision to have his private house on the Obersalzberg, pointed out that although the Nazi leader frequently admired a beautiful view, "as a rule he was more affected by the awesomeness of the abysses than by the harmony of a landscape": Speer, p. 47.
57. Van Creveld, p. 68.
58. "Er hasste immer den Begriff 'prestige' in militaerischen Handlungen." General Alfred Jodl, *European Theater Historical Interrogations–51* (hereinafter called *MS ETHINT*, hence, in this case, *MS ETHINT–51*), p. 18. "It's probable that, throughout the 1914–1918 war, some twenty thousand men were uselessly sacrificed by employing them on missions that could have been equally well accomplished by night, with less danger. How often I myself have had to face a powerful artillery barrage, in order to carry a simple post-card!" *Secret Conversations*, 13 October 1941, p. 47.
59. Ibid., 4 June 1942, p. 415.
60. *Halder Diaries*, II, 15 May 1941, p. 921. "You must give that sort of thing up in the future," Hitler told Hess in 1934 after the Deputy Führer had successfully competed in an aerial race. "There are better things in store for you. I need you, Hess": Rauschning, p. 18.
61. Van Creveld. p. 61, and Speer, p. 239. "If I had my way," Hitler said, referring to mountain climbing and Alpine skiing, "I'd forbid those sports, with all the accidents people have doing them. But of course the mountain troops draw their recruits from such fools": Speer, p. 47. During the war, Hitler condemned orchestra director Furtwängler for running the risk of skiing: *Secret Conversations*, 20 August 1942, p. 526. By that time, of course, Hitler's self-inflated image of himself provided the perfect rationale to indulge his fear of flying. "I don't take unnecessary risks in moving about by aircraft. But you know that in the heroic days I shrank from nothing": ibid., 4 January 1942, p. 145. See also ibid., 9–10 January 1942, p. 161.
62. *Halder Diaries*, I, 14 August 1939, p. 6.
63. Original emphasis. Ibid., p. 8.
64. *Mein Kampf*, p. 417. See also *Hitler Trial*, I, p. 60.
65. Rauschning, p. 17. For this paradox, see Michael I. Handel, *Perception, Deception and Surprise: The Case of the Yom Kippur War*, Jerusalem Paper 19 (Jerusalem: Hebrew University, 1976), p. 16.

66. H.R. Trevor-Roper, ed., *Blitzkrieg to Defeat. Hitler's War Directives 1939–1945* (New York: Holt, Rinehart & Winston, 1964), p. 23.

67. Raeder's 29 March 1940 meeting with Hitler: *Brassey's Naval Annual 1948*, p. 90. In 1934, Hitler had discussed in similar terms the occupation of Sweden to secure the iron-ore mines. "It will be a daring, but interesting undertaking, never before attempted in the history of the world. Protected by the fleet, and with the co-operation of the air force, I shall order a series of unexpected individual exploits": Rauschning, p. 143.

68. Emphasis in original: *Blitzkrieg to Defeat*, p. 24.

69. *Brassey's Naval Annual 1948*, p. 90. At a conference with Hitler on 10 April, Raeder emphasized that the passage to Norway would with some luck be successful, "provided the element of surprise were maintained": ibid., p. 93.

70. Quoted in Klaus Knorr, "Strategic Surprise in Four European Wars," in Klaus Knorr and Patrick Morgan, eds., *Strategic Military Surprise. Incentives and Opportunities* (New Brunswick, NJ: Transaction Books, 1984), p. 19. In referring to the Norwegian operation two years later, Hitler stated, "I cannot understand, even in retrospect, how it was that the powerful British Navy did not succeed": *Secret Conversations*, 24 April 1942, p. 335. For the official British history of the campaign, see T.K. Derry, *The Campaign in Norway* (London: HMSO, 1952).

71. *Halder Diaries*, I, 20 January 1940, pp. 189 and 358.

72. Ibid., 17 January 1940, p. 180. See also ibid., 20 January 1940, p. 188, in which Hitler lectured on the need of secrecy and concluded: "An order must not give away the underlying intention. Execution requires a large number of workers: Intentions must be known only by a very small group." Hitler need not have worried. The Allies believed the German plan to be a plant: F.H. Hinsley, *British Intelligence in the Second World War* (London: HMSO, 1979), p. 135. For measures taken to avoid any further compromises see Document 6, OKW Directive, 19 January 1940, Hans-Adolf Jacobsen, *Dokumente zur Vorgeschichte des Westfledzuges 1939–1940* (Göttingen, Berlin, Frankfurt: Musterschmidt Verlag, 1956), pp. 26–8. For "die Affare Mechelen" of 10 January, see H.A. Jacobsen, *Fall Gelb. Der Kampf um den Deutschen Operationsplan zur Westoffensive 1940* (Wiesbaden: Franz Steiner Verlag, 1957), pp. 93–9.

73. Original emphasis. *Halder Diaries*, I, 18 February 1940, p. 233. The capture of the OKH plan "may well have increased the readiness of Hitler . . . . to entertain our Army Groups' proposals later on": Erich von Manstein, *Lost Victories* (Novato, CA: Presidio Press, 1982), p. 118. Referring to Hitler, Halder later described "the soulless chill of this irresponsible gambler": Halder, *Hitler as War Lord*, p. 9. Nevertheless, there was still uncertainty on the part of Hitler and his staff as to the efficacy of achieving surprise. General Keitel referred to the tensions at the Felsennest headquarters as the western campaign began, noting "there was nobody who was not exercised by the question of whether we had succeeded in taking the enemy tactically by surprise or not": Wilhelm Keitel, *In the Service of the Reich*, trans. David Irving (New York: Stein and Day, 1979), p. 110. "I never closed an eye during the night of the 9th to 10th of May 1940," Hitler recalled in 1941: *Secret Conversations*, 17–18 October 1941, p. 58.

74. *Blitzkrieg to Defeat*, p. 34.

75. *Brassey's Naval Annual 1948*, p. 114.

76. Original emphasis. Manstein, p. 170. "He wanted to *evade* the hazard of a decisive struggle with Britain" (original emphasis), ibid. Hitler once confided to an aide: "On land I am a hero, but at sea, I am a coward." Waite, p. 463. "The arms of the future?" Hitler asked rhetorically in 1941. "In the first place, the land army, then aviation and, in the third place only, the navy." Still later in that conference, he stated that the navy "has lost its entire *raison d'être* because of the development of aviation": *Secret Conversations*, 23 August 1941, p. 23. But see Hitler's press chief who described Hitler's love of big ships from his boyhood on: Otto Dietrich, *Hitler*, trans. Richard and Clara Winston (Chicago: Regnery Company, 1955), p. 102.

77. *Brassey's Naval Annual 1948*, p. 119.
78. Ibid., pp. 120–25. The army–navy controversy revolved around the navy's insistence that it could only guarantee the defense of a narrow strip of English coast. The army, naturally enough, desired a broad invasion front. "I utterly reject the Navy's proposal," General Halder lashed out at a conference on 7 August; "from the point of view of the Army I regard their proposal as complete suicide. I might as well put the troops that have landed straight through the sausage machine": ibid., p. 125.
79. Hans-Adolf Jacobsen, ed., *Kriegstagebuch des Oberkommandos der Wehrmacht (Wehrmachtführungsstab)*, Vol. I (Frankfurt am Main: Bernard & Graefe Verlag für Wehrwesen, 1965), 19 September 1940, p. 82. Since the four German landings were to occur at four points from Ramsgate in the southeast around through to the west of Brighton in the south, the deception operation *Herbstreise* (Autumn Journey) was focused on the north. Two days before the start of Sea Lion, four German transports escorted by four cruisers were to proceed to the area between Aberdeen and Newcastle and then retreat at dusk. Other diversions were to be made toward Iceland: *Brassey's Naval Annual 1948*, pp. 128 and 133.
80. Ibid., pp. 132 and 168–9. At a conference with Hitler on 4 February 1941, Raeder insisted that there could be no further slackening in the deceptive preparations for the invasion of England. "For as operation 'Sea Lion' must be maintained as a blind, the measures cannot be further reduced; it is considered essential to continue training activities on their present scale if the deception is to be kept up." Hitler replied that "the deception must be kept up particularly in the spring": ibid, p. 179. Barton Whaley, *Codeword BARBAROSSA* (Cambridge, MA: MIT Press, 1973), pp. 172–4.
81. *Brassey's Naval Annual 1948*, p. 172.
82. Albrecht Tyrell, *Vom "Trommler" zum "Führer". Der Wandel von Hitlers Selbstverständnis zwischen 1919 und 1924 und die Entwicklung der NSDAP* (Munich: Wilhelm Funk Verlag, 1975), pp. 165–74. Dietrich Orlow, *The History of the Nazi Party: 1919–1933* (Pittsburgh: University of Pittsburgh Press, 1969), pp. 9–10.
83. Joseph Nyomarkay, *Charisma and Factionalism in the Nazi Party* (Minneapolis: University of Minnesota Press, 1967), pp. 110–44; Orlow, *History of the Nazi Party*, pp. 68–72 and 77–9; and Jeremy Noakes, "Conflict and Development in the NSDAP, 1924–1927," *Journal of Contemporary History*, I (October 1966), pp. 3–36. Hitler perceived war, as he once stated, as "a faithful copy" of conditions during the period of the struggle for power. Fest, "On Remembering Adolf Hitler," p. 31. And during the war, Hitler stated: "The experience gained while organizing the party during the Kampfzeit will stand me in good stead now." *Secret Conversations*, 24 June 1942, p. 432. The extent of Hitler's authority is illustrated by a trivial affair concerning the proposed German pavilion at the 1939 New York World's Fair. In examining the model of the fairground. Hitler noticed that an adjoining building would block from one angle a view of the proposed pavilion. Because of this, he withdrew Germany from the fair. Wiedemann, p. 224.
84. Halder, *Halder Diaries*, I, 17 March 1939, p. 277.
85. See Michael I. Handel, *The Diplomacy of Surprise: Hitler, Nixon, Sadat* (Rensselaer, NY: Hamilton, 1981), pp. 31–96, *passim*.
86. *Blitzkrieg to Defeat*, pp. 61–2. Martin van Creveld, *Hitler's Strategy 1944–1941. The Balkan Clue* (Cambridge: Cambridge University Press, 1973), pp. 139, 144 and 157.
87. John Strawson, *Hitler as Military Commander* (London: B.T. Batsford, 1971), p. 55, and Speer, p. 100. See also Harold J. Gordon's introduction to *Hitler Trial*, I, p. xxvii, who in referring to the *putsch*, points out: "No precise plans seem to have been made. Hitler in a tight situation tended to favor improvisation."
88. As recounted by Hanfstaengl to Langer: Langer, p. 72. Ernst Röhm made a similar observation about Hitler. "Usually he solves suddenly at the very last minute, a situation that has become intolerable and dangerous only because he vacillates and procrastinates": Lüdecke, p. 287.
89. Herman Kahn. *On Thermonuclear War* (Princeton, NJ: Princeton University Press, 1958), p. 257: Barton Whaley, *Stratagem, Deception and Surprise in War* (Cambridge.

MA: MIT Center for International Studies, 1969), pp. 187–8. Fest, "On Remembering Adolf Hitler," p. 29.

90. Klaus Knorr, "Strategic Surprise: The Incentive Structure," *Strategic Military Surprise*, p. 183.
91. Original emphasis. *Mein Kampf*, p. 14.
92. Rauschning, pp. 107 and 16. "Most of his thinking is carried on subconsciously, which probably accounts for his ability to penetrate difficult problems and time his moves": Langer, p. 202. See also *Mein Kampf*, p. 421, in which Hitler emphasized that "in historical instruction an abridgement of the material must be undertaken. The main value lies in recognizing the great lines of development."
93. Telford Taylor, *The March of Conquest. The German Victories in Western Europe, 1940* (New York: Simon & Schuster, 1958), p. 212; W. Melzer, *Albert-Kanal und Eben-Emael* (Heidelberg, 1957); Albert Kesselring, *A Soldier's Record* (New York: William Morrow, 1954), p. 54. Hitler, according to Rauschning, p. 30, "had the gift, like many self-taught men of breaking through the wall of prejudices and conventional theories of the experts, and in so doing, he has frequently discovered amazing truths". On the evolution of plans for the Albert Canal bridges and Eben-Emael from fall 1939 to spring 1940, see Jacobsen, *Fall Gelb*, pp. 154–70.
94. Walter Ansel, *Hitler and the Middle Sea* (Durham, NC: Duke University Press, 1972), pp. 72–3.
95. *Brassey's Naval Annual 1948*, pp. 355–7. For Operation Greif, see the Skorzeny interview in *MS ETHINT-12*, particularly pp. 2–3, and manuscript, Office of the Chief of Military History, A-862, pp. 96–100. All manuscripts from OCMH hereafter referred to as MS followed by the lettered manuscript series, the number of the manuscript within that series, and the page numbers in the manuscript. Hence, MS A-862, pp. 96–100.
96. Williamson Murray has pointed out "that it was the army above all that opposed a decision-making body equivalent to the British Chiefs of Staff." *The Change in the European Balance of Power 1938–1939. The Path of Ruin* (Princeton, NJ: Princeton University Press, 1984), p. 360. See also John M. Nolen, "JCS Reform and the Lessons of German History," *Parameters*, 3 (Autumn, 1984), pp. 12–20.
97. Manstein, p. 273.
98. Ibid., p. 113. Much of Hitler's discovery of the Manstein Plan was simply due to fortuitous chance. Even as a unitary actor, as Klaus Knorr has pointed out, "the way by which Hitler was apprised of the war-winning strategy was so much the result of uncertain chance that one is tempted to speak of a bureaucratic miracle." Knorr, "Strategic Surprise: The Incentive Structure," p. 184.
99. Langer, p. 67. See also Fest, p. 25.
100. *Blitzkrieg to Defeat*, p. 72. For Hitler's detailed involvement and obvious enjoyment in Operation Student, designed to bring down the government of Marshal Badoglio after the "resignation" of Mussolini, see Warlimont, pp. 358–73 and *Brassey's Naval Annual 1948*, pp. 355ff. When Jodl suggested that an entire airborne division be brought into Rome to overthrow the Badoglio government, Hitler asked: "But who can guarantee that they won't smell a rat? Every airfield is swarming with Italians and when a lot of parachute troops come down – what then?" Jodl reassured him that it was possible, "if we give out that the paratroops are on their way to Sicily": Warlimont, p. 367.
101. *Secret Conversations*, 27 June 1942, p. 437.
102. Langer, p. 74.
103. Michael I. Handel, "The Study of War Termination," *The Journal of Strategic Studies*, I (May 1978), p. 59, and Knorr, "Strategic Surprise in Four European Wars," p. 31.
104. Robert Jervis, *Perception and Misperception in International Politics* (Princeton NJ: Princeton University Press, 1976), p. 231. Stalin, of course, was also looking for a similar pattern before June 1941.
105. Fest, p. 21. Hitler certainly fits for the most part the personality of deceivers as outlined in Michael I. Handel, *Military Deception in Peace and War* (Jerusalem:

Magnes Press, 1985), p. 28: "They are highly individualistic and competitive: They would not fit easily into large organizations or into any type of routine work and tend to work alone. They are usually convinced of the superiority of their own opinions. In some ways they fit the supposed character of the lonely, eccentric bohemian artist, only the art they practice is different."

106. Fest, p. 24.
107. Whaley, *Barbarossa*, p. 172. For Hitler's interaction with the OKW on deception planning see MS P-044a, pp. 12–13.
108. Ibid., p. 57.
109. Ibid., pp. 67–9. A key part of the West Wall deception operation was a film on that construction shown throughout Germany and in many Allied and neutral movie houses. It used trick shots and partial views of the fortifications to make a photomontage that impressed even those who knew of the deception. Ibid., pp. 69–70.
110. Ibid., pp. 58–9.
111. 15 February 1941, OKW Instructions – "Directions for the Deception of the Enemy." Ronald Wheatley, *Operation Sea Lion. German Plans for the Invasion of England, 1939–1942* (Oxford: Oxford University Press, 1958), pp. 97–8. See also Whaley, *Barbarossa*, p. 173 and MS C-059, pp. 3–4.
112. Whaley, *Barbarossa*, pp. 41 and 147.
113. Ibid., p. 176.
114. MS P-044b, pp. 6–8.
115. See Jodl's comment: "The title 'Rundstedt Offensive' is without foundation.... Von Rundstedt neither originated it nor, was he specially concerned with it." *MS ETHINT-50*, p. l0. "Without asking the advice of C.-in-C. West – in fact without even consulting him – Hitler now laid down the fundamental principles of the offensive in conjunction with the Wehrmacht Operations Staff." Hasso von Manteuffel, "The Battle of the Ardennes 1944–5," Hans Adolf Jacobsen and J. Rohwer, eds., *Decisive Battles of World War II: The German View* (New York: G.P. Putnam's Sons, 1965), p. 394. And finally von Rundstedt's reaction: "When I received this plan early in November, I was staggered." Liddell Hart, *German Generals*, p. 275; and "I strongly object to the fact that this stupid operation in the Ardennes is sometimes called the Rundstedt offensive." Milton Shulman, *Defeat in the West* (Westport, CT.: Greenwood Press, 1975), p. 290.
116. "The states are already at loggerheads," Hitler commented referring to the Western Allies, "and their antagonisms are growing visibly from hour to hour. If Germany can now deal out a few heavy blows, this artificially united front will collapse at any moment with a tremendous thunderclap." Manteuffel, p. 401. And in his 11–12 December 1944 meetings with his commanders: "In all history there has never been a coalition composed of such heterogeneous partners with such totally divergent objectives as that of our enemies": Warlimont, p. 487. See also Richard Betts, Strategic Surprise for War Termination," *Strategic Military Surprise. Incentives and Opportunities*, pp. 166–7.
117. Russell F. Weigley, *Eisenhower's Lieutenants. The Campaign of France and Germany 1944–1945* (Bloomington: Indiana University Press, 1981), p. 715. For an historical survey of the Alpine Redoubt and US and German views on this "specter," see MS B-457 and MS B-461. See also Rodney G. Minott, *The Fortress That Never Was* (New York: Holt, Rinehart & Winston, 1964), pp. 25–6 and 135. Halder referred to the *Festung Alpenland* as nothing more "than a phantom in Hitler's brain": Halder, *Hitler as War Lord*, p. 68.
118. Kubizek, pp. 143–5, Joachim C. Fest, *The Face of the Third Reich. Portraits of the Nazi Leadership*, trans. Michael Bullock (New York: Pantheon Books, 1970), p. 7, and Smith, p. 118. Hitler once recounted how he dictated a play to his sister who commented, "your play can't be acted": *Secret Conversations*, 8–9 January 1942, p. 158.
119. It was Ludendorff who acted quickly and decisively and initiated the famous march the next day into the heart of Munich. Harold J. Gordon Jr., *Hitler and the Beer Hall*

Putsch (Princeton: Princeton University Press, 1972), pp. 332–52.
120. Manstein, p. 37.
121. Langer, p. 75. "He is afraid of logic," Otto Strasser pointed out. "Like a woman he evades the issue and ends by throwing in your face an argument entirely remote from what you are talking about": ibid., p. 201.
122. Lüdecke, p. 96. In the *Kampfzeit*, other party associates hired a secretary whose primary duty was to keep track of Hitler and get him to his appointments on time, not always successfully. "Hitler was always on the go," Lüdecke pointed out, "but rarely on time": ibid., p. 97. "Even on ordinary days in those times, it was almost impossible to keep Hitler concentrated on one point": ibid., p. 58. "Adolf is and remains a civilian," Röhm told Rauschning in 1932, "an artist, an idler": Rauschning, p. 155.
123. Dietrich, p. 137, and Speer, p. 91. For the bizarre aspects of Hitler's daily routine, see Dietrich, pp. 137–41. Speer referred to life at the Obersalzburg as "a curious vacuity": Speer, p. 94. Fest has noted how Hitler was constantly in the presence of other people. "Solitude," he concluded, "agonised him." Fest, "On Remembering Hitler," p. 20. Hitler typically romanticized the origins of his insomnia in terms of the *Kampfzeit*. "I formed the habit of staying up late during our days of struggle," he told Speer. "After rallies I would have to sit down with the old fighters, and besides my speeches usually stirred me up so much that I would not have been able to sleep before early morning": Speer, p. 44. But see Kubizek's description of Hitler in Vienna in 1908: "His work was anything but systematic. He worked practically only at night, in the morning he slept": Kubizek, p. 113.
124. Speer, p. 390.
125. Ibid., p. 210. Hitler's predilection for compartmentalization and secrecy had been demonstrated as early as the 1923 *putsch*. "The whole organization was set up so that the military leaders did not even know why they raised troops," he testified on the first day of his trial in 1924: *Hitler Trial*, I, p. 64. See also Carr, p. 26, Dietrich, pp. 117–19, Fest, "On Remembering Hitler," p. 32, and Jäckel, p. 75.
126. Michael Geyer, "National Socialist Germany: The Politics of Information," in Ernest K. May, ed., *Knowing One's Enemies. Intelligence Assessments Before the Two World Wars* (Princeton: Princeton University Press, 1984), pp. 330 and 336. The divisiveness in the armed forces worsened as the war progressed. In September 1942, for instance, General Zeitzler succeeded in persuading Hitler to agree to exclude the OKW staff from advising on anything pertaining to the east. The OKW soon lost track of the eastern theater and as a consequence the capability to advise on overall strategy and operations, much less intelligence: MS T-101,KI, p. 83. See also Michael Howard, *Studies in War and Peace* (New York: Viking Press, 1972), p. 116.
127. MS P-44a, p. 7. For the various propaganda offices, see LTC Weberstedt's description, MS P-044c. The type and quality of personnel engaged in the deception business also, of course, made a difference. The German system, which practiced deception in peacetime, had routinized this aspect by the time war came. The intelligence staffs were thus full of professionals for this function. There was no room for the gifted amateurs who were used so effectively by the British and the Americans. Handel, *Military Deception in Peace and War*, pp. 32–3.
128. Geyer, p. 340. For the myriad intelligence agencies developed in the Third Reich, see ibid., pp. 311–26. See also David Kahn, *Hitler's Spies. German Military Intelligence in World War II* (London: Macmillan, 1978), pp. 42–63. See, however, Michael I. Handel, "Intelligence and the Problem of Strategic Surprise," *The Journal of Strategic Studies*, 7 (Sept. 1984), p. 270, who makes the important point that since "the independent judgement of individual analysts at all levels cannot be guaranteed *within* each organization, the fostering of *inter*-organizational competition may enhance the diversity and freedom of the intelligence process in general." Original emphasis.
129. Speer, p. 113.
130. Rauschning, p. 183.
131. *Secret Conversations*, 1–2 August 1941, p. 15.
132. Rauschning, p. 184.

133. Kahn, *Hitler's Spies*, p. 453.
134. Manstein, p. 227.
135. Rauschning, p. 268.
136. Speer, p. 355.
137. Dietrich, p. 110.
138. Heinz Assmann, "Some Personal Recollections of Adolf Hitler," *United States Naval Institute Proceedings*, 12 (December 1953), p. 1293.
139. For the infinitely more sinister Bormann, see Jochem von Lang, *Der Sekretär: Martin Bormann, der Mann, der Hitler beherrschte* (Stuttgart: Deutsche Verlagsanstalt, 1977). General Zeitzler, who replaced Halder as OKH Chief of Staff, stated that there were "four categories of people in his immediate entourage: The blindly faithful; those with independent judgement whom he subjected to himself within a short time; those who stood by their principles but who ... could not make up their minds to break with Hitler and were therefore dismissed at any time he deemed propitious; and those who would not submit and were left the choice of suicide or resignation": MS T-111, p. 100. Speer also wrote (p. 83) that servility was "endemic among his entourage."
140. MS T-101, K1, p. 8 and Warlimont, p. 13.
141. Matthew Cooper, *The German Army 1933–1945* (New York: Stein & Day, 1978), p. 88. "He combined enthusiasm for Nazism, diligence in military administration, and an impressive stature and bearing with a complete lack of talent": Barry A. Leach, *German Strategy Against Russia 1939–1941* (Oxford: Clarendon Press, 1973), p. 34. For opinions within the army, see *Halder Diaries*, II, 6 July 1942, p. 1477: "The hardest to endure is Keitel with his undigested spoutings."
142. Speer, p. 244. Hitler had no illusions about Keitel's ability as a soldier. In the course of one of his field inspections early in the war, Keitel mistook a 75mm anti-tank gun for a light field howitzer. Hitler did not mention the mistake at the time, but on the ride back from the front, he remarked to Speer: "Did you hear that? Keitel and the antitank gun? And he's a general of the artillery": ibid., p. 235. Keitel also had no illusions about himself in this regard, once remarking "I am no Field-Marshal": Cooper, p. 88.
143. Speer, p. 389
144. Geyer, p. 342.
145. *Mein Kampf*, pp. 212–13.
146. Speer, p. 65. "Already at that early period," Rauschning commented (p. 105), "he disliked hearing anything not calculated to strengthen his own convictions."
147. Gordon A. Craig, *Germany 1866–1945* (New York: Oxford University Press, 1978), p. 714. Waite, p. 18, Cooper, p. 127, and Carr, p. 90.
148. Dietrich, p. 15, and Halder, *Hitler as War Lord*, p. 21.
149.ֹ Waite, p. 48.
150. Speer, p. 294. Speer commented later that his years of Spandau imprisonment made him realize that Hitler's life in the various war headquarters in the closing years of the war had borne a great resemblance to that of a prisoner, "particularly in terms of the psychological burden": ibid., p. 302. "I couldn't say whether my feeling that I am indispensable has been strengthened during this war. One thing is certain, that without me the decisions to which we to-day owe our existence would not have been taken": *Secret Conversations*, 13–14 October 1941, p. 48.
151. Speer, p. 238.
152. Jervis, p. 144. Such reaction to surprise was not limited to the Germans. See, for example, the classic response to a hard-pressed Soviet front-line unit on the morning after Barbarossa had begun. "We are being fired on," the Soviet unit radioed to its superior headquarters. "What shall we do?" "You must be insane," the Soviet headquarters replied. "And why is your signal not in code?" John Erickson, *The Soviet High Command* (London: Macmillan, 1962), p. 587.
153. *Mein Kampf*, p. 164. See also *Secret Conversations*, p. xvi.
154. MS B-484, p. 29. General Friedrich von Bötticher was Military and Air Attaché in the US from 1933 to 1941. Bötticher, of course, reinforced Hitler's picture of the world by

poor reporting on US military potential. Gerhard L. Weinberg, *World in the Balance. Behind the Scenes of World War II* (Hanover, NH: University Press of New England, 1981), pp. 61–2.

155. Speer, p. 246. See also Creveld, "War Lord Hitler," p. 69. "No adequate information could be provided either by our own intelligence service or that of the Italian High Command": Warlimont, p. 267. Predictably, Admiral Canaris, the intelligence chief, stated after the landings had occurred that he had forseen the invasion: *MS ETHINT-2*, p. 3. Malta was also considered an invasion possibility. William Langer, *Our Vichy Gamble* (New York: Norton, 1966), p. 365. So indirect did Hitler consider the Allied landings that he adopted a business as usual attitude the next day (8 November 1942) in his speech commemorating the 1923 *putsch*: Speer, pp. 246–7 and *MS ETHINT-20*, p. 13.

156. Lucy S. Davidowicz, *The War Against the Jews, 1933–1945* (New York: Harmondsworth, 1975), p. 21. For the quality of Hitler's pseudo-scientific racial thoughts, see *Secret Conversations*, 19–20 February 1942, p. 260: "Dirt shows on black people only when the missionaries, to teach them modesty, oblige them to put on clothes. In the state of nature, negroes are very clean."

157. Weinberg, p. 59, and Rauschning, p. 14. On the mongrelization aspect, however, see Hitler's 1928 *Secret Book*, p. 107, in which he describes the émigrés to the United States in positive Darwinistic terms as a selection of the fittest from their homelands.

158. Rauschning, p. 76. See also Speer, p. 307.

159. *Secret Conversations*, August 1941, p. 22. But see also Hitler's concern about the vast American resources *vis-à-vis* the UK and the USSR, Andreas Hillgruber, *Hitlers Strategie: Politik und Kriegsfürung 1940–1941* (Frankfurt: Bernard & Graefe, 1965), pp. 199–200 and 534, and his 1928 estimate of the US as a new power so vast that it might in the future threaten the old power order: *Secret Book*, p. 83.

160. *Secret Conversations*, 5 January 1942, p. 149. Although not mentioned, the American débâcle at the Kasserine Pass in North Africa may have prompted this comment. Hitler's opinion may have also been influenced by his experiences in the First World War. "Compared with the British and the French," he reminisced in 1933, "the Americans behaved like clumsy boys. They ran straight into the line of fire, like young rabbits. The American is no soldier": Rauschning, p. 78.

161. Speer, p. 306. "Now we have shown what we are capable of," Hitler remarked after the fall of France. "Believe me, Keitel, a campaign against Russia would be like a child's game in a sandbox by comparison": ibid., p. 173. "If we start in May 1941," Hitler stated in 31 July 1940 conference concerning an attack on the Soviet Union, "we would have five months to finish the job in": *Halder Diaries*, I, p. 534.

162. Hillgruber, *Hitlers Strategie*, p. 214. In the Soviet Army alone, the dead from the purges included three of the five marshals, 14 of the 16 army commanders, 60 of the 67 corps commanders, 136 of the 199 divisional commanders, and 221 of 397 brigade commanders. About half of the officer corps, approximately 37,000, were either shot or imprisoned: Gordon Craig, *Europe Since 1815* (New York: Holt, Rinehart & Winston, 1971), p. 532.

163. Leach, p. 202.

164. *Secret Conversations*, 5–6 January 1942, p. 150.

165. Leach, p. 202.

166. Ibid., p. 94. In December, Hitler's army adjutant had stated: "The Führer ... is ... uncertain over the Russians' strength": ibid., p. 159.

167. John Erickson, *The Road to Stalingrad* (New York: Harper & Row, 1975), p. 106.

168. Speer, p. 247. A few weeks later, Hitler's estimate of the Russians had appreciably risen. In discussing a possible Russian landing in the Crimea at a conference on 12 December 1942, Jodl remarked that if winter weather conditions were as bad as reported, the Russians would not be able to land. "The Russians will get through somehow," Hitler replied. "We wouldn't be able to land in a snow shower or anything like that. I admit that. But that doesn't go for the Russians": Warlimont, p. 292.

169. Johanna Meskill, *Hitler and Japan: The Hollow Alliance* (New York: Atherton Press,

1966), pp. 3–4. See MS P-108, pp. 275–98, the major point of which (p. 275) is that "military collaboration between Germany and Japan was at no time close enough for coordinated actions." How much of Hitler's attitude was racial is not certain. For an early negative appraisal of the Japanese, see *Mein Kampf*, pp. 290–1. But also see his positive comments in 1942: *Secret Conversations*, 1 July 1942, pp. 443–4. See also James V. Compton, *The Swastika and the Eagle: Hitler, the United States and the Origins of World War II* (London: Bodley, 1968), pp. 17–18, 31 and 60–1.

170. *Brassey's Naval Annual 1948*, p. 193. Already on 5 March, Hitler had ordered Keitel to issue specific instructions that "no hint of the *Barbarossa Operation* must be given to the Japanese." Whaley, *Barbarossa*, p. 145. Even before the Tripartite Pact, Hitler's surprises had upset the Japanese. Germany's August 1939 non-aggression pact with Russia had infuriated the Japanese military, particularly because of the serious defeat sustained by Japanese troops shortly afterwards on the Mongolian border. MS P-044a, p. 60.

171. Meskill, pp. 30–1. See also Norman Rich, *Hitler's War Aims. Ideology, the Nazi State, and the Course of Expansion* (New York: W.W. Norton, 1973), pp. 228–9, who concludes that such a simultaneous attack would have ensured a German victory and that Hitler's failure to do anything to secure Japanese co-operation before Barbarossa was incomprehensible. General von Greiffenberg has pointed out that press chief Otto Dietrich's October 1941 speech, which concluded that the war on the Eastern Front had already been decided in favor of the Germans, was designed to appeal to Japan and secure its participation in the new *Lebensraum* by timely intervention against a prostrate Russia. MS P-044a, pp. 60–1.

172. Robert J.C. Butow, *Tojo and the Coming of War* (Stanford: Stanford University Press, 1961), p. 221.

173. Murray, p. 29.

174. *MS ETHINT-2*, p. 12. For Hitler's declaration of war on the United States, see Weinberg, pp. 75–95.

175. *Secret Conversations*, 18 December 1941, p. 123. Evidently, the entire situation continued to rankle the Nazi leader. "You mustn't believe what the Japanese say," Hitler told Jodl on 5 April 1943, and later that day, referring to the Japanese, he stated, "They lie to beat the band; everything they say has always got some background motivation of deception": Warlimont, p. 211.

176. Percy Ernst Schramm, ed., *OKW KTB*, IV, p. 1719. "Italy's invasion of Greece came as a surprise": Halder Diaries, I, 29 October 1940, p. 642. At a conference on 4 November 1940, Hitler referred to the Italian invasion of Greece as a "regrettable blunder. On no occasion was authorization for such an independent action given to the Duce": *Brassey's Naval Annual 1948*, p. 146. In later, similarly ill-conceived surprises perpetrated by the Italians, there was an aspect of impatient *Schadenfreude* evident in Hitler's reactions. See, for example, Jodl's remarks in *Halder Diaries*, II, 3 March 1941, p. 818. "For the time being the Führer will not use his influence with Il Duce in the question of the Italian attack in Albania. Let them scorch their noses."

177. *Secret Conversations*, 4 January 1942, pp. 146–7.

178. *Brassey's Naval Annual 1948*, p. 346.

179. Rauschning, pp. 33–4. *Secret Conversations*, 17 February 1942, p. 257.

180. *Brassey's Naval Annual 1948*, p. 339.

181. Speer, p. 232. "Hitler possessed an astounding retentive memory and an imagination that made him quick to grasp all technical matter and problems of armaments": Manstein, p. 274.

182. *OKW KTB*, IV, pp. 1718–19.

183. *Brassey's Naval Annual 1948*, p. 152. Hitler gave up this idea by the summer of 1943 "because the few aircraft which would get there would be of no significance but would only arouse the will to resist in the population": ibid., p. 339.

184. Ansel, p. 110.

185. *Brassey's Naval Annual 1948*, p. 338.

186. MS CO65b, p. 21.

187. The six tanks were used in terrain that the army staff indicated was unsuitable for tanks. Hitler overruled these objections and when the tanks were wiped out, never referred to the incident again: Speer, p. 241.
188. Rommel's response to this amateurishness was a contemptuous, almost pitying smile: ibid., p. 254.
189. MS B-689, pp. 4–5; Speer, p. 356.
190. David Irving, *The Mare's Nest* (Boston: Little, Brown, 1964), p. 304. See also Peter G. Cooksley, *Flying Bomb: The Story of Hitler's V Weapons in World War II* (New York: Charles Scribner's Sons, 1979) and Basil Collier, *The Battle for the V-Weapons 1944–1945* (New York: William Morrow, 1965). The code name for the SAM was "Waterfall": Speer, pp. 364–5.
191. Ibid., pp. 362–3. For Allied operations against the German factories that produced the jet, see F.H. Hinsley, *British Intelligence in the Second World War*, Vol. III, p. 352.
192. Speer, p. 364, and David Irving, *The Rise and Fall of the Luftwaffe* (London, 1956), pp. 257ff.
193. Van Creveld, "War Lord Hitler," p. 71.
194. The physician was Dr Hans Karl von Hasselbach: Schramm, p. 106, and Carr, p. 80.
195. Speer, p. 228. See also David Irving, *The German Atomic Bomb* (New York: Simon & Schuster, 1967).
196. General Warlimont, *MS ETHINT–4*, p. 1.
197. *MS ETHINT–49*, p. 2. See also *MS ETHINT–4*, p. 4; Speer, pp. 352–3; Guy Hurtcup, *Code Name Mulberry: The Planning, Building, and Operation of the Normandy Harbours* (London: David & Charles, 1977) and Alfred Stanford, *Force Mulberry* (New York: William Morrow, 1951).
198. *Mein Kampf*, pp. 201–2.
199. Rauschning, p. 13.
200. Until the chemical industry was bombed in 1944, German production amounted to 3,100 tons of mustard gas and 1,000 tons of Tabun per month. Speer, p. 413. See also Stephen L. McFarland, "Preparing for What Never Came: Chemical and Biological Warfare in World War II," *Defense Analysis*, 2 (June 1986), p. 113, and John Ellis van Courtland Moon, "Chemical Weapons and Deterrence: The World War II Experience," *International Security*, 4 (Spring 1984), p. 25.
201. Speer, pp. 413–14, and McFarland, p. 114.
202. See, however, Frederick Joseph Brown, *United States Chemical Warfare Policy, 1919–1945: A Study of Restraints* (Geneva: Imprimerie Offset Blanc, 1967), pp. 175–6, 182 and 219, who contends that Hitler ordered the use of gas in 1945. Moon refutes Brown's contention as unsubstantiated. Moon, pp. 25–6. See also, in this regard, General Hermann Ochsner's testimony. "Anyone," the Army Chief of Gas Operations testified, "who suggested in Germany that chemical warfare should be initiated would have been called 'an idiot' and 'crazy'": McFarland, p. 113. See Schramm's deposition (Aussagen) for the Nuremberg Trials concerning gas warfare: "It was never planned to initiate. The deliberations in the course of the years always led to the conclusion that alone for tactical, material and other reasons it was not advisable to start a gas war": *OKW KTB*, IV, p. 1833.
203. McFarland, p. 114. In late fall 1944, Hitler ordered a special commissioner to establish a program to protect the entire population from the effects of gas warfare. Speer, pp. 413–14.
204. Moon, p. 18. Speer claimed that he tried to put poison gas in Hitler's Berlin bunker near the end of the war. But Hitler had ordered a ten-foot chimney for the ventilation shaft, knowing, as a veteran of gas warfare, that poison gas is heavier than air. Speer was stymied. Speer, p. 431.
205. Anxiety concerning Barbarossa can be sensed throughout Hitler's directive for *Merkur*, particularly his noticable hurrying of preparations for the operation against the island. Transport used in the campaign, for instance, was in no circumstances to hinder deployment for Barbarossa: Karl Grundelach, "The Battle for Crete 1941," *Decisive Battles of World War II*, p. 103, and *Blitzkrieg to Defeat*, p. 69.

206. Van Creveld, *Hitler's Strategy*, pp. 168–70; Grundelach, p. 130; MS P-105, pp. 71–81; and "Der Deutsche Feldzug in Griechenland und auf Kreta," MS C-100, *passim*.
207. Grundelach, p. 131.
208. Ibid.
209. MS P-044c, pp. 11–12.
210. Grundelach, p. 131, and Leach, p. 239. This applied, of course, to other operations as well. See, for example, Manteuffel's and Student's discussions in Liddell Hart, *German Generals*, pp. 283 and 159–60, and that author's own discussion (ibid., p. 283) of how good the opportunities were for airborne operations in the Ardennes in 1944. See also Kesselring, p. 144, in which Hitler, anxious to avoid a second costly operation like Crete, called off the "distasteful Malta venture."
211. Handel, *Perception, Deception and Surprise: The Case of the Yom Kippur War*, p. 13.
212. *Secret Conversations*, 3–4 January 1942, p. 142.
213. D.C. Watt, *Too Serious a Business: European Armed Forces and the Approach to the Second World War* (Berkeley, CA: University of California Press, 1975), p. 62; Barry R. Posen, *The Sources of Military Doctrine. France, Britain, and Germany Between the World Wars* (Ithaca and London: Cornell University Press, 1984), p. 183; Murray, p. 37; and Cooper, pp. 130–8.
214. Martin van Creveld, *Fighting Power: German Military Performance, 1914–1945* (Washington, DC: Office of Net Assessment, Office of the Secretary of Defense), p. 42. See also Lieutenant Colonel Walter von Lossow, "Mission Type Tactics versus Order-Type Tactics," *Military Review*, No. 5 (June 1977), p. 89. For the change to the infiltration tactics, see Timothy T. Lupfer, *The Dynamics of Doctrine: The Changes in German Tactical Doctrine During the First World War* (Fort Leavenworth, KS: U.S. Army Command and General Staff College), Leavenworth Papers, Number 4 (July 1981).
215. Cooper, p. 137. See also Michael Geyer, "German Strategy in the Age of Machine Warfare, 1914–1945," in Peter Paret, ed., *Makers of Modern Strategy from Machiavelli to the Nuclear Age* (Princeton NJ: Princeton University Press, 1986), p. 567.
216. Posen, pp. 206–8; Edward N. Luttwak, "The Operational Level of War," *International Security*, 3 (Winter 1980/81), 67–73; Cooper, pp. 139–59: and Alan S. Milward, *The German Economy at War* (London: Athlone Press, 1965), p. 244.
217. Luttwak, p. 69.
218. J.F.C. Fuller, *Conduct of War* (New York: Funk & Wagnalls, 1961), p. 244.
219. On the army resistance, see Posen, pp. 208–10 and 226, and Cooper, pp. 149–58.
220. Heinz Guederian, *Panzer Leader*, trans. Constantine Fitzgibbon (New York: E.P. Dutton, 1952), pp. 29–30.
221. *Secret Conversations*, 20 August 1942, p. 294.
222. Posen, pp. 226 and 212, in which Posen concludes that "to the extent that the German Wehrmacht achieved a doctrinal innovation that can be called Blitzkrieg, Hitler's intervention was decisive."
223. *Mein Kampf*, p. 356. Hitler may also have been influenced by his favorite biographer of Frederick II. "Frederick's sword was much shorter, but he unsheathed it at thrice the speed": Thomas Carlyle, *Frederick the Great* (Oxford: Clarendon Press, 1916), p. 162.
224. *Secret Conversations*, 8 July 1942, p. 471. See also ibid., p. 14.
225. Rauschning, p. 129.
226. Geyer, "German Strategy in the Age of Machine Warfare," p. 584, and Luttwak, p. 68.
227. Luttwak, pp. 70–1.
228. Rauschning, pp. 17, 158; van Creveld, "War Lord Hitler," pp. 67–8.
229. Luttwak, pp. 69, and 71–2.
230. Posen, p. 206. See, however, Cooper, p. 149, who views the Second World War as the unresolved conflict between *Blitzkrieg*, "founded on the revolutionary use of armoured, motorized and airforces engaged in a mission of paralysis," and the tradi-

tional strategy "based on mass infantry armies, with the new arms at best treated only as equal partners, the cutting edge of the old decisive manoeuvre of encirclement and annihilation." See also his Chapter 11, *passim*.

231. *Mein Kampf*, p. 192.

232. Ibid., p. 195. Commenting on the general strike that broke out just as the Ludendorff offensive began in 1918, Hitler pointed out that "at home the revolution was before the door, and not the victorious army." Ibid., p. 197. See also *Secret Book*, p. 34: "Domestic policy must secure the inner strength of a people so that it can assert itself in the sphere of foreign policy."

233. Rauschning, p. 210. See also Jodl's comments on Hitler's concern that he might have "to call the German people to arms once more in the years to come": Warlimont, p. 112.

234. Hillgruber, *Hitlers Strategie*, pp. 33–4, 254–8. This does not mean, however, that there was a very strong cause and effect connection between Germany's economic inability to fight a long war and the development of the *Blitzkrieg* doctrine. Murray, p. 37.

235. Leach, p. 95. See also Milward, pp. 9–12.

236. Van Creveld, "War Lord Hitler," p. 72.

237. Liddell Hart, *German Generals*, p. 296.

238. Manstein, p. 541. "He was just not the man to recognize the need for a far-sighted operational policy": ibid., p. 503.

239. Van Creveld, "War Lord Hitler," p. 72, and *Halder Diaries*, II, 19 December 1941, p. 49.

240. *Secret Book*, p. 148. There was for Hitler no substitute for hands-on experience. Years later, he stated: "I distrust officers who have exaggeratedly theoretical minds. I'd like to know what becomes of their theories at the moment of action": *Secret Conversations*, 20 July 1942, p. 187.

241. Michael Howard, *Studies in War & Peace* (New York: Viking Press, 1971), p. 112. See also Manstein, p. 262, who describes how Hitler questioned Halder's right to differ from him, "declaring that as a front-line infantryman of World War I he was an infinitely better judge of the matter than Halder." For the opinions of some army officers on what they referred to as Hitler's "trench perspective," see Speer, p. 241.

242. *Halder Diaries*, I, 18 June 1940, p. 471. Jodl mistakenly asserted that it was only after Hitler lost the strategic initiative in late 1942 that he began to interfere with operational matters. *OKW KTB*, IV, p. 1721.

243. Manstein, pp. 284 and 21. On the communications equipment, see Speer, p. 304. Only rarely did any of Hitler's subordinates comment to him on his tendency to command from the rear. One such incident occurred at a midday conference on 26 July 1943, when Hitler discussed the possible occupation of Rome by means of a parachute assault combined with a move by a Panzer Grenadier Division. Göring interrupted him: "But the orders must be given on the spot. You can't command that from Munich": Warlimont, p. 364.

244. Manstein, p. 284. See also MS T-101, Annex 15, pp. 9–14, particularly p. 9, which describes how after Hitler assumed command of the army in 1941, commanders of tactical units began to attend Führer conferences regularly. After the war, in discussing the Western Front after the 1944 Allied breakthroughs at Avranches, Göring stated: "The Führer decided upon all troop movements right down to regiments." *MS ETHINT-30*, p. 5.

245. Manstein, p. 285. Von Manstein, of course, was a master of the operational art who realized that initiative and a willingness to accept responsibility at the operational and tactical levels were absolutely essential. "Only when there was no other possible alternative left," he wrote, "did anyone on our side encroach upon the authority of a subordinate formation headquarters by specifically laying down the action it should take": ibid., p. 284. See also *Halder Diaries*, II, p. 1531 in which Halder noted: "It would be better to give directives broadly outlining the plan, general objectives, and resources, and leaving the rest to the best judgement of the appointed commander."

246. Handel, "Intelligence and the Problem of Strategic Surprise," p. 230.

247. Halder, *Hitler as War Lord*, p. 22. See also van Creveld, "War Lord Hitler," p. 70. It may have been due in part to his experiences in the First World War. "I can imagine the situations that the troops are called upon to face," he remarked. "And I can do that because I've been an ordinary soldier myself": *Secret Conversations*, 13 October 1941, p. 47.
248. MS C-065d, p. 23.
249. Warlimont, p. 76.
250. For the order, see Walther Hubatsch, *Die Deutsche Besetzung von Dänemark und Norwegen* (Göttingen: Musterschmidt, 1952), p. 190. See also *Halder Diaries*, I, 18 April 1940, p. 322. Apparently Hitler also considered ordering Dietl's force into internment in Sweden: Assmann, p. 1292, and MS C-065d, pp. 24–25. See LTC von Lossberg's self-serving account of his critical role in "die erste Führungskrise" which included preventing Hitler from ordering Dietl into Sweden. Bernhard von Lossberg, *Im Wehrmacht Führungsstab* (Hamburg: H.H. Noelke, 1950), pp. 67–9. See also MS C-065d, p. 27, and Buchheit, pp. 100–2, who accepts Lossberg's account.
251. Derry, p. 216.
252. Rauschning, p. 15.
253. Original emphasis. Cooper, p. 115. Liddell Hart was actually describing how he saw Germany's doctrine in operation during the war. As Cooper demonstrates, this form of pure *Blitzkrieg* was myth: ibid., pp. 115 and 139–49.
254. *Secret Conversations*, 29 October 1941, p. 77.
255. For the original 19 October 1939 *Fall Gelb* concept, see Document 10, Jacobsen, *Dokumente zur Vorgeschichte*, pp. 41–6. For von Manstein's final conceptual suggestion see his 18 December 1939 memorandum, Document 37, ibid, pp. 131–6. See also his Memorandum for Record of his 17 February 1940 meeting with Hitler in the Reichs Chancellery, Document 44, ibid., pp. 155–6 and the 24 February OKH concept, Document 19, ibid., pp. 64–8. See also Jacobsen, *Fall Gelb*, pp. 68–84, for the evolution of von Manstein's and Army Group A's concept.
256. MS C-065d, pp. 40–1 and Warlimont, p. 95. The concern had started earlier. Halder commented on Hitler's 11 May visit to the OKH headquarters: "Pleased with success, expects attack from south": *Halder Diaries*, I, 11 May 1940, p. 393. Hitler's fears were, it should be noted, shared by a large number of the German High Command. See, for instance, von Rundstedt's comments on the German southern flank in Liddell Hart, *German Generals*, p. 128: "I knew Gamelin before the war, and trying to read his mind, had anticipated that he would make a flank move from the Verdun direction with his reserves."
257. *Halder Diaries*, I, 17 and 18 May 1940, pp. 406–7. Hitler's concept: "Turn spearhead Divs. hard to southwest to protect the south flank, while holding bulk of mot. forces in readiness for a drive to the west." Ibid. Hitler never overcame his fear of flanking attacks under the *Blitzkrieg* doctrine. In his 12 August 1941 Supplement to Directive 34, he emphasized: "Only after ... threats to our flanks have been rehabilitated will it be possible to continue the offensive": *Blitzkrieg to Defeat*, p. 94. And in 1944, in devising the Ardennes counterthrust, Hitler wanted to make the base of the initial penetration even wider. "He feared that otherwise," Jodl commented later, "it ... might be driven in from the sides by the first Allied counterattacks." *MS ETHINT–50*, p. 6. This tendency may have been noticed by Göring who commented after the war to US interrogators that "General Patton is your most outstanding general .... The outstanding thing about him is that he is not afraid of his flanks." *MS ETHINT–30*, p. 20.
258. Manstein, p. 383.
259. Ibid., p. 130.
260. Cooper, p. 241.
261. Alistair Horne, *To Lose a Battle: France 1940* (Boston: Little, Brown, 1969), p. 159. At least one German general had given some thought early on to exploiting operational surprise in the West. "What do we have to rush reinforcements to Guderian in case of a breakthrough?" Jodl asked in November 1939: *Halder Diaries*, I, 14

November 1939, p. 127. See, however, Liddell Hart, *German Generals*, p. 114, who in discussing the French campaign with the German generals after the war, found that "almost every one of them admitted that he had not anticipated such a decisive victory as was actually gained." See also Halder's entry on Hitler at a 15/16 March conference in *Halder Diaries*, I, 17 March 1940, p. 227: "Decision reserved on further moves after the crossing of the Meuse."

262. *Blitzkrieg to Defeat*, p. 49.
263. On 6 November 1941. Raymond L. Garthoff, *Soviet Military Doctrine* (Glencoe, IL: Free Press, 1955), pp. 273–4.
264. Earl F. Ziemke, *Stalingrad to Berlin: The German Defeat in the East* (Washington, DC: Office of the Chief of Military History, 1984), p. 500, and *Terrain Factors in the Russian Campaign*, DA Pamphlet, No. 20–290, July 1951, p. 7.
265. Warlimont, p. 244. See also Samuel P. Huntington, *The Soldier and the State. The Theory and Politics of Civil–Military Relations* (New York: Vintage Books, 1957), p. 115, who observed: "Mystical speculation replaced considerations of time and space, and the careful calculation of the strength of one's own forces in relation to the enemy."
266. *Secret Conversations*, 6 August 1942, p. 500, and Warlimont, p. 112.
267. *Secret Conversations*, 6 September 1942, p. 563. Typically, Hitler merged his discovery of Russian space with unrealistic dreaming. "Germans must acquire the feeling for the great, open spaces .... We'll take them on trips to the Crimea and the Caucasus. There's a big difference between seeing these countries on the map and actually having visited them." Ibid., 17 September 1941, p. 29. See also 15 June 1943, p. 574: "One thing the Americans have and which we lack, is the sense of the vast open spaces." Hitler was not alone in his failure to anticipate the enormity of Russia. Even before the invasion, Halder wrote: "The imposing vastness of the spaces, in which our troops are now assembling, cannot fail to strike a deep impression": *Halder Diaries*, II, 9 June 1941, p. 951. See also General von Thoma's remarks in Liddell Hart, *German Generals*, p. 97, who stated after the war: "What we had was good enough to beat Poland and France, but not good enough to conquer Russia. The space there was so vast and the going so difficult."
268. *Secret Conversations*, 17–18 January 1942, p. 180.
269. Ibid., 12–13 January 1942, p. 164.
270. *Halder Diaries*, II, 26 July 1941, p. 1076.
271. Ibid., II, 3 July 1941, p. 1001.
272. Halder to Jodl, ibid., 7 August 1941, p. 1159.
273. Original emphasis in 5 April 1941 Directive No. 41, *Blitzkrieg to Defeat*, p. 118.
274. Manstein, p. 404.
275. Speer, p. 241.
276. *Halder Diaries*, II, endnote, p. 1586.
277. *MS ETHINT–3*, p. 5. See also Manstein, p. 278, and Erik H. Erikson, *Young Man Luther. A Study in Psychoanalysis and History* (New York: W.W Norton, 1958).
278. *MS ETHINT–3*, p. 6.
279. Schramm, *Hitler*, pp. 153–5, and Carr, p. 83. Hitler later acknowledged a conscious reversion to First World War-type defense. See his 8 September 1942 *Führerbefehl über "grundsätzliche Aufgaben der Verteidigung*," in which he stated: "Ich kehre mit dieser Auffassung bewusst zu der Art der Verteidigung zurück wie sie in den schweren Abwehrschlachten des Weltkrieges besonders bis zum Ende des Jahres 1916 mit Erfolg angewendet wurde": Andreas Hillgruber, ed., *OKW KTB*, II, p. 1293. "Hitler refused to consider the question of building defensive rear positions," General Zeitzler commented, "and even forbade the reconnaissance of such positions." MS T-111, p. 103.
280. *MS B-034a*, p. 3.
281. Warlimont, p. 561. On 17 January 1945, German forces evacuated Warsaw without Hitler's knowledge. As a consequence, he issued an order on 21 January which stipulated that divisional commanders or higher were to report to him every decision

204 Churchill and Hitler

concerning any operational movement whether it be an attack, a disengagement or withdrawal: *Blitzkrieg to Defeat*, p. 203.

282. See, for example, Ziemke, p. 15, who points out that "even his harshest critics later had to concede that the German armies, caught as they were without prepared positions to fall back on or even adequate winter equipment and clothing, might well have disintegrated if the command had conceded a necessity to retreat." "Halten und kämpfen bis zum Äussersten," the 21 December order stipulated: *OKW KTB*, I, p. 1085. "In der *Verteidigung*," Hitler continued in his order of 26 December, "ist um jeden fussbreit Boden mit letztem Einsatz zu kämpfen": ibid., p. 1086.

283. Guderian, p. 305. "It is characteristic of Hitler's lack of strategic understanding," Halder noted, "that he never showed the slightest interest in the disposition and movement of the reserves, although minor occurrences at the front could occupy him for hours." Halder, *Hitler as War Lord*, p. 31. Referring to the Eastern campaign, von Manstein wrote: "Hitler never made the right forces available at the right time." Manstein, p. 162. See also MS T-111, in which General Zeitzler referred to Hitler as being "busy shifting replacement battalions" – hardly operational level units. See also H.J. Hopfgarten, "The Movement of Operational Reserves," *Military Review*, 3 (April 1954), pp. 103–7.

284. MS T-101, K1, p. 81. Manstein, p. 278.

285. MS T-26, p. 8; Manstein, pp. 445–6; and Cooper, p. 456.

286. "Mir ist bei dem Gedanken an diesen Angriff auch immer ganz mulmig im Bauch": Buchheit, p. 367.

287. Manstein, p. 448.

288. Cooper, p. 458; Manstein, pp. 447–9; and MS T-26, p. 9.

289. Richard Betts, "Strategic Surprise for War Termination: Inchon, Dienbienphu, and Tet," *Strategic Military Surprise*, p. 165.

290. Ibid., pp. 166–7. "The forces available for the offensive on December 16th ... bore no proper relation to the far-reaching objectives which had been set for it." Manteuffel, p. 414. SS General Dietrich, commander of the key and newly created 66th (SS) Panzer Army, looked back with irony on the ends–means disparity. "All I had to do was to cross a river, capture Brussels, and then go on and take the port of Antwerp. And all this in ... the worst three months of the year, through the Ardennes where the snow was waist-deep and where there wasn't room to deploy four tanks abreast, let alone six armoured divisions": Cooper, p. 521.

291. Halder, *Hitler as War Lord*, p. 66. Manteuffel, p. 144, and MS B-235, pp. 50–1. For the effectiveness of deception measures, see MS P-044a, pp. 43–5.

292. Manteuffel, p. 415.

293. Milward, *German Economy*, pp. 3–5, and Alan S. Milward, *War, Economy and Society, 1939–1945* (Berkeley, CA: University of California Press, 1977), pp. 24, 29 and 31.

294. Posen, p. 196.

295. Original emphasis. Cooper, p. 128; Waite, p. 460.

296. *Secret Conversations*, 17 September 1941, p. 28. See also T.W. Mason, "Some Origins of the Second World War," *Past and Present*, 29 (Dec. 1964), p. 69.

297. *Secret Book*, p. 43.

298. Telford Taylor, *The Breaking Wave* (New York: Simon & Schuster, 1967), p. 298.

4
Churchill:
The Victorian Man of Action

The most common public image of Winston Churchill is the wartime picture taken by the famed photographer, Yousuf Karsh. The British leader glowers out from that photograph, truculent and combative. Never mind that the menacing look was reportedly caused by Karsh's insistence that Churchill remove his ever-present cigar from his mouth. What remains is that quintessential aura of resistance and defiance against all odds that came to symbolize the spirit of the nation Churchill led throughout the Second World War. Paradoxically, this image of a modern warlord in the greatest of all twentieth century conflicts owes its existence to the late Victorian era into which Churchill was born in 1874. For it was during those years in the Indian summer of Queen Victoria's reign that the future British Prime Minister developed his singular traits of character and formed his concepts of war and personal leadership that were to endure throughout his long life.

Foremost among the Victorian influences on the young Churchill was a pervasive sense of historical continuity that stretched beyond the Victorian years. To begin with, there was Blenheim Palace with its obelisks of victory, its grand vistas that created a sense of drama, and the great achievements ubiquitously carved in stone, woven in tapestries and painted on canvas. A monument, in short, to one man, John Churchill, first Duke of Marlborough, whose exploits fed into the unique Whig legend devised by the British in the intervening centuries to underpin their imperial ambitions. It was in that great mansion that Churchill was born, and it was among the patrician descendants of the great Whig aristocracy from Stuart and Georgian England that he spent his formative years. It was thus no accident that he never deviated throughout his life from what the British historian J.H. Plumb has described as "that curious ideology of the Whigs, half truth, half fiction; half noble, half base."[1]

In pursuing that course, Churchill was doing no more than accepting the historical assumptions of his fellow patricians in the late nineteenth-century. For them, English history was an evolutionary development by trial and error in which the Englishman's inherent national characteristics such as love of liberty and justice were gradually matched by the appropriate institutions of government. There were problems throughout this process, of course, ranging from the Stuarts and Civil War to the loss of America and the threats posed to institutions by industrialization. But the Empire that emerged from those travails was the greatest and most just in history, founded on the richest and freest democracy the world had ever known. In this interpretation, to which Churchill fully subscribed, it was the play of time working on natural genius that produced Great Britain and its institutions. Finally, it was the landed Whig squirearchy, the "great Oaks" as Edmund Burke referred to them, who had been through the centuries England's natural rulers, the guardians of her destiny, and who had brought that miraculous historical development to fruition.[2]

British imperialism, in the later Victorian era, was an extension of the Whig version of England's development. Two years before Churchill's birth, Disraeli had confirmed this in his Crystal Palace address, in which he denounced the Liberals for viewing colonies simply from an economic viewpoint, ignoring "those moral and political considerations which make nations great, and by the influence of which alone men are distinguished from animals."[3] It was these higher considerations that caused Victorians to venerate the soldiers on the Imperial frontier, those men of action who maintained the British Empire, which, by the time of Churchill's formative years, was an all-engulfing red splash on the world map, three times the size of the Roman Empire.

The young Churchill was extremely susceptible to such hero worship. On 14 February 1885, he wrote to his father in India commenting on the death of Colonel Frederick Gustavius Burnaby, Royal Horse Guards, who had been killed in action the previous month "sword in hand, while resisting the desperate charge of the Arabs at the battle of Abu Klea."[4] The letter was indicative of the name recognition concerning the heroic Victorian men of action. Burnaby had ridden through Asia Minor to Persia, served as a war correspondent for *The Times*, and had undertaken a solo balloon flight from Dover to Normandy. It never occurred to the 11-year-old Churchill that his father would not have heard of the colonel.

But if the daily exploits of such heroes were not enough, there was always the prolific pen of George Alfred Henty. In 1876, Henty published the first of his 80 novels on English and Imperial history. Whether it was with Wolfe at Quebec or with Clive in India, young Victorians like Churchill could relive vicariously every British triumph throughout the Empire. In 1898, the year that Churchill participated in Kitchener's victory over the Mahdi at Omdurman, it was estimated that Henty's annual sales were as high as 250,000.[5] Added to this was the wide variety of nonfiction dealing with the same subject. Between 1852 and 1882, the increasingly literate British masses purchased 31 editions of Creasy's *Fifteen Decisive Battles of the World*, at least partially, according to Herbert Spencer, to "revel in accounts of slaughter."[6] Equal success awaited the Macmillan series entitled "The English Men of Action," each story of 250 pages being immediately sold out. By 1891, such stories were the staple of the popular press. That year, a new series entitled "Story of the VC: Told by those who have won it" appeared in the new *Strand Magazine* and enjoyed as much success as the Sherlock Holmes short stories that also began in the magazine that year.[7]

From these influences, Churchill created an inner historical world in which there was only the grand and the grandiose. Progress was measured through politics and war, rarely in terms of economic, intellectual and social issues. Throughout his long life, he was always conscious of the continuity in this world and of his place in it. At one point in June 1940, for instance, General Ismay, his Chief of Staff, urged him to delay sending troops to organize a redoubt in Brittany. "Certainly not," was the Prime Minister's immediate reply. "It would look very bad in history if we were to do any such thing."[8] And in December 1943, while recovering from pneumonia at Eisenhower's villa at Tunis, Churchill's physician reported that the British leader was well enough to mutter with his lifelong lisp: "I shpposhe it ish fitting I should die beshide Carthage."[9]

Like the great heroes of old, Churchill was at stage center in his inner world, at all times, as he had written of Pitt the Elder, "a projection on to a vast screen of his own aggressive dominating personality."[10] Harry Hopkins, Roosevelt's envoy in the Second World War, recognized this early in the war. "Churchill ... always seemed to be at his Command Post on the precarious beachhead ..," he wrote; "wherever he was, there was the battlefront – and he was involved in the battles not only of the current war but of the whole past, from Cannae to Gallipoli."[11] That romantic outlook was captured in 1913 in an astonishingly prescient biographical sketch of Churchill in A.G. Gardiner's *Pillars of Society*:

> He is always unconsciously playing a part – an heroic part. And he is himself his most astonished spectator. He sees himself moving through the smoke of battle – triumphant, terrible, his brow clothed with thunder, his legions looking to him for victory, and not looking in vain. He thinks of Napoleon; he

thinks of his great ancestor. Thus did they bear themselves; thus, in this rugged and most awful crisis, will he bear himself. It is not make-believe, it is not insincerity; it is that in that fervid and picturesque imagination there are always great deeds afoot with himself cast by destiny in the Agamemnon role.[12]

School life only accentuated these tendencies. The British public schools were an integral part of the nineteenth-century English patrician life. The spirit of those select institutions was captured in Thomas Hughes' *Tom Brown's Schooldays*, the 1856 fictional account of life at Rugby under that school's famous headmaster, Dr. Arnold. The book provided an ideal of life for two generations of British schoolboys, best summed up in Squire Brown's parting thoughts concerning Tom: "If he'll only turn out a brave, helpful, truth-telling Englishman, and a gentleman and a Christian, that's all I want."[13]

Underlying that ideal in Tom Brown was a tradition of manliness from the English squirearchy, with its cult of games and field sports and its emphasis on physical strength and prowess. It was also a morally righteous manliness to be used against bullies – usually older, if not stronger – in defense of the small and the weak, the downtrodden fags that seemed to populate Hughes' Rugby. "I want to leave behind," Tom states in formulating his ambitions at Rugby, "... the name of a fellow who never bullied a little boy, or turned his back on a big one." Allied to this theme, but even more fundamental to the manly tradition, was the concept of combativeness, the love of a good fight. "After all," Tom conjectures, "what would life be without fighting, I should like to know! From the cradle to the grave, fighting, rightly understood, is the business, the real, highest, honestest business of every son of man."[14]

By the time Churchill entered Harrow in 1888, the manliness cult in the public schools had been augmented by the amateur ideal in the gamesmanship formulation of the new imperialism, as new generations of military men of action and civilian ruler administrators were produced for the Empire. The essential linkage of sport and the wars to support that Empire was summed up in Sir Henry Newbolt's classic public school poem, "Vitai Lampada."

> There's a breathless hush in the close tonight –
>   Ten to make and the match to win –
> A bumping pitch and a blinding light,
>   An hour to play and the last man in.
> And it's not for the sake of a ribboned coat,

> Or the selfish hope of a season's fame,
> But his Captain's hand on his shoulder smote –
> "Play up! play up! and play the game!'

Years later, the former cricket player, under native attack, exhorts his colonial troops to greater deeds:

> The sand of the desert is sodden red –
> Red with the wreck of a square that broke;
> The Gatling's jammed and the Colonel dead,
> And the regiment blind with dust and smoke;
> The river of death has brimmed his banks,
> And England's far, and Honour a name;
> But the voice of a schoolboy rallies the ranks:
> "Play up! play up! and play the game!"[15]

At Harrow, Churchill shed his egocentric world view to some degree and began "to play the game." Most importantly, the school experience reinforced his determination to make himself physically and mentally tough, to mold himself in more courageous, heroic and manly terms than were naturally his in physique and temperament. "I am cursed with so feeble a body," Churchill wrote to his mother from Sandhurst in 1893, "that I can scarcely support the fatigues of the day."[16] His frustration was understandable. He stood five feet, six and a half inches at the time, with a chest measurement of 31 inches, inadequate by Sandhurst standards. He had extremely sensitive skin and, as has been noted, suffered all his life from a difficulty, like his father, in pronouncing the letter "s." As a young man, he would walk up and down attempting to remedy this problem by rehearsing such phrases as: "The Spanish ships I cannot see for they are not in sight." Later on the lecture circuit, he began to cure his lisp and to lose the inhibitions that it had caused. "Those who heard him talk in middle and old age," his son commented later, "may conclude that he mastered the inhibition better than he did the impediment."[17]

Despite these physical disadvantages and a temperament that was not naturally courageous, Churchill emerged as a mentally tough, physically brave man. In fact, it was precisely because he lacked the very mental and physical traits that were the quintessential staples of the British public schoolboy and the Victorian man of action, that Churchill persevered, forcing himself to go against his inner nature. It would be a lifelong and successful effort to compensate, to keep, as he termed it, from "falling below the level of events."[18]

As a consequence, the history of Churchill's involvement in the late Victorian wars was one of continual search for physical danger, whether it was with the Malakand Field Force on India's northwest frontier in 1897, with Kitchener's forces in the Sudan in 1898, or with Lord Roberts' troops in the Boer War at the turn of the century. "I am more ambitious for a reputation for personal courage than [for] anything else in the world," he wrote early on in the Malakand Field Force campaign, and the remainder of his first combat experience was involved in achieving that goal.[19] It required constant test and examination. A matter, in other words, of finding situations which afforded "opportunities for the most sublime forms of heroism and devotion." "I am glad," he wrote to his mother, "to be able to tell you ... that I never found a better than myself as far as behaviour went;" no one, he pointed out a month later, would "be able to say that vulgar consideration of personal safety ever influenced me."[20]

Underlying this attitude was also the ideal of chivalry. Within that ideal and given the relatively small number of British casualties in the limited Victorian conflicts, Churchill could indulge an almost Arthurian concern with the manner of death. In general, he could mourn the fact that "golden lads and girls all must, as chimney-sweepers, come to dust." But if the death was honorable, or in an honorable cause, he could still celebrate it as part of a chivalric necessity. "Lord Ava is seriously wounded," he wrote in February 1900 about that nobleman, fatally wounded in the Boer War, "a sad item, for which the only consolidation is that the Empire is worth the blood of its noblest citizens."[21]

Churchill's experience in nineteenth-century wars also confirmed a ruthless rationality and pragmatism in Victorian combat. Under the new imperialism, there was in all classes almost a religious faith in Britain as the great force for good in the world. That England could be in the wrong in any one of the countries splashed with red on the world map was almost inconceivable, particularly against itinerant natives. Those tribesmen would often mutilate the British wounded and dead, as Churchill discovered in India. In return, he noted, the British "do not hesitate to finish their wounded off .... I have not soiled my hands with any dirty work – though I recognize the necessity of some things."[22] It was also a necessity recognized by the British public. Thus, Churchill could describe in his best-selling account of the Indian campaign how the 11th Bengal Lancers formed a line across a plain on the northwest frontier and began a merciless pursuit of the enemy up a narrow valley: "No quarter was asked or given, and every tribesman caught was speared or cut down at once. Their bodies lay thickly strewn about the fields, spotting with black and white patches, the bright green of the rice crop. It was a terrible lesson and one which the inhabitants ... will never forget. Since then the terror of Lancers has been extraordinary."[23]

The terror could also be extended in that type of warfare to noncombatants. "Of course, it is cruel and barbarous," Churchill acknowledged concerning the burning of native shacks in India as punishment for raids, "as is everything else in war, but it is only an unphilosophic mind that will hold it legitimate to take a man's life and illegitimate to destroy his property."[24] It was a rationale that could also be applied to weapons, such as the new Dum-Dum bullet fired from the Lee-Metford rifle. Churchill had nothing but praise for the expansive character of the new round, "a wonderful and from the technical point of view a beautiful machine," since it "tears and splinters everything before it, causing wounds which in the body must be generally mortal and in any limb necessitate amputation." Results and effectiveness were the ultimate criteria. "I would observe," Churchill concluded on the Dum-Dum, "that bullets are primarily intended to kill, and these bullets do their duty most effectively without causing any more pain to those struck by them than the ordinary lead variety."[25]

That pragmatic approach to weapons and technology gained further ascendancy as the young Victorian continued to encounter the realities of military life on the Imperial frontier. In late 1897, Churchill and a small group of British and Indian troops from the Malakand Field Force were being pursued by a band of Swati tribesmen. The lead warrior paused to slash at one of the British wounded, and Churchill, as he later recounted, decided to kill him.

> I wore my long Cavalry sword well sharpened. After all, I won the Public School fencing medal. I resolved on personal combat *à l'arme blanche*. The savage saw me coming. I was not more than twenty yards away. He ... awaited me, brandishing his sword. There were others waiting not far behind him. I changed my mind about the cold steel. I pulled out my revolver, took ... most careful aim, and fired.[26]

In a similar manner, at Omdurman in 1898 during the initial charge of the 21st Lancers, Churchill used a 10 shot Mauser pistol instead of a saber. The choice was dictated by the necessity to strap his arm to his side because of his easily dislocated shoulder. Nevertheless, Churchill was grateful for the choice of the pistol, "the best thing in the world," which probably saved his life. As a result of the charge, "the most dangerous 2 minutes I shall live to see," the Dervish lines fell back

"A.O.T. [arse over tip]." In those two minutes, 70 officers and men and 119 horses had been killed or wounded in a totally unnecessary charge. Once again, at least momentarily, pragmatism replaced romanticism, as Churchill recounted two days later to his mother.

> I was very anxious for the regiment to charge back because it would have been a very fine performance and men and officers could easily have done it while they were warm. But the dismounted fire was more useful, though I would have liked the charge – "pour la gloire" – and to buck up British cavalry. We got a little cold an hour afterwards and I was quite relieved to see that "heroics" were "off" for the day at least.[27]

After Omdurman, Churchill walked among the thousands of Dervish bodies stacked on the battlefield and found "nothing of the dignity of unconquerable manhood."[28] Those feelings were reinforced by a steadily mounting British casualty list that included many of his closest friends. "The realization came home to me with awful force," he wrote later, "that war, disguise it as you may, is but dirty, shoddy business which only a fool would undertake. Nor was it until the night that I again recognized that there are some things that have to be done, no matter what the cost."[29] Duty was something that late Victorians could understand. And with duty would come the romanticizing of what had to be done. The brave deeds of Omdurman, Churchill told his readers in the *Morning Post*, "brighten the picture of war with beautiful colours, till from a distance it looks almost magnificent, and the dark background and dirty brown canvas are scarcely seen."[30]

Such sentiments did not survive the First World War. That conflict was a gradually evolving shock not only to Churchill, but to the British public who also approached it with Victorian idealism and optimism compounded by romantic public school notions of chivalry and combat. "War declared by England," a schoolboy destined to die in that conflict wrote in his diary on 5 August 1914; "Intense relief, as there was an awful feeling that we might dishonour ourselves."[31] Disillusion began to creep in as the carnage mounted. But the horror of modern warfare was generally concealed well into the conflict from the British public by a conspiracy of silence in the form of stiff upper, if not sealed, lips. That tendency was supported by the popular literature of the time. There was, for instance, a rear guard fictional movement for most of the war, in which a band of brothers continued to protect the weak and vanquish the villains under the leadership of such heroes as Bulldog Drummond and Major-General Richard

Hannay. And in 1917, Conan Doyle ended *His Last Bow*, the final volume of Sherlock Holmes, by observing that after the war, "a cleaner, better, stronger land will lie in the sunshine."[32]

For Churchill, the full impact of the conflict came midway through the war, after his resignation from the Asquith government when he moved to the western front, where from January to May 1916 he commanded the 6th Battalion, Royal Scots Fusiliers located at the village of Ploegsteert on the Ypres–Armentières road. Even then, Churchill's command experience at "Plug Street" only confirmed his ambivalence about war in the modern era. On the one hand, no matter how grim the troglodyte world of the trenches, there was the visceral, combat exultation that had not changed since the days of the Malakand Field Force. "My beloved," he wrote to his wife in January, "... I have just come back from the line, having had a jolly day."[33] In that context, even the grimness of attritive warfare could be viewed through the romantic prism of death's grandeur in a Victorian "last stand" for men of action. In a letter written while an offensive was in progress, Churchill referred to "the bloody & blasted squalor of the battlefield," noting of a battalion that had lost 420 men out of 550 in that battle: "I shd feel vy proud if I had gone through such a cataclysm."[34]

On the other hand, there was the daily proof offered by the ongoing carnage of Verdun. "Do you think we should succeed in an offensive," he wrote to his wife in April, "if the Germans cannot do it with all their skill & science?" And that same month, he wrote to Lord Birkenhead that Verdun seemed "to vindicate all I have ever said or written about the offensives by either side in the West."[35] Churchill returned to the subject of Verdun in May 1916 after his release from the army. In a speech to the House of Commons, he warned against a philosophy that within two months would result in the Somme campaign. "The argument which is used that 'it is our turn now,'" he said, "has no place in military thought." And later that month, he reminded his listeners in Commons that every 24 hours, nearly 1,000 men, "English, Britishers, men of our own race, are knocked into bundles of bloody rags."[36]

The effect of all this for the men of action, Churchill observed as he looked back during the first decade after the First World War, was the "obliteration of the personal factor in war, the stripping from high commanders of all the drama of the battlefield, the reducing of their highest function to pure office work." For him, the modern commander had become "entirely divorced from the heroic aspect by the physical conditions which have overwhelmed his art."

No longer will Hannibal and Caesar, Turenne and Marlborough, Frederick and Napoleon, set their horses on the battlefield and by their words and gestures direct and dominate ... the course of a supreme event. No longer will their fame and presence cheer their struggling soldiers. No longer will they share their perils, rekindle their spirits and restore the day. They will not be there. They have been banished from the fighting scene, together with their plumes, standards and breast-plates.[37]

The general in such an environment was for Churchill no more than a "high-souled speculator," who would in the future "sit surrounded by clerks in offices, as safe, as quiet and dreary as Government departments, while the fighting men in scores of thousands are slaughtered or stifled over the telephone by machinery." It would be efficient, but not heroic. "My gardener last spring," he commented in that regard, "exterminated seven wasp's nests .... It was his duty and he performed it well. But I am not going to regard him as a hero." It would be, as Churchill, envisioned it, a pale, lifeless, unromantic, unemotional world of the masses, without the splash of color and the verve of great deeds and individual heroism.

The heroes of modern war lie out in the cratered fields, mangled, stifled, scarred; and there are too many of them for exceptional honours. It is mass suffering, mass sacrifice, mass victory. The glory which plays upon the immense scenes of carnage is diffused. No more the blaze of triumph irradiates the helmets of the chiefs. There is only the pale light of a rainy dawn by which forty miles of batteries recommence their fire, and another score of divisions flounder to their death in mud and poison .... The wars of the future will be even less romantic and picturesque.[38]

War, in short, had "been stripped of glitter and glamour." If war should come again, Churchill reflected bitterly, "poets will not sing nor sculptors chisel the deeds of conquerors. It may well be that chemists will carry off what credit can be found," in a competition "to kill women and children and the civil population generally, and victory will give herself in sorry nuptials to the diligent hero who organizes it on the largest scale."[39] Once again it was the scale. Mass, impersonal death meant the end of the very basis of Victorian wars. Looking back on the battle of Omdurman after the First World War, Churchill commented in sorrow on the passing of that "sporting game" for the men of action under the new rules of total war:

This kind of war was full of fascinating thrills. It was not like the Great War. Nobody expected to be killed. Here and there in every regiment or battalion, half a dozen, a score, at the worst thirty or forty, would pay the forfeit; but to the great mass of those who took part in the little wars of Britain in those vanished light-hearted days, this was only a sporting element in a splendid game. Most of us were fated to see a war where the hazards were reversed, where death was the general expectation and severe wounds were counted as lucky escapes, where whole brigades were shorn away under the steel flail of artillery and machine-guns, where the survivors of one tornado knew that they would certainly be consumed in the next or the next after that. [40]

Nevertheless, the Great War could not completely destroy Churchill's reverence for such abstractions as glory, honor and courage, which remained for him permanent and reliable, no matter what had transpired in the grim world of the Western Front. One day, after that conflict, someone remarked in his presence that nothing was worse than war. "Dishonor," Churchill immediately replied in full voice, "is worse than war – slavery is worse than war."[41] It was not that Churchill failed to see the conditions of modern war, as he experienced them. "Never for a moment," he could write to his wife in 1917, "does the thought of this carnage & ruin escape my mind."[42] But he would not allow the squalor to penetrate fully his inner Victorian core. His romantic perception of what conflict had been for men of action and therefore what it should be remained a dominant counter-weight to his realistic assessment of total war. It was a perception so powerful that it influenced even the disillusioned like Robert Graves and Siegfried Sassoon. Sassoon later recounted how he wondered if Churchill, during their September 1918 meeting, had been entirely serious when he said that "war is the normal occupation of man." Churchill had gone on to add "war – and gardening" as a qualifier. "But it had been unmistakable," Sassoon concluded, "that for him war was the finest activity on earth."[43]

It was, of course, not as simple as Sassoon described. Churchill's portrayal, for instance, of the French General Mangin reflected the ambiguity of his feelings. On the one hand, there was the incredibly brave and resourceful general personally leading the men at Verdun and along the Chemin des Dames "like a hungry leopard." On the other hand, there was "Mangin the Butcher," relieved temporarily from his command for the losses his leadership had inflicted on his own troops. In a similar manner, there was Churchill's mixed analysis of General Hubert Gough, the Fifth Army Commander. "He was a typical cavalry officer, with a strong personality and a gay and boyish charm of manner," Churchill wrote. "A man who never spared himself or his troops, the instrument of costly and forlorn attacks, he emerged from the Passchendaele tragedy pursued by many fierce resentments."[44]

Amidst this ambiguity, Churchill could still find the heroic men of action as he looked back on the Great War from the late 1920s. There was, for instance, Bernard Freyberg, the New Zealander whom he had befriended as a sub-lieutenant at the beginning of the war, commanding elements of four divisions in 1918 while successfully holding a front of 4,000 yards. And there was General Tudor and his Ninth Division, whom he visited just before the Ludendorff offensive in March 1918. "The impression I had of Tudor," Churchill wrote, "was of an iron peg hammered into the frozen ground, immovable. And so indeed it proved." Before he left the battlefield that day, Churchill turned and once again looked back on the men of the Ninth Division. "I see them now, serene as the Spartans of Leonidas on the eve of Thermopylae."[45]

Strength, in other words, even in total war was not enough for Churchill. There must also be the valor and steadfastness of the men of action that he had known in his early years in a previous era. Marshal Foch, in this view, despite disastrous errors was redeemed by his "obstinate combativeness." "He was fighting all the time," Churchill wrote of Foch, "whether he had armies to launch or only thoughts."[46] Such characteristics, Churchill came to believe, were even more important at the political level of total war, when national survival was the stake. That lesson was provided by Georges Clemenceau, whom he met many times during the war and who, he considered, "embodied and expressed France. As much as any single human being, miraculously magnified, can ever be a nation, he was France." It was the fiery French Premier's indomitability and willingness to take any measures on both the Home Front and the fighting front in order to emerge triumphant that most impressed Churchill. "Happy the nation," he wrote of Clemenceau, "which when its fate quivers in the balance can find such a tyrant and such a champion." And in a passage that presaged his own emergence in the spring of 1940, Churchill described Clemenceau's final call to public life, which marked the beginning of the end for France's misfortunes: "He returned to power as Marius had returned to Rome; doubted by many, dreaded by all, but doom-sent, inevitable."[47]

In the 1930s as he researched and wrote his history of Marlborough,

Churchill returned again and again to his ancestor's "combination of mental, moral, and physical qualities adapted to action which were so lifted above the common run as to seem almost godlike." His studies renewed his faith in the man of action, whose every word "was decisive. Victory often depended upon whether he rode half a mile this way or that."[48] Such a man could make a difference in any type of conflict, particularly if he combined valor with common sense. In the fourth volume of his biography, Churchill lingered over the aftermath of the battle of Elixem in which Marlborough had pierced the Lines of Brabant with almost no Confederation casualties. Even as the battle neared its end, his grateful troops responded with spontaneous mass affection. As he rode up sword in hand to take his place in the final cavalry charge, the soldiers and their officers broke into cheering, extremely unusual considering the formal military etiquette of the time. And afterwards, when Marlborough moved along the front of his army, the veterans of Blenheim, as Churchill described it, "cast discipline to the winds and hailed him everywhere with proud delight."[49]

Surely there was still room in modern warfare for men like Marlborough who, in order to seal the victory at Ramillies, personally led a cavalry charge on the left wing, in Churchill's words, "transported by the energy of his war vision and passion."[50] In October 1940, that was still a concern as the Prime Minister attempted to bring Major-General P.C.S. Hobart back to active duty against the resistance of the General Staff. "I am not all impressed by the prejudices against him in certain quarters," he minuted. "Such prejudices attach frequently to persons of strong personality and original view." This was a time, the British leader added, "to try men of force and vision and not to be exclusively confined to those who are judged thoroughly safe by conventional standards." The minute continued:

> We are now at war, fighting for our lives, and we cannot afford to confine Army appointments to persons who have excited no hostile comment in their career. The catalogue of General Hobart's qualities and defects might almost exactly have been attributed to most of the great commanders of British history. Marlborough was very much not the conventional soldier, carrying with him the goodwill of the Service. Cromwell, Wolfe, Clive, Gordon, and in a different sphere Lawrence, all had very close resemblance to the characteristics set down as defects.[51]

It was a perceptive use of history for a man whose belief in historical continuity had been momentarily suspended by the advent of total war. Already in 1929, however, Churchill was referring to history "as a guide to present difficulties. How strange it is," he wrote "that the past is so little understood and so quickly forgotten. We live in the most thoughtless of ages. Every day headlines and short views."[52] This renewed faith in historical continuity also caused Churchill to face up pragmatically to the prospect of future total wars. "The story of the human race is War," he wrote, "...But the modern developments surely require severe and active attention."[53] In looking back on the First World War in that regard, Churchill acknowledged that any romantic predilections in total war had to be supported by large doses of realistic, tangible support. "Foch's famous war cries, 'Allez à la bataille,' 'Tout le monde à la bataille,'" he commented, "would have caused no more meaning to history than a timely cheer, but for the series of tremendous drives and punches with which the British armies ... trampled down the ... German military might."[54]

But a world without emotional romanticism, without heroic men of action, did not have to be the fate of total war. "There is a sense of vacancy and of fatuity, of incompleteness," he wrote as he observed disillusioned Britain in the inter-war years. "We miss our giants." If the British people, Churchill warned, quit "the stern, narrow high-roads which alone lead to glorious destinies and survival," there would be nothing left but a "blundering on together in myriad companies, like innumerable swarms of locusts, chirping and devouring towards the salt sea."[55] Despite modern forces and trends, despite the stark realities of modern life and warfare, there must still be the indefinable, romantic aspirations of Victorian times. Britain had emerged victorious from the Great War on a new and higher plateau, Churchill concluded,

> but the scenery is unimpressive. We mourn the towering grandeur which surrounded and cheered our long painful ascent. Ah! if we could only find some new enormous berg rising towards the heavens as high above our plateau as those old mountains down below rose above the plains and marshes! We want a monarch peak, with base enormous, whose summit is for ever hidden from our eyes by clouds, and down whose precipices cataracts of sparkling waters thunder.[56]

That mountain peak appeared when Churchill kissed hands in May 1940. Immediately, the new Prime Minister's rich, romantic historical sense of continuity dominated the scene, allowing him to proceed with what appeared to many an obstinate irrationality against overwhelm-

ing forces in the darkest days of the war. But he knew with an absolute
certainty that it had been done before, not the least successfully by the
first Duke of Marlborough.

This perception of historical continuity also fueled Churchill's sense
of personal destiny as a man of action, a key ingredient of his success
as a leader in the Second World War. "The statesman ... must behold
himself," Hans Morgenthau has pointed out in this regard, "not as the
infallible arbiter of the destiny of men, but the handmaiden of
something which he may use but cannot control."[57] And so it was with
Churchill, who believed that he was the servant of an historical entity
called England, and that he was destined to maintain that entity and
its Empire on the upward path that reached back to Alfred the Great.
It was this belief, this inner certainty, that could inspire the masses in
general, and his civilian and military subordinates in particular. One
scientist described the effect whenever he met Churchill during the
war as "the feeling of being recharged by contact with a source of
living power."[58] On another occasion, the Permanent Under-Secre-
tary of War in 1940 urged the Prime Minister to meet with a general
about to leave on an urgent arms purchase mission to the United
States, "in order that he may have the glow of Mount Sinai still on him
when he reaches Washington."[59]

To his feeling of destiny, Churchill brought an absolute sense of
combativeness from his Victorian heritage as a man of action.
General Smuts, for instance, who was on General Botha's staff during
the Boer War, described Churchill after his capture in that war as "a
scrubby, squat figure of a man, unshaved. He was furious, venomous,
just like a viper."[60] And when the King of Sweden wrote to George VI
on 1 August 1940 to propose a conference to examine possible peace
ties, Churchill would have no part of it, pointing out to the Foreign
Office "that the intrusion of the ignominious King of Sweden as a
peacemaker, after his desertion of Finland and Norway, and while he
is absolutely in the German grip, though not without its encouraging
aspects, is singularly distasteful."[61]

This type of combativeness was as natural as breathing to Churchill,
often governed by his visceral reaction of the moment. On 10 June
1940, as an example, in another one of the increasingly dismal Anglo-
French meetings, French Premier Reynaud asked the Prime Minister
what would happen if France capitulated and all of German strength
were concentrated upon invading England. Churchill replied instantly
that he had not thought out his response in detail, but that basically he
would drown as many as possible of the invaders on their way over to
England, leaving it only to "*frapper sur la tête*" anyone who managed

to crawl ashore. At the end of that meeting, the increasingly emotional British leader once again reassured his French counterpart that Britain "would fight on and on, *toujours*, all the time, everywhere, *pas tout, pas de grâce*, no mercy. *Puis la victoire!*"[62] That such emotions were also governed by his Victorian concept of the heroic last stand was illustrated in a conversation the British leader had with President Roosevelt's Special Envoy, Averell Harriman, while sailing to the United States on the *Queen Mary* in spring 1943. When that conversation turned to the U-boat menace, the Prime Minister informed Harriman that he had arranged for a machine gun to be added to his own lifeboat, should it be necessary to abandon ship. "I won't be captured," he concluded. "The finest way to die is in the excitement of fighting the enemy."[63]

But it was in Hitler that Churchill found a perfect outlet for his combative nature – a threat on which he could lavish all his contempt. On 19 July 1940, three days after he had issued the Sea Lion Directive to prepare for the invasion of England, Hitler offered Britain, in a Reichstag address that mocked Churchill, the choice between peace with Germany, or "unending suffering and misery." When asked if he wished to respond to the speech, Churchill replied: "I do not propose to say anything to Herr Hitler's speech, not being on speaking terms with him."[64] Nevertheless, he delighted in bringing to bear against the threat the full brunt of his command of the English language. When, for instance, he spoke of the "Na–sies," the very lengthening of the vowel carried a stunning message of his contempt. Moreover, there were always the visual images invoked by his vivid descriptions of the enemy. Von Ribbentrop was "that prodigious contortionist" and Mussolini was a "whipped jackal, frisking at the side of the German tiger – this absurd imposter." And when Barbarossa was unleashed on Russia, he brought the event, which Hitler considered would cause the world to hold its breath, down to its basic level. "Now this bloodthirsty guttersnipe," Churchill announced, "must launch his mechanized armies upon new fields of slaughter."[65]

Against this threat and despite the realities of a newer, even more complicated total conflict, the British leader returned in the Second World War to his Victorian concept of the heroic man of action. The RAF, in particular, excited his admiration, arousing in him, a Private Secretary recalled, "a latent schoolboy instinct of hero-worship" for what he considered "the cavalry of modern war."[66] It was a type of emotion that had led him throughout his civilian and military careers to seek out the recipients of Britain's highest award for valor. "You can generally tell what he finds attractive in his friends," a close friend commented late in Churchill's life; "a V.C. is in his good graces before he begins."[67]

Closely allied to Churchill's ingrained hero worship was his Victorian sense of honor. How men conducted themselves in crises was all important to him. The Czechoslovak legionaries after the First World War, for instance, "forsook the stage of history" in their dishonorable treatment of the White Russian leader, Admiral Kolchak. In the Second World War, Pétain was a similar example. Admiral Darlan, on the other hand, redeemed himself in Churchill's eyes with the 1942 scuttling of the French fleet at Toulon.[68] It was a conception that penetrated the remoteness he needed in order to send mass heroic men of action to their deaths in total war. The German sinking of the *Royal Oak* in Scapa Flow in October 1939, for instance, triggered his overactive imagination concerning the 800 "heroes" who had lost their lives. "Poor fellows, poor fellows," he muttered after receiving the news, "trapped in those black depths."[69] And in the later stages of the war, General Eisenhower witnessed a meeting at Chequers when a logistics briefer struck a nerve in the normally imperturbable Prime Minister by using the phrase "so many thousand bodies" in referring to British reinforcements. "Sir," Churchill broke in with great indignation, "you will not refer to the personnel of His Majesty's Forces in such terms as 'bodies'. They're not corpses. They are live men, that's what they are."[70]

This sensitivity to the mass men of action was complicated by the ambivalence about conflict that had dogged Churchill since the era of total war began. "War is a game played with a smiling face," he told his daughter Sarah at Tehran in 1943, "but do you think there is laughter in my heart?"[71] And yet he was still fascinated by it. Captured German combat films, for instance, often marked the evening's entertainment. He also loved newsreels of the war and took particular delight if he was featured, often shouting to General Ismay: "Look Pug, there we are."[72] And, finally, there was his pride in *Desert Victory*, the film history of the Eighth Army, which he viewed over and over again, even sending a copy to Stalin.

J.F.C. Fuller touched upon a major reason for Churchill's ambivalence in a description that he eventually excised from his classic *The Conduct of War*. "The truth would appear to be," he wrote of the British leader, "that throughout his turbulent life he never quite grew up, and like a boy, loved big bangs and playing at soldiers."[73] Certainly, Churchill felt more intense exhilaration in battle than most professional soldiers. At one point early in the Second World War, enemy bombing commenced as he was being conducted around anti-aircraft sites in Richmond Park. Only after great difficulty and many protests, could the commander persuade the Prime Minister to take cover. "This exhilarates me," Churchill gleefully explained. "The sound of these cannons gives me a tremendous feeling."[74]

It was a pattern that was to be repeated many times in the war. Only George VI's intervention, for instance, kept Churchill from sailing with the assault forces on D-Day. "There is nothing I would like better than to go to sea," the King wrote his Prime Minister, "but I have agreed to stay at home; is it fair that you should then do exactly what I should have liked to do myself?"[75] Such restraint could only last a short time. A week later, Churchill crossed the Channel and had, as he wrote to Roosevelt, "a jolly day ... on the beaches and inland."[76] In another example, Churchill also described in his memoirs how he had gone to view a railroad bridge over the Rhine in March 1945 and how incoming artillery rounds had forced him and his party, escorted by the American General Simpson, to move off the bridge. General Alanbrooke also described the scene, detailing how urgently Simpson had requested that Churchill evacuate the bridge. "The look on Winston's face was just like that of a small boy being called away from his sand-castles on the beach by his nurse!" he wrote. "He put both his arms around one of the twisted girders on the bridge and looked over his shoulder at Simpson with pouting mouth and angry eyes .... It was a sad wrench for him; he was enjoying himself immensely."[77]

Churchill's enjoyment and exhilaration in these incidents were closely tied to a Victorian schoolboy's perception of the danger and the drama involved. "A German came over," General Montgomery recalled of an air raid while Churchill was visiting the front lines in 1944. "There was an air battle. Everyone was rather alarmed. Winston was rather pleased."[78] In a similar manner, the British leader loved the thrill of sailing through the submarine-infested waters of the Atlantic to a dramatic meeting place. But the drama, as he knew full well, was only a momentary escape from the business of total war. In the last spring of what he realized would be his last conflict, as he stood on a hillside watching British troops methodically cross the Rhine, he returned to the dramatic days of earlier wars. "I should have liked," he said wistfully, "to have deployed my men in red coats on the plain down there and ordered them to charge."[79]

Churchill's presence at the Rhine crossings demonstrated a key advantage offered him by the British Constitution which ideally suited his temperament and views on leadership as a man of action. Unlike Roosevelt, constrained because of his special position as President, or Hitler who elected to isolate himself increasingly in command posts, the British Prime Minister traveled freely within the war zones. This allowed him to solve major military issues by face-to-face contact with his operational commanders. Moreover, the fact that his constitutional role did not prevent him from visiting the front lines meant that he could fulfill his romantic conception of a war leader at the scene of action. Wherever he went, whether in the fighter control rooms of 1940, in the Egyptian desert, on the beaches at Normandy, or at the Rhine crossings, Churchill's visible, inspirational presence in the most outrageous of ad hoc uniforms was a key factor that contributed not only to the prosecution of the war, but to the genuine affection in which he was held by the officers and men throughout the services.[80]

Such visits also renewed Churchill, allowing him to escape from the pressures of his office and exercise the degree of personal leadership that he associated with the great men of action from previous eras. Writing to his wife in August 1942 from Egypt, he recounted his visit to the front lines where he "was everywhere greeted with rapture by the troops," the same words he had used in his Marlborough biography to describe the great commander in 1705 after the battle of Elixem.[81] And on 3 February 1943, Churchill flew to the forces just outside of Tripoli. In a small natural amphitheater he told the assembled soldiers and airmen that "after the war when a man is asked what he did it will be quite sufficient for him to say, 'I marched and fought with the Desert Army'." The next day, he drove in an armored car into Tripoli, moving past the assembled forces, amazed to see the Prime Minister among them, but recovering sufficiently to remove their helmets and give three cheers. A short time later in Tripoli's main square, surrounded by veterans of the Eighth Army, Churchill took the march past of one of the Desert divisions, the tears streaming down his face.

But leadership in total war goes beyond that exercised on the battlefield. Ultimately, it depends on the people. And it was here that Churchill's background as a Victorian man of action made its most lasting contribution to the Second World War. For it was primarily because of national will that Britain survived that conflict. And that national will owed its existence to a nineteenth-century man in his seventh decade, who in his dealings with the British people returned to his Victorian inheritance and allowed his emotional, romantic picture of his country and its citizens full rein. It was a picture that did not reflect the contemporary world of 1940. Instead, Churchill created an imaginary world of action steeped in Victorian visions with such power and coherence and imposed it on the external world with such irresistible force that for a short time it became reality. Imagination can be a revolutionary force that destroys and alters concepts. But as Churchill demonstrated, imagination can also fuse previously isolated beliefs, insights and mental habits from an earlier time into strongly unified systems. In those systems he created romantic ideal models in which, by dint of his energy, force of will and fantasy, facts were so ordered in the collective mind as to transform the outlook of the entire British population.

Those facts were firmly grounded in the British leader's sense of historical continuity which had always engendered in him high expectations of the British people. "I hope that if evil days should come upon our country," he wrote after contemplating the thousands of Dervish dead at Omdurman, "and the last army which a collapsing Empire could interpose between London and the invader was dissolving in rout and ruin, that there would be some ... who would not care to accustom themselves to the new order of things and tamely survive the disaster."[82] And so it was in 1940 when he molded the people's aspirations to fit his by recognizing no other mood in them than what he felt. During the Blitz, while walking with Churchill in the garden at Chequers one evening after dinner, General Ismay remarked how the Prime Minister's speeches had inspired the nation. "Not at all," was the almost angry retort from Churchill who could see the glow of London burning in the distance. "It was given to me to express what was in the hearts of the British people. If I had said anything else, they would have hurled me from office."[83]

This was the essence of Churchill's power. If his fellow citizens were not initially with this man of action in their hour of danger, that soon changed. Because he idealized them with such fevered intensity, in the end they approached his ideal and began to view themselves as he saw them with their "buoyant and imperturbable temper." It was the intense eloquence in his speeches that caught the British people in his spell until it seemed to them that he was indeed expressing what was in their hearts and minds. As a consequence, Churchill created in 1940 a heroic mood in which his countrymen conceived a new image of themselves as acting in a larger litany of great deeds ranging from Thermopylae to the Spanish Main. He imposed those responses through his speeches and through his expectations of the people, which in turn caused the British people to impose upon the present, however momentary, the simple virtues they believed had prevailed in the past. The combination of his personality and powerful imagery focused through the medium of radio invested the squalid and fearful circumstances of those days with overtones of glory.[84]

In the end, Churchill accomplished all this, not by catching the mood of his country, which in Isaiah Berlin's estimate was "somewhat confused; stout-hearted but unorganized," but by being obstinately impervious to it, as he had always been to the details, to the passing shades and tones of ordinary life. For him, the Battle of Britain was "a time when it was equally good to live or die."[85] His busy imagination, imposed on his countrymen, lifted them to abnormal heights in their nation's supreme crisis and allowed Churchill to enjoy a Periclean reign. But it could only last a short time. It was a climate in which people ordinarily do not want to live, demanding as it does a violent tension, which if not soon ended, destroys normal perspectives, overdramatizes personal relationships and distorts normal values to an intolerable extent. Nevertheless, for a time in the 1940s, by dramatizing their lives and making them seem to themselves and to each other as acting appropriately for a great historic moment, Churchill transformed the British people into a collective, romantic and heroic whole – a supreme optimization for total war.[86]

It was natural, then, that, at the moment of victory, Churchill should turn again to the people whose faith, which he had unconsciously brought forth, had done so much to sustain him. "This is your victory!" he told the vast VE Day crowds assembled before where he stood on the Ministry of Health balcony overlooking Whitehall. The crowd immediately roared back: "No – it is yours." Later that night, Churchill addressed another crowd stretching far up Whitehall to Parliament Square. "My dear friends, this is your hour .... It is a victory of the great British nation as a whole. We were the first ... to draw the sword against tyranny .... There we stood alone. Did anyone want to give in?" "No," the crowd shouted. "Were we downhearted?" "No," was the response to the greatest of all Victorian men of action.[87]

NOTES

A shorter version of this paper was first published in the US Army War College Qurterly *Parameters*, Vol. XXIII, No. 4 (Winter 1993–94), pp. 74–90.

1. J.H. Plumb, "The Historian," in *Churchill Revised. A Critical Assessment*, A.J.P. Taylor, ed., (New York: Dial Press, 1969), p. 133. See also H. Butterfield, *The Whig Interpretation of History* (New York: Charles Scribner's Sons, 1951).
2. Plumb, p. 135.
3. Herman Ausubel, *The Late Victorians* (New York: Van Nostrand, 1955), p. 135.
4. Randolph S. Churchill, *Winston S. Churchill, Volume I, Youth 1874–1900* (Boston: Houghton Mifflin, 1966), p. 61.
5. Janet and Peter Phillips, *Victorians at Home and Away* (London: Croom Helm, 1978), p. 166.

6. Walter E. Houghton, *The Victorian Frame of Mind, 1830–1870* (New Haven: Yale University Press, 1957), pp. 210–11.
7. Phillips, p. 166.
8. Lionel Hastings Ismay, *The Memoirs of General Lord Ismay* (New York: Viking Press, 1960), p. 142.
9. S.W. Roskill, *Churchill and the Admirals* (New York: William Morrow, 1978), p. 224.
10. Maurice Ashley, *Churchill as Historian* (New York: Charles Scribner's Sons, 1968), pp. 216–17.
11. Robert E. Sherwood, *Roosevelt and Hopkins. An Intimate History* (New York: Harper & Brothers, 1950), p. 241.
12. Martin Gilbert, *Winston Churchill, Volume III, The Challenge of War 1914–1916* (Boston: Houghton Mifflin, 1971), p. 132.
13. Bernard Darwin, *The English Public School* (London: Longmans, Green, 1929), p. 26.
14. Houghton, p. 203. Herbert Tingsten, *Victoria and the Victorians*, trans. David Grey and Eva Leckstroem (London: Allen & Unwin, 1972), p. 51.
15. Paul Fussell, *The Great War and Modern Memory* (New York: Oxford University Press, 1975), pp. 25–6.
16. R. Churchill, Vol. I, p. 203.
17. Randolph S. Churchill, *Winston Churchill, Volume II, Young Statesman 1901–1914* (Boston: Houghton Mifflin, 1967), p. 28. See also ibid., Vol. I, p. 282.
18. Churchill's phrase in his 6 March 1915 letter to Sir Edward Grey. Winston S. Churchill, *The World Crisis 1915* (New York: Charles Scribner's Sons, 1929), p. 205; Anthony Storr, "The Man," in *Churchill Revised*, pp. 235–236.
19. 4 December 1897 letter. Randolph S. Churchill, *Winston S. Churchill, Companion Volume I, Part 2, 1896–1900* (Boston: Houghton Mifflin, 1967), pp. 833–4.
20. 2 November and 22 December 1897 letters. Ibid., pp. 815 and 839.
21. Frederick Woods, *Young Winston's Wars. The Original Dispatches of Winston S. Churchill War Correspondent 1897–1900* (New York: Viking Press, 1972), pp. xxv–xxvi.
22. 2 October 1897 letter, R. Churchill, Companion Volume I, Pt. 2, p. 797.
23. Winston S. Churchill, *The Story of the Malakand Field Force. An Episode of Frontier War* (New York and Bombay: Longmans, Green, 1901), p. 88. See also Churchill's fictional account in his only novel, *Savrola*, when a subaltern in the Lauranian army recalls a campaign in Africa: "It was damned good fun.... My squadron had a five-mile pursuit. The lance is a beautiful weapon." Winston S. Churchill, *Savrola. A Tale of the Revolution in Laurania* (New York: Random House, 1956), p. 73.
24. *Malakand Field Force*, p. 241.
25. Ibid., p. 288. In *Savrola*, when told that it will be necessary for government troops to fire on a crowd in order to create a diversion, the officer in charge replies: "Excellent, it will enable us to conclude those experiments in penetration, which we have been trying with the soft nosed bullet." *Savrola*, p. 11.
26. "Whether I hit him or not, I cannot tell. At any rate he ran back two or three yards and plumped down behind a rock." Winston S. Churchill, *A Roving Commission* (New York: Charles Scribner's Sons, 1951), p. 142.
27. All quotes from Churchill's 4 September 1898 letter to his mother, R. Churchill, Vol. I, pp. 400–4.
28. Woods, p. xxiv.
29. Winston S. Churchill, *The River War. An Historical Account of the Reconquest of the Soudan* (London: Longmans, Green, 1902), p. 360.
30. Woods, p. 122.
31. Mark Girouard, "When Chivalry Died," *The New Republic* 13 (30 Sept. 1981) p. 28.
32. Ibid., p. 29, and Phillips, p. 189.
33. Gilbert, Vol. III, p. 643.
34. Martin Gilbert, *Winston Churchill, Volume VI, Finest Hour 1939–1941* (Boston: Houghton Mifflin, 1983), footnote, p. 521.

35. Ibid., Vol. III, pp. 755 and 754.
36. Ibid., Vol. VI, p. 521.
37. Winston S. Churchill, *Thoughts and Adventures* (London: Odhams Press, 1949), pp. 198–9.
38. Ibid., pp. 198 and 200.
39. Ibid., p. 201, and Winston S. Churchill, *The Aftermath* (New York: Charles Scribner's Sons, 1929), p. 479.
40. *Roving Commission*, p. 180.
41. Collin Brooks, "Churchill the Conversationalist," in *Churchill By His Contemporaries*, Charles Eade, ed., (New York: Simon and Schuster, 1954), p. 302. See also Fussell, p. 21.
42. Martin Gilbert, *Winston Churchill, Volume IV, The Stricken World 1916–1922* (Boston: Houghton Mifflin, 1975), p. 111.
43. Ibid., p. 151.
44. All quotes in the paragraph, Winston S. Churchill, *The World Crisis 1916–1918*, Part II (London: Thornton Butterworth, 1927), pp. 496, 501 and 426.
45. Ibid., pp. 410, 412 and 435. Freyberg, a VC winner in the First World War, remained a hero to Churchill, who was to place him in command of Crete in the next war. One day in the inter-war years while visiting Freyberg, Churchill asked the New Zealander to show him his wounds. "He stripped himself," Churchill recounted, "and I counted twenty-seven separate scars and gashes." Winston S. Churchill, *The Second World War. Volume III, The Grand Alliance* (Boston: Houghton Mifflin, 1950), p. 273.
46. Winston S. Churchill, *Great Contemporaries* (London: Thornton Butterworth, 1937), p. 190.
47. Ibid., pp. 310, 302 and 312.
48. Winston S. Churchill, *Marlborough. His Life and Times, Volume III, 1702–1704* (New York: Charles Scribner's Sons, 1935), p. 116.
49. Winston S. Churchill, *Marlborough. His Life and Times, Volume IV, 1704–1705* (New York: Charles Scribner's Sons, 1935), p. 215.
50. Ashley, p. 152.
51. Winston S. Churchill, *The Second World War, Volume II, Their Finest Hour* (Boston: Houghton Mifflin, 1949), p. 678.
52. Martin Gilbert, *Winston Churchill, Volume V, The Prophet of Truth 1922–1939* (Boston: Houghton Mifflin Company, 1977), p. 319.
53. *Aftermath*, p. 479.
54. *Great Contemporaries*, p. 323.
55. *Thoughts and Adventures*, pp. 201 and 196.
56. Ibid., p. 202.
57. Hans J. Morgenthan, "Henry Kissinger, Secretary of State," *Encounter* 5 (Nov. 1974), p. 61.
58. R. V. Jones, *Most Secret War* (London: Hamish Hamilton, 1978), p. 107.
59. Gilbert, Vol. VI, p. 697.
60. John Kennedy, *The Business of War*, Bernard Fergusson, ed. (London: Hutchinson, 1957), p. 316.
61. Gilbert, Vol. VI, pp. 694–695.
62. Ismay, p. 141, and Gilbert, Vol. VI, p. 507.
63. W. Averell Harriman and Elie Abel, *Special Envoy to Churchill and Stalin, 1941–1946* (New York: Random House, 1975), p. 205.
64. Gilbert, Vol. VI, p. 672.
65. Leslie Hore-Belisha, "How Churchill Influences and Persuades," in Eade, p. 338, and Richard Dimbleby, "Churchill the Broadcaster," in ibid., p. 353.
66. John Colville, *The Churchillians* (London: Weidenfeld & Nicolson, 1981), p. 143.
67. Lord Moran, *Churchill. Taken from the Diaries of Lord Moran: The Struggle for Survival, 1940–1965* (Boston: Houghton Mifflin, 1966), p. 796.
68. Ashley, p. 230.
69. Gilbert, Vol. VI, p. 62.

70. James Nelson, ed., *General Eisenhower on the Military Churchill* (New York: W.W. Norton, 1970), p. 19.
71. William Manchester, *The Last Lion. Winston Spencer Churchill: Visions of Glory: 1874–1932* (Boston: Little, Brown & Company, 1983), p. 10.
72. Martin Gilbert, *Winston Churchill, Volume VII, Road to Victory 1941–1945* (Boston: Houghton Mifflin, 1986), p. 354.
73. Excised section of Fuller's manuscript provided by Professor Jay Luvaas, U.S. Army War College.
74. Gilbert, Vol. VI, p. 828.
75. Ronald Lewin, *Churchill as Warlord* (New York: Stein and Day, 1973), p. 247. When Churchill asked Eisenhower if he could go in with the initial wave at Normandy, Eisenhower turned him down, adding: "And I think it's rather unfair to give me another burden on a day like this." Nelson, p. 38.
76. Winston S. Churchill, *The Second World War, Volume VI, Triumph and Tragedy* (Boston: Houghton Mifflin, 1953), p. 15.
77. Arthur Bryant, *Triumph in the West* (Westport, CT: Greenwood Press, 1959), p. 334, and *Second World War*, Vol. VI, pp. 416–17.
78. Gilbert, Vol. VII, p. 806.
79. Manchester, p. 10.
80. Lewin, pp. 21–2.
81. Gilbert, Vol. VII, p. 167. For the description of his welcome in Tripoli, see ibid., p. 330.
82. Ashley, p. 49.
83. Ismay, p. 156.
84. Isaiah Berlin, *Mr. Churchill in 1940* (Boston: Houghton Mifflin, n.d.), pp. 26–7.
85. *Second World War*, Vol. II, p. 279.
86. Berlin, pp. 28–9.
87. Gilbert, Vol. VII, pp. 1347–8.

5

Strategic Rationality is not enough: Hitler and the Concept of Crazy States

There is an historical tendency to measure international behavior against familiar styles and norms. One reason, for example, why the threat posed by Napoleon was only gradually recognized was that previous events, even Louis XIV's attempt at European hegemony, had accustomed policy-makers to international actors who desired only to modify, not overthrow, the existing system.[1] In this century, there was a similar failure of perception in terms of the revolutionary character of Adolf Hitler and his regime. That failure, followed by the most ferocious conflict in history, has remained a cause and effect *idée fixe* for Western policy-makers ever since. President Truman's intervention in Korea, for instance, was due in part to his perception of how easy it would have been to suppress Nazism in 1936 if there had been some reaction to the German remilitarization of the Rhineland. And Anthony Eden, a veteran of the diplomatic battles in the interwar years, acted in the 1956 Suez crisis on the basis of his memories of the Munich Conference and his conviction that Nasser was another Hitler.[2] Finally, there was President Johnson, rejecting a proposal to pull US troops out of Vietnam "because we learned from Hitler at Munich that success only feeds the appetite of aggression."[3]

The durability of the Hitler analogy in international affairs continues. As the Iraq–Kuwait crisis evolved, Hitler's Germany re-emerged as the symbol of a nation gone wild. Some analysts described parallels between the Nazi party and Saddam Hussein's emphasis on the Ba'th party as a carrier of pan-Arabism. Most importantly, top US policy-makers were not averse to invoking the Nazi analogy, with President Bush at one point suggesting that Hussein's use of Americans as shields against attack was even beneath the Nazi leader. "Hitler did not stake people out against potential military targets," the President stated, "and he did ... respect the legitimacy of the embassies."[4] All that caused one columnist, an obviously exhausted veteran of the Containment years, to comment in exasperation:

> I don't know why people keep saying we don't have a good reason for going to war with Iraq. Of course we do; the same one

we've had for every war we've fought in the past 45 years: We're
going to war to stop Hitler. Ever since we missed our first chance
to stop Hitler 50 years ago, we've been trying to make up for it;
in Korea, in Cuba, in Vietnam, in Nicaragua, in El Salvador – in
Grenada, for crying out loud – and now in Iraq.[5]

That observation notwithstanding, Hitler and his 12-year regime is
a valuable starting point in examining what have been called "crazy
states," an increasingly important concept in the post-cold war era
involving both the personal characteristics of national policy-makers
as well as the decision-making structure and processes of their states.[6]
Historically, nation states like Gadhafi's Libya have emerged from
time to time. But the majority remained isolated, local phenomena.
Modern technology has changed this. The communications revolution now allows these states to achieve notoriety, if not status, in the
international arena. Equally important, technology offers the potential for these countries to build up significant power to influence
regional, if not global, events. Added to this is the breakdown of the
superpower bipolar nexus which, while reducing East–West tensions,
has also mitigated the pattern of client state stability. In such a
multipolar, interdependent environment marked by the proliferation
of conventional and nuclear weaponry, the capabilities and potential
impact of a rogue or "crazy" state will be of increasing importance.

But what constitutes "craziness"? What variables can be applied for
what will be, at least in part, culturally biased value judgments based
on Western norms? In such a context, as will be demonstrated, pure
rationality in the instrumental sense of being able to make means
commensurate with ends is not enough. And yet by simultaneously
holding up an ideal type of means-ends rational decision-making as
the basis for comparison, and by opening up the "black box" of the
unitary nation state in order to focus on the structure of government
and the personal characteristics of the decision-makers, other criteria
can emerge as a basis for evaluation.

The pure rationality model in Figure 1 is a "synoptic conception" of
decision-making, by which the policy-maker places all available alternatives before him and measures, against his scale of preferred values,
all the possible consequences of the various courses of action under
consideration.[7]

The model is what Max Weber called an "ideal type," which, with
few exceptions, is impossible to achieve since it presupposes omniscience and a capability for comprehensive analysis that time, cost and
other factors simply do not permit. Nevertheless, as Weber pointed
out, by using such types, "it is possible to understand the ways in

which actual action is influenced by irrational factors of all sorts ... that ... account for the deviation from the line of conduct which would be expected on the hypothesis that the actions were purely rational."[8]

The so-called "black box" geopolitical strategic model superimposes the nation state on this pure rational typology. Key to this conceptualization is the idea that governments are rational, value-

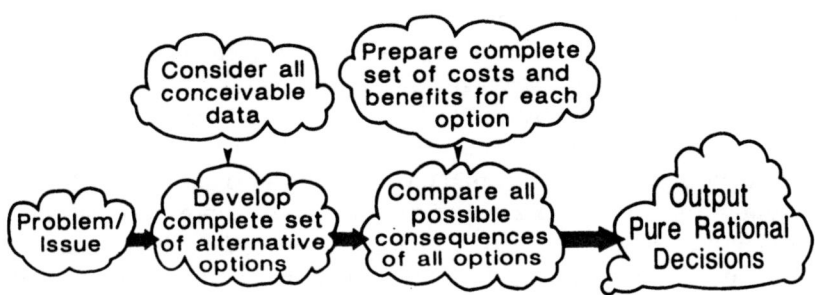

Figure 1
Pure Rational Decision-Making

maximizing actors in international affairs – black boxes with goals, whose behavior reflects purpose and intention. Scholars have pursued variations of the concept. Some have explored ideology, culture and national characteristics as explanations for the difference in state behavior.[9] Others have enhanced the usefulness of the "black box" concept by focusing on the elements of national power, with variations determined by geostrategic theories.[10] The boxes, in other words, are not just black, but of varying colors depending on such elements as power, ideology and strategic position. But they remain boxes in the sense that the analysts do not have to know what goes on inside them, deducing from the actions of the boxes in the global arena the nature of their intentions.

Other scholars, however, have not been content with this model and its variants and have opened up the boxes in search of better explanations for the behavior of states, normally focusing on the personal characteristics of policy-makers as well as the structure and concomitant decision-making processes of the government.[11] This chapter will take a similar approach with the black box of Nazi Germany. The inner decision-making workings of that 12-year regime will be compared with the pure rationality model as a basis for a first cut at inductive validation of general patterns and relationships from that era for use in evaluating the potential for emergence of "crazy" states and possible methods for dealing with them in an increasingly complex world.

Rationality and the Means-Ends Connection

The classic description of strategy as the calculated relationship of means to ends is in keeping with the ideal pure rationality model. Clausewitz, in fact, defined rationality in warfare as the commensurate relationship between violence as means and politics as ends in which war, governed by political calculation and reasoning, became not "a mere act of policy; but a true political instrument, a continuation of political activity by other means."[12] Basic to this approach is the assumption that means are not arbitrary in the sense of constituting subjective behavior that has no relation to any ends such as essentially random or strictly cathartic action. Instead, means are related to ends in a manner that can be justified in terms of positivistic instrumentality. Astrology, for instance, may be claimed subjectively to relate means to ends, but would be perceived objectively as unjustified and, therefore, irrational. Thus, irrationality can be said objectively to have governed the strategic decisions of Julius Caesar to the extent that those decisions were based on the patterns of chicken entrails.[13]

This abstract concept of instrumental rationality is acultural, but as one theorist has noted, "at a human level one man's god is another man's heresy, and manifestations of rationality, like those of religion, are a matter of geographical accidents and cultural heredity.[14] In fact, cultural or, as in the case of Hitler, mental differences may prevent one nation from accepting or even perceiving certain options for another nation which in an objective instrumental sense may be fully rational. Clausewitz's dictum that war is a means to serve political ends, for instance, is open to varying interpretations. For the West, those political ends normally can only be served through military victory. As a consequence, a state acting in accordance with Western experience will never resort to war if it has little or no chance of winning militarily. Political goals, in short, cannot be achieved by defeat.

For many non-Western countries, on the other hand, war is considered a worthwhile means even if military success is not assured, so long as political and psychological goals can be achieved, to include such intangibles as the preservation of national honor and pride. The rational instrumentality of this "success through failure" approach is not easily grasped by Western states, which see it as a tendency in non-Western societies to assume much greater risks than those considered by Western standards to be rational or profitable. Thus, in the 1973 Yom Kippur war, Egyptian President Sadat was willing to risk military defeat to improve his political status, both of which came to pass. For the Israelis, the majority of whom perceive that there is no substitute for victory, there was little comprehension of a military offensive that *a priori* could not result in military victory. By projecting this different concept of instrumental rationality onto the Arab culture, the Israelis were unable to anticipate Egyptian behavior.[15]

In a similar manner, the Second World War began for the United States because, in part, of an inability to understand the Japanese readiness to accept risks that were unacceptable by American standards. The United States in 1941, one author noted, could not believe "that a power as small as Japan would make the first strike against a power as big as the United States . . . . Japanese sanity cannot be measured by our own standards" which "reckoned the risks to the Japanese as too large and therefore not likely to be taken."[16] And in the final months of that war, the divine wind sacrifice of the kamikaze pilots was perceived by the US military and public as a horrendously irrational tactic. For the hard-pressed Japanese elites, however, it seemed a rational means to sink enemy warships.[17]

Cultural subjectivity, then, can hinder the objective evaluations of means and ends commensurability that essentially defines "rationality." For some theorists, this simply adds to the argument that such a definition is too narrow since it omits a discussion of values which, at the very least, influence the decision-maker's appreciation of the situation and, as a consequence, the means-to-ends process associated with instrumental rationality.

> The choice of means, however these are conceived, is as much dictated by value-related ends as by strictly limited notions of capability. Rationality thus becomes normative in character and is related to justifiable action in terms of some ethical code or set of rules as much as to standards of effectiveness. If strategy is treated ... as action directed to the fulfillment of political purposes, their instrumentality has to be related to the nature of these purposes. It is not a mechanical or teleological relationship between means and ends.[18]

Rationality, in other words, is not simply limited to a choice among means, but also includes the judgment of what values are worth pursuing. "The paranoid who waits till dark before turning on his persecutors may be a master strategist," Abraham Kaplan has pointed out in this regard; "he is surely not a paragon of rationality."[19] To admit otherwise, Kaplan adds, is to concede that Satan has a fine mind and is lacking only in heart. Goals must be examined beyond mere instrumentality in order to demonstrate that Satan "is a fool from beginning to end."[20]

Most theorists, however, believe that the term "rational" can be used legitimately only to describe judgments or beliefs about matters of fact or logical relation, such as whether given means are adequate for the fulfillment of given ends. The term cannot be applied to ends themselves, since they are neither rational nor irrational. Instead, they represent values which are not the type of entity to which the conception of rationality is applicable since it involves the realm of moral judgment without an empirical referent. Thus, even in the pure rationality model, the formulation and prioritization of final values can only be determined by value judgments, not by rational processes, which creates, as Felix Oppenheim has pointed out, a different perspective on the Prince of Darkness: "If Satan has a fine mind, he is no fool, but a rational actor, however diabolic his goals may appear to us. This is precisely why he is so dangerous, as are those wielders of absolute power who incarnate him. Their scale of values may be abhorrent to us, but reason is of no avail to prove that they should act otherwise."[21]

The two approaches mean that rationality is a necessary but not sufficient factor upon which to base an examination of the concept of crazy states. On the one hand, the use of "morality" in the "rational" evaluation of goals, what Weber called *Zweckrationalität*, can, like the use of "realism," lead to doubtful judgments. In the early years of Hitler's rule, for instance, the anti-appeasers made a case for military intervention in Germany before the Nazi leader had consolidated his power. At that time, it was generally considered immoral to sacrifice lives for such an effort, although, as one of the leading anti-appeasers has pointed out, 30,000 casualties might have sufficed. In 1939, however, intervention was considered moral; but by that time, the action involved some 30,000,000 lives.[22]

On the other hand, to accept only the instrumental definition implicit in the pure rationality model can lead to moral relativism which, as Ivan Karamazov observed, will inevitably lead to an "anything goes" position. Isaiah Berlin has noted in this regard that "when we are told that it is foolish to judge Charlemagne, or Napoleon or Ghengis Kahn, or Hitler, or Stalin [or that such judgments are beyond the point, we can only answer that] ... to accept this doctrine is to do violence to the basic notions of our morality, to misrepresent our sense of our past, and to ignore the most general concepts and categories of normal thought."[23]

The answer is to accept the objective, more limited definition of rationality while applying moral judgments at the same time as separate subjective, value-laden criteria. In this context, a state may behave rationally in an instrumental sense of effectively achieving its ends or goals which in themselves may be "crazy."[24]

Operational Code and the Cognitive Trail

Decision-making structures within the governmental black box can range from single individuals to large bureaucracies, each invoking different types of processes for dealing with uncertainty and conflict among participants as well as for making and implementing choices. The size of these structures as well as the distribution of power and the roles of group members within the organization are among properties present in all decisions which insure the probability that certain decision processes are more likely to be used than others. Variations in these processes influence the foreign policy behavior selected by the decision-maker(s). This in turn allows the strategic analyst to examine the impact of the personal characteristics of national leaders as well as the organizational processes on the rational decision-making model.

The duality is important, because there will inevitably be tendencies to focus on one to the detriment of the other. There has been, in fact, a great deal of controversy concerning the effect that a leader's personal characteristics can have on foreign policy. Some analysts contend that individual actors are limited in their effect on events by social forces – the so-called "great man" versus *Zeitgeist* debate. Others point to organizational constraints and the fact that political leaders are merely "agents" or representatives, reflecting the views, beliefs and ideologies of their constituencies and agencies. "Names and faces may change," one analyst has concluded in this regard, "interests and policies do not."[25] Most cases actually fall between the extremes, with theorists taking into account the total situation in which the decision-makers act as well as their own traits. Political leadership in this context is "the interaction of personality, role, organization, task, values, and setting as expressed in the behavior of salient individuals."[26]

In the case of Nazi Germany, there is a broad historical school, the so-called "intentionalists," that views Hitler as the principal force in the Third Reich whose personality and ideology drove events, a perception summarized succinctly in one historian's conclusion that "the point cannot be stressed too strongly: Hitler was master in the Third Reich."[27] As opposed to those who see the absolute centrality of Hitler in the Third Reich, the "structuralist" or "functionalist" school focuses more on the structure of Nazi rule and the functional nature of policy decision-making. As early as the 1940s, analysts had begun to challenge the concept of the rational "monolithic" Nazi state, pointing out that beyond the unitary facade of the Third Reich, the power structure seethed with personal and organizational rivalries, reminding a counsel at the Nuremberg trials of "the minor courts of the Italian Renaissance."[28] In more recent times, analysts have renewed the examination of how organizational processes and bureaucratic politics influenced decision making in Nazi Germany by focusing on the civil service, Party–State relations, the regional power bases of the *Gauleiter*, and policy implementation at local levels.[29]

An analysis of crazy states must also recognize this duality. Beginning with the political leader, his *Weltanschauung,* depending on the degree of his political control, will shape the strategies that his government will employ in its foreign policy. Because the personal characteristics that feed into this world view can structure the policy-maker's interpretation of the environment, he will be likely to ensure, again depending on his control, that the government acts consistent with this image. A policy-maker "acts upon his 'image' of the situation," one analyst has noted in this regard, "rather than upon 'objective' reality."[30] His *Weltanschauung,* in other words, forms the basis for the political leader's cognitive map that charts his course in overcoming the type of conditions which, as Alexander George has pointed out, normally accompany the rational decision-making process:

> (1) The political actor's information about situations with which he must deal is usually incomplete; (2) his knowledge of ends-means relationships is generally inadequate to predict reliably the consequences of choosing one or another course of action; and (3) it is often difficult for him to formulate a single criterion by means of which to choose which alternative course of action is "best."[31]

Understanding this cognitive map – the so-called "Operational Code" – of a political leader can be an important tool for dealing with crazy states, since it sets the boundaries within which the leader will act. The code would include the actor's decision-making style as well as his philosophical and instrumental beliefs, which might be better labeled, in George's estimation, "approaches to political calculation."[32] These can range from orientation on opponents and notions about chance or risk to ideological goals and the ability and commitment to implement those goals – all important inputs needed for behavioral analysis of political decision-making and leadership. The operational code is not a simple key to explanation and prediction, but as George has pointed out, "it can help . . . to 'bound' the alternative ways in which the subject may perceive different types of situations and approach the task of making a rational assessment of alternative courses of action."[33]

The Personal Link

Any examination of decision-making in a crazy state should begin with an investigation of the leader's beliefs about politics, which can be postulated as part of his operational code on the basis of generally accessible data. Hitler's thoughts on the "Jewish problem" and the acquisition of *Lebensraum* in the east, for instance, were two fundamental constants in his world view that were available in written form through *Mein Kampf* as early as the mid-1920s and were normally used in the Nazi leader's speeches both during the *Kampfzeit* and after he assumed power.[34] The emergence of such a belief system may be affected by developmental problems encountered in personality formulation, which should also be examined when possible. Walter Langer's wartime psychological study of Hitler, in this regard, was remarkably accurate in some respects, even to the conjecture of monorchism (one testicle) on the part of the Nazi leader; and his analysis of Hitler's decision-making cognition pattern remains an insightful point of departure.

> He does not think things out in a logical and consistent fashion, gathering all available information pertinent to the problem, mapping out alternative courses of action, and then weighing the evidence pro and con for each of them before reaching a decision. His mental processes operate in reverse. Instead of studying the problem, as an intellectual would do, he avoids it and occupies himself with other things until unconscious processes furnish him with a solution. Having the solution he then begins to look for facts that will prove that it is correct. In this procedure he is very clever, and by the time he presents it to his associates, it has the appearance of a rational judgment. Never-

theless, his thought processes proceed from the emotional to the factual instead of starting with the facts as an intellectual normally does. It is this characteristic of his thinking process that makes it difficult for ordinary people to understand Hitler or to predict his future actions. His orientation in this respect is that of an artist and not that of a statesman.[35]

Psychoanalytical examinations of crazy-state leaders, however, should also be handled carefully. The concept of motivation, for example, is not only typically Western in its purposive behavior and degrees of freedom, but is extremely complex, often with any prioritization, when that can be even ascertained, being only fortuitous. In this regard, the German annexation of the Sudetenland, according to one former associate of the Nazi leader, was at least partially inspired at the 1937 Breslau festival, when participants from that province marched past the reviewing stand shouting demands for liberation from Czechoslovakia.[36] This points to two major problems in dealing with explanations of the phenomenon of crazy-state leaders which depend for their validity on psychohistorical insights. To begin with, the conjectural element in any analysis of this kind will, of necessity, be extremely high, simply because patient diagnosis is at best secondhand. More importantly, however interesting an analysis of a national leader's psychopathology can be for understanding why he acts, it cannot explain or predict how the psychic tensions in that leader, even when diagnosed properly, are or will be translated into political action. How, in other words, is one to take Langer's conclusion that Hitler was "a neurotic psychopath bordering on schizophrenia" and relate it to real world correctives in terms of crazy states.[37]

The Cultural Link

One way to move beyond the psychoanalytical approach is to recognize that the structure of expectations by policy-makers as well as their perceptual interpretations, motivations and behavioral norms are directly affected by their society and culture. Certainly, decision-makers, even those in crazy states, tend like everybody else to organize their cognition and perception of reality in terms of cultural meanings and values. "If triangles had a god," Montesquieu once remarked in this regard, "he would have three sides."[38] Common or similar values, in other words, are not necessary for understanding or attempting to predict the actions of crazy-state leaders. But understanding the difference in values is absolutely essential, otherwise the strategist of one nation will project his own values and sense of priorities on those of another state. "If my opponent is rational," the thought process goes in this circumstance, "he will do what any rational man would do in this situation. I am rational. Therefore he will do what I would do in his shoes. If I were in his shoes I would . . . ." By this process, ethnocentrism sabotages one of the central thought processes in strategy – that of knowing the enemy. There is a need, in this regard, to realize as André Gide has pointed out, that "grey is the colour of truth."[39]

The distortion of threat perception, of course, can be deliberate on the part of a crazy-state leader, further complicating the process. It is, in fact, not unusual for such leaders to use ethnocentrism to demonize or dehumanize outsiders, to create a foreign "Great Satan" – all in furtherance of Bertrand Russell's dictum that "few people can be happy unless they hate some other person, nation or creed."[40] Hitler played upon this concept throughout his career in order to intensify in-group/out-group feelings within German society and to make the population sensitive about German rights and interests. "Identifying enemies helps define who we are *not*," he wrote as early as 1925, "which is a necessary part of defining who we *are*."[41] And years later, when asked if the Jews should be destroyed, Hitler replied in the negative, adding that otherwise "we should have then to invent him. It is essential to have a tangible enemy, not merely an abstract one."[42]

In addition to threat perception, cultural relativism is particularly important in negotiating with crazy states since that process involves a psychological relationship in terms of bargaining and is therefore inevitably prone to the problems of ethnocentric perceptions. In the inter-war years, for example, Allied diplomats were not prepared for the different styles, strategies and tactics of their counterparts in the totalitarian regimes. Western diplomacy had long put a premium on honesty and compromise. Hitler, on the other hand, viewed compromise as a sign of weakness, appreciated the instrumentality of the "big lie" over the small one, and believed that negotiating in good faith was, at the very least, a silly principle. In this context, appeasement could only be a sign of weakness for the Nazi leader. "Our enemies are little worms," he reassured his military leaders on 22 August 1939, "I saw them at Munich."[43]

Hitler's contempt for his adversaries was due in part to the ease with which he manipulated the perceptions of the Allies by masking his intentions to fit the beliefs of most Western European nations that any defeated, proud and powerful nation state would attempt to change an enforced peace treaty in order to regain a legitimate, but not dominating, position in world politics. What Hitler did not realize,

however, was that the March 1939 annexation of rump Czechoslovakia convinced the Western Powers that what had appeared to be
a policy of revisionist status quo in terms of the Versailles Treaty had
really been from the start a policy of imperialism, of continental, if not
world-wide dimensions. The result was that until the outbreak of war,
the Nazi leader and the Western Powers were locked in a game of
"chicken" by which either side could avoid collision by turning the
steering wheels of their national vehicles headed at each other. The
problem was that, by then, one side had metaphorically been
conditioned to drive on the left-hand side of the road, the other on the
right. Under those conditions, attempts to engage in "rational" war
avoiding strategic behaviors only helped precipitate confrontations. [44]

All this was illustrated on 23 August 1939 when the British Ambassador in Berlin personally delivered a letter from the British Prime
Minister, Neville Chamberlain, to Hitler which began: "It has been
alleged that if H.M. Government had made their position more clear
in 1914 the great catastrophe would have been avoided."[45] The letter
then went on to emphasize Britain's determination and resolve to
honor the guarantee to Poland. Unfortunately, the Prime Minister
also continued by pointing out that all issues between Germany and
Poland could be settled by negotiations and suggested ways such a
process could be initiated. For the Nazi leader, the British warning
was vitiated by Chamberlain's familiar offer to continue his search for
a peaceful solution. Added to this was his confidence that the actual
conclusion of the German–Russian non-aggression pact would shake
the resolution of the Western Allies. Acting on these misperceptions,
on the evening of his interview with the British Ambassador, Hitler
scheduled the German attack on Poland.

Cognitive Dissonance and Consistency

This type of diplomatic denouement also reveals how far the cognitive
trail of a policymaker can stray from the pure rational decision-
making model. In that ideal type, all aspects of the problem are fully
understood by the decision-maker. In actuality, however, as several
theorists have postulated, because most problems are too complex to
allow a total or synoptic rationality, policy-makers are forced to begin
with existing problems and take incremental decision steps when
issues arise, as shown in Quadrant 3 of Figure 2.[46] Thus, even in Nazi
Germany, the first tentative steps in foreign affairs were nothing more
than incremental "muddling through," with Hitler assuring visitors
that there would "be no change in the policy ... laid down in 1932."[47]
These assurances were followed by Germany's continued member-

ship in the European disarmament conference as well as the Nazi leader's so-called *Friedensrede* to the Reichstag, the conciliatory aspects of which "could scarcely have been equalled by Stresemann or Bruening."[48]

It is almost an historical cliché, of course, that German foreign policy in the inter-war years moved increasingly into Quadrant 4 where decisions of ever greater import were made with incomplete information. For the Nazi dictator, however, always intolerant of ambiguity and moving with each foreign policy success in those years further into his own world of hubristic infallability, there were cognitive adjustments that allowed him in his own perception to move into Quadrant 2, "the realm of superhuman decision-makers."[49] The most common of such adjustments was Hitler's consistent tendency through the years to fit incoming information into his pre-existing beliefs and hopes. In terms of his unfavorable opinion of Soviet capabilities, for example, when the two armies met in October 1939 at the German–Russian demarcation line in what had been Poland, Hitler was particularly interested in the reports of the bedraggled state of the Russian troops.[50]

One result of such developments, on the one hand, is cognitive dissonance in which decision-makers like Hitler seek to increase their comfort level with strategies already implemented by minimizing any evidence that might lead to different conclusions. This tendency to make the enemy "fit" into a strategy that has been decided upon was

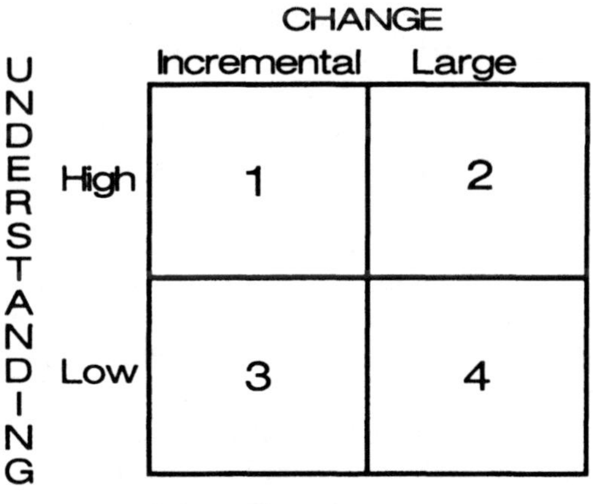

Figure 2
Change and Understanding in Decision-Making

demonstrated not only by Hitler, but by two generations of German leaders who convinced themselves that Great Britain would not enter into a continental war or that it would not be a major factor in such a war.

In Hitler's case, the process began with his conviction that the Allied guarantee of Poland was not serious, a conviction bolstered by organizational and political deviations from the unitary rational actor model. Most noticeable, in this regard, was Joachim von Ribbentrop, the Nazi foreign minister, who provided a constant flow of information on a weakened, indecisive Britain to Hitler who, in turn, was happy to receive reinforcement of his own beliefs. If any member of the Wilhelmstrasse deviated from this position, Ribbentrop warned his subordinates at the time, he "would kill him myself in his office and take responsibility before the Fuehrer for it."[51] Based on such input, the British guarantee was perceived by Hitler as a bluff; and in the coming months, the Tripartite Alliance with Italy and Japan and the Nazi–Soviet non-aggression pact were his counters to that bluff. How limited the Nazi leader's perception was of the danger in calling this bluff was demonstrated when he heard the news of the British ultimatum on 3 September 1939. "What now?" he demanded savagely of the hapless Ribbentrop.[52]

From Hitler's perception, the German–Polish war presented no threat to British and French vital interests and could have been treated as a limited conflict. From the standpoint of military technology, as well as Poland's geographical position, the target for German aggression was far removed from both countries. Moreover, the German thrust into Poland eliminated that state's buffer role with the USSR, thus increasing the possibility of military confrontation between Germany and the Soviet Union. Britain and France, however, viewed the German invasion of Polish territory as part of a larger war in which their national security was at stake. As a consequence, they entered into a conflict in which, as the astounded Hitler noted, the means adopted were both inadequate and incommensurate with the ends desired – a study in Clausewitzian instrumental irrationality. "(T)hink of the declaration of war in 1939!" he commented, still incredulous three years later. "They had no armaments at all – and yet they declared war!"[53]

Such manifestations of cognitive dissonance naturally increased in number and severity as the flow of events became more unacceptable. "The myth of our vulnerability, in the events of the war becoming prolonged, must be resolutely discarded," Hitler declared in 1941. "It's impermissible to believe that time is working against us."[54]

Towards the end of the war, this had become the outright refusal to acknowledge the hopelessness of the situation. On 15 March 1945, for instance, Albert Speer prepared a memorandum for the Nazi leader that outlined the situation in stark terms. Hitler began to read the report but when he came across the words "The War is lost," he refused to read another line. And on another occasion during that period, General Guderian described Hitler's plaintive refusal of a request by Speer to meet with the Nazi leader alone.

> All he wants is to tell me again that the war is lost and that I should bring it to an end. Now you can understand why it is that I refuse to see anyone alone any more. Any man who asks to talk to me alone always does so because he has something unpleasant to say to me. I can't bear that.[55]

On the other hand, a case can be made for rational cognitive consistency on the part of the Nazi leader for much of his career, since his actions were based on what he perceived as constants in the environment and because within that construct, he did not violate generally agreed-upon rules for dealing with evidence.[56] How rational Hitler's cognitive pattern could be was illustrated by Operation Barbarossa. At the strategic level, the German underevaluation of Soviet strength not only made means and ends apparently more commensurate in an instrumental sense, but served to further the Nazi leader's logical conclusion that Russian military power limitations militated against deferring an invasion. To begin with, there was the inevitability of war with the Soviet Union which had permeated his goal of eastern *Lebensraum* since the beginning of the *Kampfzeit*. "Therefore," he concluded, "it's better to stave off the danger now, while we can still trust in our own strength."[57] Moreover, by 1941 the Nazi leader believed that Britain could not be conquered in the near future and that London's hope of Soviet support only served to prevent an Anglo-German settlement. At the same time, US entry into the conflict seemed inevitable – a disastrous possibility if the Soviet Union remained intact, able to threaten Germany's rear or to cut off supplies it was providing the Third Reich. For Hitler, "the road to London passed through Moscow."[58]

This was particularly true since there was at least a year and a half's breathing space before an Allied invasion of the Continent would be possible. The armored might of the *Wehrmacht*, sitting idly in the west since the conquest of France, could thus be used in this period to eliminate the Soviet Union and its rapidly growing military strength in order to avoid the traditional German nemesis of a two-front war.

"Britain's hope lies in Russia and the United States," Hitler concluded. "If Russia drops out of the picture, America, too, is lost for Britain, because the elimination of Russia would greatly increase Japan's power in the Far East."[59]

Such rational grand strategic calculations notwithstanding, Hitler's ethnocentrically-based, ideologically-tinged underestimation of his Slavic opponent permeated down to the theater strategic and operational levels where it had disastrously irrational logistical results in terms of means-end commensurability. Using calculations of ammunition expenditure from the western campaigns, for example, the army planners swallowed any misgivings they might have had as they made new estimates conform with quantities German troops could carry. Even the original assessment of five months for conquering the Soviet Union was scaled down. By the time Barbarossa was launched, there was no buffer in the planning for flexible solutions to unexpected problems. "Rather than culling down their goals to suit their limited means," Martin van Creveld has pointed out concerning the German planners, "they persuaded themselves that their original calculations were overcautious."[60]

Style

The style of policy-makers encompasses personal methods of making decisions based on operational codes and might include preferences for compromising and optimizing as well as for such components as confidence and openness to new information. Or it might include, at an interpersonal level, how a leader views his environment in dealing characteristically with other policy-makers, which in turn might entail an examination of his use of threats and praise to persuade, his sense of political timing or his sensitivity to other leaders. Above all, in terms of crazy state analysis, the examination of a leader's style should focus on deviations from what is regarded as normal behavior in international actions – that is, the propensity and preference (conscious or unconscious) for stylistic innovations which are not constrained by accepted patterns such as the hijacking of aircraft and the seizure of diplomats as hostages.[61]

In Hitler's case, the first indications of such deviations were in domestic politics, beginning with the systematic use of terror after the February 1933 Emergency Decrees and the "Blood Purge" of Röhm and the SA in June 1934. For the Nazi leader, normal rules and conventions must be swept away when, at "critical periods in history all the tinsel falls away and the great rhythm of life alone rules the hour."[62] At such a point, he had no choice. "I must do things that cannot be measured with the yardstick of bourgeois squeamishness."[63]

> If these gentlemen, with their outworn ideas, imagine that they can go on pursuing policy like the honest merchant with his business, in accordance with precedent and convention, let them go on. But I am concerned with power politics – that is to say, I make use of all means that seem to me to be of service, without the slightest concern for the proprieties or for codes of honor . . . . I certainly have an advantage . . . in my freedom from pedantic and sentimental inhibitions. Am I to be so generous as to throw away this advantage, simply because my opponents have not progressed so far? If anyone is prepared to be deceived, he must not be surprised that he is . . . . It is characteristic of the narrowness of these outlived classes that they should be indignant with me, indignant at our contempt for past customs and assumptions in political life. I recognize no moral law in politics. Politics is a game, in which every sort of trick is permissible, and in which the rules are constantly being changed by the players to suit themselves.[64]

Pragmatism, Technique and Hubris

These statements notwithstanding, there was a pragmatic side to Hitler which can often mark the leaders of crazy states, with their authoritarian ability to subordinate ideology to national interests. Thus, there was the August 1939 Nazi–Soviet non-aggression pact in which Hitler's diplomatic surprise prepared the way for a military surprise. In *Mein Kampf*, he had pointed out the "tactical considerations" for such an alliance. And a decade later, he returned to the theme in his discussions of the need to conquer the great spaces of the Soviet Union. "That does not mean that I will refuse to walk part of the road together with the Russians, if that will help us," he concluded. "But it will be only in order to return the more swiftly to our true aims."[65]

Pragmatism also includes an ability to learn from failure. The classic case with Hitler was his abortive 1923 *putsch* in Munich. From that failure, the Nazi leader not only learned that he could not prepare for battle with the power of the German state on a purely Bavarian plane, but that *Konflikt* itself was not a suitable means to his end. "It will take longer, to be sure," he declared, "to outvote our opponents than to outshoot them, but in the end their own Constitution will give us success."[66]

After he achieved that success, Hitler continued to employ the dual approach as a rational and conscious means to overcome all opposition. On the domestic front there was the process of *Gleichschaltung*, or coordination, in which every breach of the law or the Constitution was camouflaged and accompanied by vociferous claims to respect legality. It was, in fact, a classic case of a rational, totalitarian capture of democratic machinery with the assistance of, not in opposition to, the state. The key to the process was the linkage of legal and revolutionary actions which produced an overarching screen of legality, dubious in individual instances and yet convincing enough as a whole to keep the essential illegal construct of the regime hidden. As a consequence, certain areas of public life like civil law were initially untouched, leaving, in this example, reassuring preserves which made it difficult to assess the regime's legality and, as a consequence, whether it should be supported or not. Many people, in this regard, hoped to domesticate the revolutionary side of the revolution – an illusion that was fostered by a nationalistic smoke screen that ultimately persuaded the civil service, the army, the political parties, the trade unions and the legal profession to support totalitarian goals.[67]

In a similar manner in foreign affairs, force was always accompanied by expressions of scruples and protestations of peaceful intent. From 1933 to 1936, for example, a time of relative weakness for Germany in the international arena, Hitler's diplomacy of *fait accompli* increased his adopted country's strength while carefully and rationally avoiding retaliation. In order to accomplish this, he inaugurated a pattern of diplomacy completely in keeping with his parvenu and revolutionary background and his disrespect for bourgeois values, but so alien to diplomatic norms that it shrouded, at least temporarily, the Nazi leader's intentions.

Each *fait accompli* would start and end with firm declarations of Germany's desire for peace and for friendly collaboration followed by new proposals for disarmament and non-aggression treaties. And because sanctions were still a possibility in the early years, Hitler issued a solemn promise after each diplomatic surprise that there would be no more such actions and that he would personally guarantee each of Germany's treaty obligations, particularly the one that was next on his repudiation list. In these promises, however, Hitler added the important condition that he would stand behind Germany's treaty obligations and remain peaceful only so long as the other international players followed suit. To this condition, the Nazi leader then invariably set other conditions that he knew were unacceptable to the nations involved, thus leaving each post-*fait accompli* speech with the normally unnoticed justification for the next *fait accompli.*[68]

For the leader of a crazy state, such deviations from diplomatic norms may be the rule. As with Hitler, surprise and shock tactics, in some form, will have been the most likely tools in the leader's rise to power and will thus continue to play a role in foreign policy after assumption of power. "Our present struggle is merely a continuation on the international level," Hitler declared in this regard, "of the struggle we waged on the national level."[69] Moreover, the very centrality of such a leader's power will ensure a large measure of control and thus diplomatic surprise. In the 7 March 1936 Rhineland crisis, for example, Hitler issued all orders on short notice. As a consequence, the operations divisions of the *Wehrmacht* services had less than a day to plan and then issue orders to their relevant departments. At the most, the Nazi leader only confided in nine people in late February and early March as the decision process was underway. Most of his Cabinet did not learn about the operation until late on the night of 6 March; and most of the participating troops did not realize the nature of their true objective until they crossed into the demilitarized zone.[70]

That same crisis also illustrated how the ethics of "old school" diplomacy played a major role in the general inability to anticipate the actions of the Nazi leader. Professional diplomats simply found it difficult to accept the fact of Hitler's revolutionary diplomacy in which lies and deceit were basic to an approach that accepted no conventional obligations and that constantly used diplomatic instruments and language for deception. "I shall shrink from nothing," the Nazi leader confided in this regard. "No so-called international law, no agreements will prevent me from making use of any advantage that offers."[71] As a consequence, in the weeks preceding the Rhineland coup, the German Foreign Ministry issued no less than nine assurances that Germany had no intention of repudiating the Locarno Pact. Moreover, there was no rise in diplomatic tensions of the type that normally precede the breaking of a diplomatic pact, no formal ultimatums or demands, and, in fact, Hitler even did his best to maintain normal relations with France during the period.[72]

Added to this was Hitler's unique sense of timing. His move into the Rhineland was originally planned for the spring of 1937, but the international conditions in the winter of 1936 caused him to change his mind. Like his March surprise of the previous year, the Rhineland move took place on a weekend when Western diplomats were normally absent from their capitols, thus providing initial insurance against prompt diplomatic reaction. The move also occurred at a time when relations between Britain and Italy, two of the Locarno guarantors, were at a particularly low ebb over Ethiopia. Moreover, it was a period when French public opinion was firmly opposed to any military action beyond the borders of France. And as 7 March was only a few days before the French general election (not to mention the Easter holiday), French leaders were not ready to take decisive action in anticipation or reaction to the German move.[73]

After 1936, this "diplomacy by challenge" soon reached its natural limits, a development noted by an official of the Wilhelmstrasse who called it as early as 1937 a policy of "accelerating the Last Judgment."[74] Others at home and abroad were not so prescient. By that time, the vitriolic style of discourse that marked the diplomacy of Nazi Germany and other totalitarian regimes had become so normal that messages which in former years would have meant hostilities, if not war, were by 1941 accepted as demonstrating mild protest. As a consequence, when the Japanese foreign minister made a blunt and hostile statement in May 1941 to the American ambassador, "in Washington no one made much ado about his words. Hitler had hardened statesmen to the whole vocabulary of abuse."[75]

Long before 1941, however, Hitler's diplomatic coups had come to an end in accordance with the adage that nothing fails like success. Paradoxically in this regard, hubris is based on an excessive belief in reason – a penalty, in other words, for success due to a reliance on successful techniques that eventually fall to a new challenge. Thus, as we have seen, when it came to the Polish issue in 1939, Hitler did not realize in his preoccupation with his successful strategy that his success had undermined the conditions that made it possible. Because Britain and France had not fought before, when Germany was weaker, he believed in the summer of that year that they would not fight for Poland. What the Nazi leader failed to realize was that his opponents who had succumbed at Munich were by that time ready to behave differently for many reasons, not the least of which was that the uninterrupted chain of Hitler's triumphs in foreign affairs had convinced them that his ambitions were unlimited.[76]

The nemesis of having wishes completely fulfilled applies as much to politics as to personal life, as Hitler repeatedly demonstrated. For the Nazi leader, reason alone did not suffice to guard against its own excesses. "That is the miracle of our age," he stated, "that you have found me, that you have found me among so many millions. And that I have found you, that is Germany's good fortune." In the end, dazzled by his successes and corrupted by arrogance and impatience fused in a hubristic infallibility, Hitler returned to the extra-legal solution discarded so many years before as a means to his ends: a *putsch* on a monstrous scale.[77]

Personal Styles and the Structure of Decision-making

To the more general style of interacting in the international arena, the analysts of crazy states should add the personal decision-making styles of the leaders, which can provide insights not only into deviations from the pure rationality model, but also into the very nature of the governments themselves. The popular image of Hitler, for example, as an energetic and decisive policy-maker does not stand up to scrutiny. In fact, as Karl Dietrich Bracher has pointed out, all the great decisions in the Nazi leader's life were actually acts of avoidance, whether it was leaving school and moving to Vienna, entering politics almost as a last resort, or launching the Second World War.[78] In this context, Hitler was a study in irresoluteness, normally allowing chance to govern developments and making decisions only when circumstances or opponents provided him no other choice. What he termed *Schicksal* (destiny) or *Vorsehung* (province), in this regard, was nothing more than rationalization of his unwillingness to make decisions.

The discipline involved in regular work had always been anathema to the Nazi leader who believed that "a single idea of genius is worth more than a whole lifetime of conscientious office work."[79] As a consequence, after becoming Chancellor, Hitler reverted to form, returning to the idle bohemian style of the Vienna cafés and rejecting any administrative duties that smacked of what he had contemptuously termed, as an 18-year-old, *Brotberuf*, a "bread and butter" trade. One result was increasingly longer absences from Berlin in order to avoid official duties. Soon he settled in as Chancellor to a daily pattern of leaden inactivity, only occasionally breaking into manic restlessness, creating the lasting impression of breathless effort from these spurts of abrupt frantic activity. When war came, he could only look longingly back on these peacetime work habits. "When peace has returned," he remarked in 1942, "I'll begin by spending three months without doing anything."[80] The work forced upon Hitler as warlord took a physical and mental toll, the latter, as Joachim Fest has pointed out, because the work schedule "did violence to his nature and was in deliberate opposition to his inveterate yearning for passivity and indolence."[81]

These non-bureaucratic habits and the idiosyncratic style of governing also contributed to the chaotic nature of the Third Reich. For example, Hitler was averse to putting anything down on paper, and his lengthy absences from the capitol meant that he was increasingly inaccessible even to his top ministers. Added to this were his continued impatience with the complex details of intricate problems and his tendency to seize compulsively upon stray pieces of information or ill-considered analysis from the paladins and court favorites in his inner circle. "Ministerial skill," it has been pointed out in this regard, "consisted in making the most of a favourable hour or minute when Hitler made a decision, this often taking the form thrown out casually, which then went its way as an 'Order of the Führer.'"[82]

All this, of course, was a far cry from the pure rationality model for decision-making, with the effective transfer to state administration of the Nazi Party's (NSDAP) basic social Darwinistic principle of letting things develop until the strongest had won. As a consequence, by the mid-1930s, influence in key state decisions had passed to the rotating cast of cronies in Hitler's inner circle, with governmental ministries effectively cut off from the process. In this inner circle, Martin Borman became the "Brown Eminence" in the declining years of the regime, a man of "darkness and concealment," as Richelieu called Père Joseph, the sinister prototype of anonymous power-seekers, a functionary who derived his power solely from his office. Borman filled all the key posts in the party with men who owed these powers to him personally, not because of past service or qualifications. Within a short time, his intimate knowledge of Hitler's personal peculiarities and weaknesses gave him decided advantages over his rivals, ensuring that he was keeper of the mystic gate right up to the moment of Hitler's suicide.[83]

To some degree, the institutional chaos in the Third Reich was a result of Hitler's calculated policy of "divide and rule." This does not mean, however, that there was a consistent and systematic strategy on his part in terms of that policy.[84] In some cases, for example, the Nazi leader promoted the establishment of huge power bases, the most notable being the tremendous accretion of political strength which Himmler and Göring enjoyed with Hitler's active support. In addition, as we have seen, there was the case of Martin Borman, who accumulated unprecedented power during the war without any anxiety being evidenced by the Führer. Finally, there was the intense danger posed by Ernst Röhm and the SA in the early phase of the dictatorship, which Hitler eliminated only after intense pressure from the army as well as Himmler and Göring.

The domestic chaos was also due, in part, to Hitler's charismatic form of leadership which, in essence, rejected the institutional and bureaucratic norms required for the "rational" governing of a modern state in favor of dependence on personal loyalty to the Führer as the basis of authority. This transference of the NSDAP ethos from the *Kampfzeit* to a modern government also led to the Nazi leader's almost pathological hypersensitivity to any attempts to impose the slightest institutional or legal restrictions upon his authority, which in the *Kampfzeit* had to be completely untrammelled and, in theory, absolute. "Constitutional law in the Third Reich," the head of the Nazi Lawyers' Association stated in 1938, "is the legal formulation of the historic will of the Führer, but the historic will of the Führer is not the fulfillment of legal preconditions for his activity."[85] As a consequence, Hitler grew increasingly distrustful of any form of institutional loyalty and authority, whether it was demonstrated by army officers and civil servants or by lawyers, judges and church leaders.[86]

The corollary to this distrust of institutional links was Hitler's re-emphasis on the personal loyalty, which had marked the basis of his charismatic authority from early *Kampfzeit* days, until it was elevated to a dominant governmental principle. As long as that loyalty remained intact, the Nazi leader, as we have seen, had no problem with power bases emerging from his own chosen knights in the inner circle based on his Führer authority. But when it failed, there was the corresponding distress, as demonstrated by his behavior in the *Tiefbunker* at the end when notified of the treachery by Himmler, his "loyal Heinrich."[87]

Nevertheless, the loyalty principle remained the bond between all followers and the person of the leader, bringing an almost neo-feudal aspect to the Reich. In fact, however, as Ian Kershaw has pointed out,

> the bonds of personal loyalty – a pure element of "charismatic" rule – did not replace but were rather superimposed upon complex bureaucratic structures. The result was not complete destruction as much as parasitic corrosion. The avoidance of institutional restraints and the free rein given to the power ambitions of loyal paladins offered clear potential for the unfolding of dynamic, but unchannelled, energies – energies, moreover, which were inevitably destructive of rational government order.[88]

Risk

Risk propensity on the part of decision-makers is an important dimension in the analysis of crazy states. There are major problems in evaluation, however, many already encountered in the cognitive deviations from the pure rationality model. To begin with, the concept is directly related to instrumental rationality, a connection expressly made in John Collins' definition of risk as the "danger of disadvantage, defeat, or destruction that results from a gap between ends and means."[89] And yet crazy-state leaders may not perceive the gap as dangerous or even that it exists. In terms of the former, as we have seen, in some countries where there are concepts of martyrdom or nobility in failure, the element of risk may hardly apply in some instances. When that is translated to foreign policy moves on the part of a particular country, US policy-makers, who generally prefer low risk alternatives in strategic options, often label such moves as "reckless behavior, brinksmanship and adventurism."[90]

RISK

		HIGH (MAX)	LOW (MIN)
G A I N	HIGH (MAX)	1 MAXIMAX	2 MAXIMIN
	LOW (MIN)	3 MINIMAX	4 MINIMIN

Figure 3
Strategic Risk and Gain

More common are the cases where some states that use force to alter the status quo may differ from others less in the willingness to take perceived risks than in the perception of low risks where others perceive high ones. In many of Hitler's "Saturday surprises" in the 1930s, for instance, the Nazi leader may have been "reckless," not because he willingly tolerated a high probability of conflict but because he was certain that the other side would back down. When the German military opposed such policies as the Rhineland coup and the *Anschluss* on the basis that they were too dangerous, Hitler did not argue that the risks were worth the prizes, but that, instead, the risks were negligible – in other words, in terms of Figure 3, the MAXIMIN approach of Quadrant 2, not that of MAXIMAX in Quadrant 1.[91]

For example, in the Rhineland episode of March 7, 1936, the correlation of forces was quantifiably against Germany, as Hitler was well aware. "We had no army worth mentioning," he reflected later;

"at that time it would not even have had the fighting strength to maintain itself against the Poles."[92] But, unlike military advisers focused firmly on French capabilities, the Nazi leader examined all the instruments of power, particularly national will, and concluded that France had no intention of responding militarily, thus decreasing in his mind a risk already mitigated, as we have seen, by the timing of the reoccupation.[93] The difference between Hitler's focus on intentions and the German military's necessary attention to capabilities only increased tensions as the crisis heightened. On 9 March, the *Wehrmacht* commander received warning of impending French military countermoves and asked to withdraw troops from major cities in the Rhineland. But the Nazi leader, still taking an essentially MAXIMIN (Quadrant 2) approach, discounted the possibility of French intervention.[94]

The Rhineland coup also illustrated how Hitler deliberately fostered a MAXIMAX (Quadrant 1) picture of himself in international relations throughout the 1930s, accomplishing the extremely difficult job of projecting an image of a leader willing to pay a high cost to prevail in a specific dispute, but not willing to contest higher issues. The trick, as the Nazi leader realized, was to convince others of his willingness to go to war over a relatively minor issue by tying his stand to principles with more general applicability – in his case, those provided by the Versailles Treaty. By couching his demands in terms of the generally perceived legitimate right to reverse that treaty rather than the general righting of previous wrongs, Hitler convinced others that the vehemence of these demands was not a harbinger of unlimited aggressiveness.[95]

The Allied guarantee of Poland and subsequent declaration of war over the issue illustrate two important aspects in terms of risk propensity and crazy states. First, the lack of rationally instrumental means on the part of the Allies, as we have seen, caused Hitler to discount the guarantee in MINIMAX (Quadrant 3) if not MAXIMAX (Quadrant 1) terms. "The men of Munich will not take the risk," he told his commanders at Obersalzberg on 14 August 1939.[96] Second, Hitler's actions illustrate that there is always the possibility that a crazy state may not perceive the extent to which its actions will change the status quo. A. J. P. Taylor has argued in this regard that the Nazi leader was quite possibly attempting to achieve "international equality" for Germany without comprehending that "the inevitable consequence of fulfilling this wish was that Germany would become the dominant state in Europe."[97] For the nations that had to deal with the Nazi state in close proximity, however, as Robert Jervis has pointed out, "this distinction was important only if they could alert Germany to the consequences of her actions and Germany would then modify her policies. If this was not possible, it mattered little whether Germany was attempting to dominate out of inadvertence or design."[98]

Risk propensity for a leader like Hitler is also dependent on the flow of accurate information. German preparations for Barbarossa, for instance, were marked by atrocious intelligence, resulting in gross underestimations of Soviet power by the German military and Hitler, who less than a year later remarked that "if someone had told me that the Russians had ten thousand tanks, I'd have answered: 'You're completely mad!'"[99] Even so, the Nazi leader hovered nervously between the upper quadrants of the gain–risk matrix. "I never closed an eye during the night of ... the 21st to 22nd of June 1941," he reported later, also remarking that if he had known how large the Soviet forces were, he would never have initiated the operation.[100] Nevertheless, it should not be forgotten that Barbarossa was a very near thing – a point that suggests, as Richard Betts has indicated, "how thin the line may be between foolhardiness and masterstroke and how deceptive hindsight can be about actual risks before the fact."[101]

The interplay of intelligence and risk is also a reminder of how important the concepts of preemptive attack and preventive war are in terms of the rational gain–risk matrix. A preemptive attack is concerned with immediate threats and is designed to take the initiative once strategic warning of enemy attack preparations are received. Even for an actor as blatantly aggressive as Hitler, the defensive motivation inherent in such attacks cannot be discounted. The German attack on Norway, for example, was a form of preemption, since the Nazi leader was aware that Britain was about to launch a strike into that country. And a few weeks later, Hitler's impatience to attack into the Lowlands can be explained, in part, by his fear that the Western Allies might move first.[102]

Preventive war, on the other hand, is conducted in anticipation of future vulnerability on the part of the aggressor and is designed to take on an adversary before that adversary's capabilities can be improved. With Hitler and Barbarossa, for example, there is abundant evidence that preventive war in terms of his anxieties about growing German vulnerability reinforced his inclination to take risks for his long-held aggressive goals in the east. "What confirmed me in my decision to attack without delay," he recalled in 1942, "was the information ... that a single Russian factory was providing by itself more tanks than all our factories together."[103] Time, in this construct, could only work against Germany. "If Stalin had been given another ten or fifteen years, Russia would have become the mightiest State in the world."[104]

Added to this was the Nazi leader's perception that Stalin was increasingly intolerant of German conquests in the West and his own intolerance concerning Soviet annexations in the Balkans and Eastern Europe. In November 1940, Hitler offered Molotov, the Soviet foreign minister, a chance for the USSR to share in the booty and expand southward towards the Persian Gulf and India as the British Empire retreated. Molotov replied by indicating Soviet intentions to occupy all of Finland; and later, as he and Ribbentrop sat out a British bombing raid in a Berlin shelter, he revealed to his German counterpart that the Soviets were interested in the western approaches to the Baltic. "He demanded that we give him military bases on Danish soil on the outlets to the North Sea," Hitler still recalled with incredulity in the last week of his life. "He had already staked a claim to them. He demanded Constantinople, Romania, Bulgaria, and Finland – and we were supposed to be the victors."[105]

In addition, during this period, there was Hitler's continued lack of accurate intelligence on the Soviets that fed, as we have seen, into his rational strategic calculation that the road to London led through Moscow. Risk thus became acceptable for Barbarossa when the cost of not attacking the Soviet Union became unacceptable – a rational enough calculus, the "irrationality" of which in Stalin's perception ensured German military surprise. As a consequence of all this, the preventive war rationalization was strongly present the following month in Hitler's directive which stipulated that necessary measures were to be carried out "as a precaution against the possibility of the Russians adopting an attitude towards us rather than what it had been up to now."[106] For Stalin, this aspect of the Nazi leader's motivation was lost in his preoccupation with the danger of a German preemptive attack – a danger that, in the Soviet leader's estimation, would be increased by any improvements in Soviet readiness, since such actions would provoke rather than deter. As a result, Stalin never considered countermobilization, one of the worst ways to deter preemption, but the best way to deter preventive war.[107]

The "irrationality" of Barbarossa in the Soviet leader's perception reemphasizes the fact that crazy-state leaders may logically try to close the instrumental risk gap between means and ends by various techniques. Hitler, in this regard, was certainly aware that military surprise was one such way. The invasion of the Lowlands in the spring of 1940, for instance, demonstrated the Nazi leader's appreciation of the paradox that the greater the risk, the greater the surprise, producing as a result less risk – a move by means of theater operational technique from MAXIMAX (Quadrant 1) to MAXIMIN (Quadrant 2). On the other hand, when the means-end gap was extremely wide and could not be closed in an operation like the invasion of England, Hitler backed off. Thus there is the picture of the Nazi leader sitting through interminable conferences on Sea Lion in the summer of 1940, sifting logically through Service squabbles and fluctuating estimates of the correlation of forces. Whether he was operating at that time from the MINIMAX (Quadrant 3) position because, as Admiral Raeder concluded, "in Hitler's opinion the war was already won," or the MAXIMAX position (Quadrant 1), Hitler made the logical decision for postponement based on his staff's final estimate on 10 September 1940, the day before he was to give the executive order:

> The weather conditions which for the time of year are completely abnormal and unstable, greatly impair transport movements and mine-sweeping activities for "Sea Lion." It is of decisive importance for the judgment of the situation, that no claim can be made to the destruction of the enemy air force over Southern England and the Channel area .... The English bombers ... and the minelaying forces of the British Air Force, as the experiences of the last few days show, are still at full operational strength, and it must be confirmed, that the activity of the British forces has undoubtedly been successful .... The operational state, which the Naval War Staff ... gave as the most important prerequisite for the operation, has so far not been achieved, i.e. clear air superiority in the Channel area and the extinction of all possibilities of enemy Air Force action in the assembly areas of the Naval Force, auxiliary vessels, and transports.[108]

As the war dragged on, however, the risks associated with the means-end gap at the grand strategic and theater strategic levels grew too great for the Nazi leader to cope with. Consequently, he "tacticalized" strategy, increasingly confining himself to the operational and tactical levels of war where he could still find some measure of means-ends commensurability. Thus, in contrast to Churchill's use of the Chiefs of Staff system, Hitler created the OKW to deal with grand strategic matters while he played operational commander of the army in the east without having to face the increasingly grim strategic problems. "The other day I called off an attack that was to procure us a territorial gain of four kilometres," Hitler recounted with obvious relish midway through the war, "because the practical benefit of the operation didn't seem to me to be worth the price it would have cost."[109] How far this descent to rationality had gone by the final year of the war was described in one instance by General Blumentritt:

> The plan came to us ... in the most minute detail. It set out the specific divisions that were to be used .... The sector in which the attack was to take place was specifically identified and the very roads and villages through which the forces were to advance were all included. All this planning had been done in Berlin from large-scale maps and the advice of the generals ... was not asked for, nor was it encouraged.[110]

The Extrarational Factor

All policy-makers rely on extrarational means to some extent because of limited resources, uncertain conditions and lack of knowledge – the normal barriers to pure rational decision-making. There appears to be no way to make a valid normative model from the extrarational process, primarily because there is no way to compare the quality of a policy derived by means of subconscious processes such as intuition and judgment or through the interplay of charisma and ideology. What is certain, however, is that the extrarational factor is an important dimension of analysis in dealing with leaders of crazy states that can also account partly for the difficulty in predicting the foreign policy decisions of those actors. Moreover, the efficacy of extra-rational decision-making should not be discounted when examining these leaders. For example, in the two person, non-zero-sum, non-cooperative version of the "Prisoner's Dilemma" shown in Figure 4, it can be demonstrated that extrarational decision-making is better than that of pure rationality.[111]

In this version, two robbers are arrested in a stolen car and jailed separately. The police have evidence that they stole the car, but not that they committed the robbery. If both remain silent, each will be convicted only of the car robbery and each will receive a sentence of three years in prison. The prosecutor approaches the prisoners separately, promising each one that if he also confesses to the robbery and is the only one to make this full confession by a given time, he will be acquitted, while the other prisoner will be convicted on both counts and receive a 15-year sentence. When each prisoner then asks what happens if both of them confess, the prosecutor replies that the two prisoners will be convicted on both counts, but will receive relatively lighter sentences of ten years because of their confessions.

   If both prisoners use pure rationality to make their decisions, their
considerations would follow this pattern: "If the other guy confesses
(or keeps quiet), should I confess or keep quiet? I should confess,
never mind what he does, because if he keeps quiet, zero years in
prison is better than three years; and if he confesses, ten years in
prison is better than 15." As a consequence of this rational decision-
making, both prisoners talk and spend ten years in prison contemplating the limitations of pure rationality.

Prisoner
B

		No Confession	Confession
Prisoner A	No Confession	A - 3 Years B - 3 Years	A - 15 Years B - 0 Years
	Confession	A - 0 Years B - 15 Years	A - 10 Years B - 10 Years

Figure 4
Prisoner's Dilemma

The Great Simplifier

There is a normal bias on the part of policy-makers towards "intuition" rather than "information" and towards "guess" rather than
"estimate." Ideally, the optimum solution is to strengthen the rational
components as much as possible in order to achieve some form of
"informed intuition" or "guesstimate." The degree to which policy-
makers disregard such components can have a direct effect on the
degree of "craziness" in their states, particularly if their decisions are
bound up in a self-image of what Thomas Carlyle classified as the
intuitive heroic figure who:

   brings events to pass that emanate directly from a rare ability to
   see through appearance and the plethora of detail, to discount
   the false and trivial, and to highlight the great and the tragic. His

source of wisdom does not come from empirical knowledge. He has an intuitive sense of reality that allows him to feel and grasp new and unusual possibilities that otherwise are hidden to the senses. He cannot objectively prove what he so strongly feels. But when the event which was shaped by the force of the hero comes to pass, his people recognize the need if not the logical fitness of his deed.[112]

Hitler clearly saw himself in this mold – as one who could pierce the intricate complexities of the modern world. "I have the gift," he remarked in 1932, "of reducing all problems to their simplest foundations."[113] In that process, he was guided by a "divine Providence" that protected him throughout his career. *Mein Kampf* is filled with such references. It was fate that caused him to be born so close to the German frontier; that sent him to Vienna to suffer with the masses; that spared him in the First World War, etc. – all for a larger purpose. "Divine Providence," he concluded, "has willed it that I carry through the fulfillment of the German task."[114]

Allied to this was the inner intuitive voice provided in "the commands that Providence has laid upon me," which governed his decisions. "No matter what you attempt," Hitler stated in this regard,

> if an idea is not yet mature, you will not be able to realize it. I know that as an artist, and I know it as a statesman. Then there is only one thing to do: have patience, wait, try again, wait again. In the subconscious, the work goes on. It matures, sometimes it dies. Unless I have the inner, incorruptible conviction, *this is the solution*, do nothing. Not even if the whole party tries to drive me to action. I will not act; I will wait, no matter what happens. But if the voice speaks, then I know the time has come to act.[115]

The intuitive process reinforced a natural tendency of the Nazi leader to procrastinate, evident in his refusal to make up his mind in 1932 to stand as a presidential candidate and in his initial attempt to defer action against Röhm in 1934. Such inaction drove his advisers to distraction. But Hitler was adamant. "Trust your instincts, your feelings, or whatever you like to call them," he admonished. "Never trust your knowledge."[116] And when pressed for an explanation of his intuitive process, the Nazi leader fell back on his earlier vocation. "Do you know how an artist creates? In the same way the statesman must allow ... his own thoughts to mature"[117]

On the other hand, Hitler's "inner voice" at the subconscious level certainly accounted, in part, for his ability to penetrate difficult problems and to time his moves. Thus, there was his correct insistence against all advice throughout the latter part of 1932 on holding out for the Chancellorship. And, as we have seen, there was his perceptive, intuitive analysis of the situation during the Rhineland crisis. "I follow my course," he stated at that time, "with the precision and security of a sleepwalker."[118] This was the essence of Carlyle's hero figure who with his cry of "Act or you may never" transcends the power of reason and sense experience with his intuitive awareness. Apparent hesitations are swept away with one instantaneous, decisive blow. "The spirit of decision," Hitler observed, "consists simply in not hesitating when an inner conviction commands you to act."[119]

Ideology and Charisma

In *The Brothers Karamazov*, Dostoyevsky pointed out that the Church showed no tendency to become a "state," but that the state did all it could to become a church. Faith in this context is a major point of examination when analyzing a crazy state, particularly if it relegates reason to a subordinate role as was the case with medieval Christianity and with Hitlerism. The result can be state fanaticism which not only ignores values incompatible with those fanatically pursued, but, by ignoring the gravity of potential problems and obstacles, also removes national decision-making from anything approaching the instrumental rational process. In such cases, an increasingly interdependent world may no longer be willing to tolerate the principle *cuius regio; eius religio*.[120]

In the case of the Third Reich, there was nothing original in Hitler's ideology – a basic admixture of Social Darwinian thought and *Völkisch* themes tied in with Pan-Germanism, all espousing anti-Semitism and German superiority linked through concepts of "race" and "blood." Spread throughout were certain emotional elements focused on hostility to civilization, among which were the German Romantics as well as such luminaries as Wagner and Nietzsche. Finally, reflecting the mood of the times, were the strong undercurrents of nationalist and socialist ideas, tied in with anything that could be thrust into the grab bag that was National Socialism. "We gathered our ideas from all the bushes along side our life's road," Hitler once stated, "and we no longer know where they come from."[121]

For Hitler, this ideology, apart from the overpowering drive for eastern *Lebensraum* and concomitant racial dominance, was nothing more than the "great landscape painted on the background of our stage." As a result, contradictions and inconsistencies in the articles of faith were unimportant to him so long as they did not interfere with success, "the sole arbiter of right and wrong." This outlook was evident throughout his career when, on innumerable occasions, he jettisoned the so-called "granite" principles of National Socialism as soon as they became impediments to tactical considerations. There was, in fact, nothing the Nazi leader was not prepared to proclaim or abandon for the sake of gaining power even if it meant that he had to "swear six false oaths a day." "Any idea, even the best," he concluded in *Mein Kampf*, "becomes a danger when it imagines it is an end in itself, whereas in reality it represents merely a means to such."[122]

Underlying all this was the ideological link to the Nazi leader's "authority" or "rule," defined by Max Weber as power that is recognized by other people in any one of three ways. The first authority is that of the "eternal yesterday," derived from ancient custom and tradition. "My father was the king . . . . He is now dead. I am therefore King and you must obey me." Rational-legal, the second authority, is basically expressed as: "I was elected . . . by legal and constitutional procedures, and therefore you must obey me." Finally, there is the authority derived from an "extraordinary and personal gift of grace," which Weber called charisma, derived from the Greek word for spirit. "God . . . has laid His Hand upon me to make me the leader and therefore you must obey."[123]

It was this third authority that developed within the NSDAP during the *Kampfzeit*. And after the assumption of power by rational–legal means in 1933, National Socialism was revealed for what it had been since the emergence of the *Führerprinzip* in that period of struggle: the ideological justification of the Führer as the ultimate source of truth and power with the concomitant requirement for absolute obedience, the basis of charismatic domination. This merger of ideology into the charismatic leader created a faith solely focused on Hitler.[124]

Key to this concept of charismatic *Führerprinzip* was the assertion of Hitler's infallibility which fed, as we have seen, into an ever-expanding sense of destiny and hubris. By the late 1930s, the Nazi leader began to believe that he was actually free from human error, his goals supported by the will of Providence. "When I look back upon the five years that lie behind us," he stated in the summer of 1937, "I can say, this was not the work of human hands alone."[125] The following year at the Nuremberg Party Rally, his followers proclaimed this to the world. For one high party official, the Nazi leader was the only human who had never made a mistake, while another compared him to God. But it was left to SS Gruppenführer Schulz from Pomerania to add the final touch by asserting that Hitler was greater than Christ,

since the latter had been followed by only 12 disloyal disciples, while
the former had a nation of 70 million loyal citizens behind him.[126]

This type of charismatic authority is also a reminder how such rule
can further the "craziness" of states. For there is a need with such
authority to maintain both in the ruling elites and among the people
themselves the myth of the leader's unerring correct judgment and
independence from factional disputes. Thus, in contrast to the
massive unpopularity of the NSDAP and the daily greyness of life
under Nazism as the Third Reich wore on, there was Hitler's soaring
popularity that stemmed, in part, from his image of a leader aloof
from the daily realities of political and domestic life. It was an image,
as we have seen, to which Hitler to some extent had to adjust, thus
adding to a leadership style of aloof non-interference and a tendency
always to side with "the big battalions."[127] This, in turn, fostered the
need on the part of the Nazi leader to produce even greater achievements in order to bind the masses closer to the Führer and to prevent
the "vitality" of the Third Reich from slipping into stagnation,
disenchantment and probable collapse. This aspect played a major
role in impelling the Nazi regime toward craziness, always impeding
the establishment of a "state of normalcy"; promoting instead the
same radical, essentially negative dynamism that had led to the social
integration of the Nazi movement, but when applied to the international scene, could only lead to disaster.[128]

The Rationality of Irrationality

"Irrational" reactions from leaders of crazy states are a common
phenomenon. Thus, there is the picture of Libya's Gadhafi working
himself into a paroxysm of rage in a television interview; or Saddam
Hussein calmly outlining how he will defeat the United States. In
many cases, however, such displays are nothing more than rational
combinations of fanaticism and calculation. For example, if general,
conventional war is seen to be disastrous and essentially irrational, as
it was in the 1930s by most states still reflecting on experience in the
First World War, then the state most able, like Hitler's Germany, to
demonstrate a willingness to move closer to such a war is more likely
to succeed in intimidation. "There is a rational advantage," Herman
Kahn has pointed out in this regard, "to be gained from irrational
conduct or from the expectations of irrational conduct."[129]

This so-called "rationality of the irrational" can thus be applied in
an ends-means instrumental sense to the pure, value-maximizing
decision model. At the most basic level, as Thomas Schelling has
noted, some inmates in mental institutions often seem to cultivate,

deliberately or instinctively, value systems that cause them to be less susceptible to disciplinary threats and more capable of exercising coercion. Examples include a deliberately induced inability to hear or understand, or acquiring in the way of small children a reputation for frequent lapses of self-control that reduce or eliminate the deterrent effect of punishment. Even a careless or self-destructive attitude in terms of injury moves toward the rational construct of the model with the threat "I'll cut a vein in my arm if you don't let me ..." constituting a genuine strategic advantage. In Joseph Conrad's *The Secret Agent*, in this regard, a known anarchist with a container of nitroglycerin in his pocket is left unmolested by the London Police because he has threatened to blow himself up. A companion wonders why the police would believe anything so preposterous, to which the anarchist calmly replies: "In the last instance it is character alone that makes for one's safety .... I have the means to make myself deadly, but that by itself, you understand, is absolutely nothing in the way of protection. What is effective is the belief those people have in my will to use the means. That's their impression. It is absolute. Therefore, I am deadly."[130]

Hitler had a firm understanding of this approach. After his father's death, for example, his mother attempted to have him continue his education at the *Realschule*. Not for the last time, the 13-year-old successfully imposed his will by means of an hysterical reaction in the form of an illness. "The goal for which I had so long silently yearned, for which I had always fought," he wrote of the incident in *Mein Kampf*, "had through this event suddenly become reality almost of its own accord."[131]

In a similar manner in later years, Hitler played upon his reputation as a *Teppichfresser*, a rug chewer given to ungovernable rages. This was particularly evident in his tirades against the Austrian chancellor in February 1938 and the Czechoslovakian president in March of the following year, the latter almost succumbing to a weak heart in response to the verbal onslaught. At no time was the calculated irrationality of the Nazi leader better illustrated than in his 23 August 1939 meeting with the British ambassador, at which, as we have seen, he was presented with a note from Chamberlain indicating Britain's readiness to honor its Polish guarantees while still holding out hope for negotiations. Hitler responded by working himself into a frenzy, launching a violent tirade against the British, whom he held responsible for the crisis. "To all appearance," Alan Bullock has noted, "Hitler was a man whom anger had drawn beyond the reach of rational argument." And yet, as an official from the Wilhelmstrasse recorded that day: "Hardly had the door shut behind the Ambassador

than Hitler slapped himself on the thigh, laughed and said: 'Chamberlain won't survive that conversation; his Cabinet will fall this
evening.'"[132]

Goals and Organizations

Foreign policy goals are the key ingredient in determining or examining crazy states. These goals can be revealed in many ways, ranging
from a leader's official and semi-official pronouncements to actual
implementation. In terms of strategic analysis of crazy states, the
primary focus is on those national goals which involve external
aggression. These could vary from slight border incursions to an
ideological crusade for converting entire regions, if not the world. An
important corollary to all this, of course, is goal commitment. With
what intensity and consistency, in other words, do policy-makers
pursue these goals? The answer can be found in the combination of
the leader, who establishes the goals in accordance with his world
view, and the governmental structure in which every leader, no matter
how authoritarian the system, must play a functional role in terms of
organizations and politics.[133]

Crazy Goals

In examining the goals of any leader as a means of finding some
consistency in the course of that leader's foreign policy, it is necessary
to guard against teleologically squaring the circle in interpreting
results only in terms of those goals. Nevertheless, early goals or
indications of a world view on the part of a political leader are good
startpoints for any analysis. In Hitler's case, for instance, his ideological goals of *Lebensraum* and racial domination were clearly
spelled out in *Mein Kampf* and in myriad speeches. How those goals
were translated into reality forms the basis of most controversy on this
subject, which in turn provides valuable insights into the progressive
radicalization process that leads to the formation of a crazy state.

The "intentionalists" see the conquest of *Lebensraum* and racial
domination as intrinsically related, programmatic components of
Hitler's *Weltanschauung*, which formed the essence of the Nazi
leader's politics. The "programmatic" concept has been complemented by studies on Hitler's so-called *Stufenplan* to expand German
power in rationally calculated stages, beginning with the restoration

of the Reich to great power status; followed by achieving a position of preeminence in Central Europe for Germany; leading then to continental hegemony and the acquisition of *Lebensraum* in the east by subjugating the USSR; and finally the triumphant march of this "Greater German Reich" on the road to global dominance.

None of these views is intended to indicate a definite timetable on the part of the Nazi leader, but to demonstrate, instead, the consistent driving force of these components, while acknowledging Hitler's ability to improvise and his great tactical flexibility.[134] Added to this picture of relentless consistency are studies of Hitler's role in domestic policy, in which the Nazi leader, working with tactical adroitness, moved in a logical and internally rational series of moves toward the attainment of total power in order to implement the ideological goals of his world view.[135]

The "functionalists" or "structuralists," on the other hand, focus on a picture of Darwinian rivalries preventing any full coordination by fractured governmental machinery to bureaucratize the Führer's charismatic authority. As a result, the Nazi leader's personal world view served, at least in part, a functional role by providing "directions for actions" (*Aktionsrichtungen*), with Hitler not so much creating policy, as sanctioning pressures operating from different forces within the regime. In this construct, the pursuit of *Lebensraum* was nothing more than Hitler's need to maintain the dynamic momentum he had unleashed, which in turn created the need for complete freedom of action, only obtainable by breaking all diplomatic norms. In any event, the Nazi leader's foreign policy goals before 1939 were nebulous, unspecific and basically utopian with the goal of attaining *Lebensraum* being an ideological, metaphorical symbol that explained incessant foreign policy activity. That activity was used as a means of integrating and diverting antagonistic forces in the Reich brought about by the increased involvement of the masses. In this manner, the plebiscitary social dynamic in foreign policy exerted increasing pressure on Hitler and his regime to transform the *Lebensraum* metaphor into reality.[136]

In terms of the Jewish component, both schools are agreed that Hitler maintained throughout his political career a deep, pathologically violent hatred of Jews and that this paranoid obsession was a major factor in determining the climate within which the spiraling radicalization of anti-Jewish actions occurred. In this regard, it was Hitler in his capacity as head of the Third Reich whose fanaticism on the "Jewish Question" provided impulse, sanction and legitimation for the escalating horrors that eventually culminated in the "Final

Solution." The "intentionalists," however, go beyond this to see unwavering continuity in Hitler's aim to destroy the Jews, his dominance in shaping anti-Jewish policy, and his decisive role in the initiation and implementation of the Holocaust. Lucy Dawidowicz, for instance, views the entire process as one long rational search on the part of the Nazi leader for the means to fulfill his goal of destroying Jewry – a goal which he had formed as early as 1918 and which was openly espoused as a program of annihilation in *Mein Kampf*. This program was to become "a blueprint" for his policies when he came to power. "There never had been any ideological deviation," Dawidowicz concludes. "In the end only the questions of opportunity mattered."[137]

The "functionalists" continue to see at least part of the Holocaust in organizational terms, focusing on the oddly fragmented decision-making process in Nazi Germany which resulted in improvised bureaucratic initiatives with their own internal momentum that led to the dynamics of cumulative radicalization. In this context, the key lay in the local exterminations in the east, stemming from the unexpected failure of the *Blitzkreig* invasion of the Soviet Union and the con-comitant inability, after fall 1941, of *Gauleiter*, SS chiefs and other officials in the Occupied Territories to cope with the growing number of Jews in their domains as a result of breakdowns in the deportation plans. "It thus seems that the liquidation of the Jews began not solely as the result of an ostensible will for extermination," Martin Broszat has pointed out in this regard, "but also as a 'way out' of a blind alley into which the National Socialists had manoeuvred themselves. The practice of liquidations, once initiated and established, gained predominance and evolved in the end into a comprehensive 'pro-gramme.'"[138]

Crazy Reality

Any examination of deviations from the pure rationality model has to deal with the personal characteristics and intentions of the policy-maker as well as the structure and process of decision-making. In most cases there will not be a single focus. For example, in terms of Hitler's part in shaping anti-Jewish and foreign policy, the "intentions" of the Nazi leaders as well as the impersonal "structures" of the regime were both critical elements, though in varying degrees of mix. In the case of foreign policy, Hitler shaped initiatives and made the key decisions. In this regard, his ideological goals were important factors in the course that policy was to follow, but are often difficult to separate from strategic-power considerations and economic interests. Nevertheless, the force of Hitler's personality and his pragmatic goals can be seen throughout the foreign policy of the Third Reich.

At the same time, the charismatic function of the Führer role was also key to a foreign policy moving inexorably toward war, since it provided rational legitimization of the means towards the ends it was presumed were desired by the Nazi leader. This legitimization extended to the self-interest of the army leadership, to the ambitions of the diplomats at the Wilhelmstrasse, and to the greed and ruthlessness of the industrialists. Equally important, the charismatic legitimization provided in the realm of foreign policy a chauvinistic and imperialistic basis for the party masses to clamor for the restoration of Germany's power and glory. In the end, it is the complex mixture of Hitler's ideological intentions, combined with the conditions and forces which structured those intentions, that explains the foreign policy of the Third Reich.

In the case of the "Jewish Question," Hitler's primary input was establishing the long-range goal, shaping the climate and sanctioning the increasingly radical actions of those around him who were acting in the name of his "heroic" charismatic intentions. The overall effect should not be underrated. The Nazi leader's obsession with *Lebensraum*, for instance, was tied to a reemergence of the Ottonion and Hohenstaufen feudal empires populated by a new master race of Ayrian blood – all of which would begin with the resumption of medieval colonization. In this context, the east was to be considered initially an area like the "Wild West," outside the bounds of law and order where a new "nobility of the sword," the Gauleiter, would be expected after ten years, according to Hitler, "to be in a position to inform me that these regions have once again become German."[139]

In addition, there were the Nazi leader's public tirades of hate and vague threats against the Jews, the most notorious being his 30 January 1939 Reichstag speech in which he "prophesied" that the war would result in the "annihilation (*Vernichtung*) of the Jewish race in Europe." This was a prophecy to which he frequently referred during the war and which he postdated to the opening day of the conflict, reflecting his mental merger of that war and his "mission" to destroy the Jews.[140]

At the same time, the "Final Solution" was not just a product of the myriad "local initiatives" in the east which gained retroactive sanction "from above." There was also direction from the centers of power, primarily the Reich Security Head Office, but not from Hitler, although "undoubtedly the most important steps had his general approval and sanction."[141] What emerges from this is a picture of the

Nazi leader who, despite his undeviating hatred of the Jews over the
years, took little part in the overt formulation of anti-Jewish policy.
Instead, his major role, as we have seen and as Ian Kershaw has
noted,

> consisted of setting the vicious tone within which the persecution
> took place and providing the sanction and legitimation of initia-
> tives which came mainly from others. More was not necessary.
> The vagaries of anti-Jewish policy both before the war and in the
> period 1939–41, out of which the "Final Solution" evolved, belie
> any notion of "plan" or "programme." The radicalization could
> occur without any decisive steerage by Hitler. But his dogmatic,
> unwavering assertion of a vague ideological imperative – "get-
> ting rid of the Jews" from Germany, then finding a "final
> solution to the Jewish question" – which had to be translated into
> bureaucratic and executive action, was nevertheless the indis-
> pensable prerequisite for the escalating barbarity and the
> gradual transition into full-scale genocide.[142]

External Action Capabilities

In order for a crazy state to become a strategic problem, it must
possess external action capabilities. Thus a deranged Russia would
have far greater significance than a small Third World country
similarly affected. Figure 5 is a three-factor model with just some of
the variables that shape such capabilities.[143] The initial factor is
concerned with a nation's infrastructure bearing on external action
capabilities.. The second factor includes realization variables which
transform the infrastructure variables into external action capabilities, the dimensions of which form the third factor in the model and
provide a breakdown into external action instruments. Each of the
factors, of course, does not exist in a vacuum. Realization variables,
for example, depend on and to some extent, in some situations, are
even a function of the infrastructure variables. And infrastructure
elements also depend in the long run on such realization variables as
political capacity and national will. On the other hand, nuclear,
biological and chemical weapons or terrorist groups may provide
external action capabilities out of proportion to the infrastructure of a
relatively small state.[144]

In a pure rational model, the realization variables would operate
perfectly, with the full potential of the external action infrastructure
being realized. In reality, of course, in most nations there are

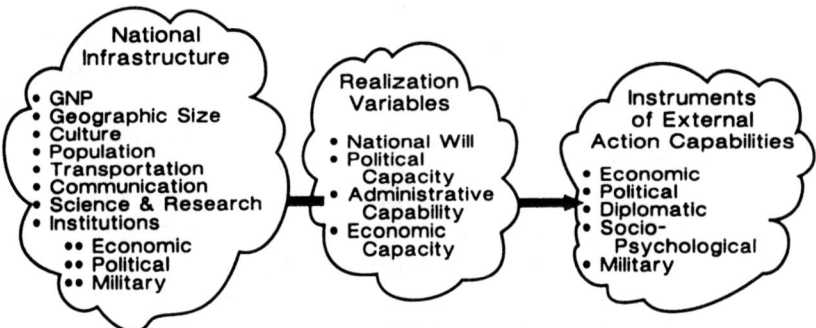

Figure 5
Capabilities for External Action

significant deviations from the model. There may be an even greater tendency in this regard with crazy states if Nazi Germany is any indication. For as the political and ideological demands of the Nazi leaders came to play a dominant role in their alliance with Germany's military–industrial complex, instrumental rationality declined accordingly with a concomitant loss in ability to generate external action capabilities.[145]

Realization Variables – Irrationality in Preparation

At the outset of the Nazi regime, the political decision was made to give absolute priority to rearmament. And yet there emerged over the years of peace many factors which inhibited the rational instrumental realization process that would have transformed resources efficiently into armaments. To begin with, there were the various interest groups with direct connections to Hitler in the charismatic chaos of the *Führerstaadt*. Party members were particularly onerous, often making it difficult for Hjalmar Schacht, the Minister of Economics, to cut public expenditures not associated with armaments. For example, in almost every instance when Schacht attempted to cut funds earmarked for municipal improvements, he was opposed by prominent Gauleiter and other party officials, all of whom could take the issues directly to the Nazi leader who invariably sided with them.[146]

Equally important in terms of rearmament was the regime's unwillingness to increase public expenditures and thus incur larger deficits which, it was believed, would destroy confidence in the government and lead to inflation. The fear of inflation, of course, was a holdover from previous non-Nazi governments that affected everybody from

Hitler to Schacht, the latter a paragon of financial conservatism. For the Nazi leader, it was all simply a matter of control. "Inflation is lack of discipline – lack of discipline in the buyers and lack of discipline in the soldiers," he stated early on. "I will see to it that prices remain stable. That is what my S.A. is for."[147] In the end, the specter of inflation was a major restraining influence on both the development of raw material industries and total military expenditures.

Added to all this was the inefficiency of the National Socialist regime in planning and implementing the war production program. There was, for instance, no central agency before the war for coordinating the material demands of the services of the armed forces. The result in terms of rearmament was that the services created and then pursued their own weapons and equipment programs with almost no coordination and, as a result, with interservice rivalry for scarce raw materials often being intense, if not vicious.

Göring's "Four-Year Plan" in 1936 did not alleviate the situation; and Hitler did not improve matters by sanctioning armaments programs that were far removed from reality. Thus, the programs proposed and approved by the Nazi leader in 1938–39 included a massive expansion of the Luftwaffe and the so-called "Z-Plan" for creating a large-ship navy – both of which, if implemented, would have required an amount of crude oil reserves in excess of the total annual production in the world. How far removed from reality such planning could be for an individual service was demonstrated in the initial army force for Barbarossa of approximately 150 divisions, only 46 of which were fully provided with German equipment and arms, the remainder either operating with captured Czech and French materiel or deficient in such key items as anti-tank guns. "In the light of such evidence," Burton Klein has concluded in this regard, "it would be difficult to deny that a more rational and better executed program would have given the National Socialists larger rearmament."[148]

Finally, in terms of impediments to rearmament, there was the unwillingness of Hitler and his regime to ask civilians to lower consumption, demonstrated by the government's reluctance in 1937 to cut food imports in order to increase those of raw materials. In actuality, no choice had to be made until 1936 on the question of *Kannon oder Butter*, since increased civilian consumption as well as rearmament could be achieved together by using unemployed resources. After that, however, the German economy was operating at almost full employment; and a large increase of expenditures for armaments would have required, at the very least, a sharp curtailment of some types of civilian goods production. That did not happen. And although shortages of such vital raw materials as iron and other ores repeatedly delayed armaments production in the late 1930s, the volume of imported foodstuffs and textiles actually rose steadily during that period – all permitted by the regime in order to satisfy public demand, upon which, despite its use of terror, the government believed its stability and popularity rested. In these circumstances, major armament increases in the latter part of the decade overburdened economic capabilities while producing bottlenecks in all sectors of the economy as well as latent inflation. The armament achievements were considerable, but they were obtained at the expense of growing economic instability, while still falling short of the goals in Hitler's strategic plans.[149]

The concern with public well-being is a paradoxical development in an authoritarian, non-democratic government, which, depending on the background of the leader and the culture of the population, could prove to be an important focus in examining a crazy state's ability to generate external action capabilities. Certainly it is key to understanding the Nazi leader's impact on the realization variables that affected such capabilities in the Third Reich. As a consequence, the domestic policies of the Nazi regime swung inconsistently back and forth between vicious attacks on personal and political rights, on the one hand, and fearful circumspection inspired by the need to gain and maintain popularity with the masses on the other. At the same time, there was a disinclination on the part of the regime to require sacrifices from the population, demonstrated by the government's rejection of higher taxes as an alternative to deficit spending, its unwillingness in 1937 to cut food imports in order to increase raw material imports, and its failure to move workers out of non-essential occupations. All this culminated in 1938 when, despite pleas from the Ministry of Food and Agriculture, Hitler would not raise food prices for fear of the effect on living standards as well as worker morale. "How can I expect to wage war," he asked, "if I drive the masses into the same state of apathy that they were in during 1917–1918?"[150]

Realization Variables – Irrationality in Execution

Hitler's concern with any threat to "social peace" continued to affect the realization of external action capabilities throughout the Second World War. During the first months of that conflict, for instance, he withdrew plans for labor mobilization after workers protested the impact that such a move would have on wages, working conditions and living standards. In a similar manner as the war went on, other aspects of "business as usual" included the high level of consumer goods until 1943; the personal allowances paid to the wives of the military, which were so generous that most married women could not be persuaded to take up industrial work when that was belatedly sanctioned for women; and finally the incessant propaganda stressing that final victory was just around the corner, making sacrifice and effort appear less necessary. In all this, Hitler's concerns were the dominant factor. "For there is no doubt about it," he stated in 1942 while emphasizing the need to bring captured grain from the Soviet Union, "the morale of the people is dependent to quite a considerable degree on a sympathetic understanding of and catering for the little things that make life more pleasant for them."[151]

The Nazi leader's doubts about the solidarity of his regime continued up to the end to affect adversely Germany's ability to generate external action capabilities. In the spring of 1945, all ministers agreed that a drastic increase in taxation was necessary for combatting inflation and reducing civilian purchasing power. Hitler signed an order for this increase, but insisted that it should only be enacted when the German military won a victory. The tax increase was never put into effect.[152]

Another aspect of the wartime realization variables for Germany involved the twin goals of *Lebensraum* and the destruction of the Jews. That these goals were "crazy" is an almost universal judgment. But there is another component of this issue that brings it full circle to the question of whether rationality can go beyond instrumentality and also apply to goals. The final solution against the Jews was initiated on a large scale in the summer of 1941 at the same time that the final solution of Hitler's grand design in foreign policy, the war against the Soviet Union, was undertaken. At that same moment, as the blueprints of Hitler's foreign and racial policies reached their climaxes, the inevitable struggle for priorities between the two began. To have discontinued temporarily the persecution of the Jews in the summer of 1941 and concentrated all means on the conquest of living space, or vice versa, was not possible for the Nazi leader whose policies, as Eberhard Jäckel has pointed out, "took their course with an obstinate, brutal, and final self-destructive consistency."[153] Thus, Hitler's goals were irrational in the limited meaning of choosing mutually conflicting ends – in the sense, for instance, that mankind's insistence on using up the earth's resources as quickly as possible in order to maximize self-interest is irrational.[154]

Hitler's Jewish policy was also profoundly irrational in an instrumental sense in terms of external action capability. For the most part in the early stages of that policy, there was no contradiction between the objectives of the relatively autonomous SS and the interests of the German capitalists. The latter were quite willing to take advantage of the Jewish labor concentrated in the Polish ghettos and the lack of restrictions concerning the total exploitation of this labor for minimal cost. But with the huge extension of the "Jewish Question" in the Occupied Territories and the move into the final solution accompanying the increased labor shortage of the Third Reich, the radical nihilism of National Socialism emerged triumphant over "rational" economic interest. This was illustrated by the initial extermination of Polish Jews, thousands of whom were skilled metal workers from Polish arms factories. And although the Nazis continued to use some Jewish labor almost to the war's conclusion, the regime also used on a continuous basis the scarce transport equipment to move human cargo across Europe to extermination points – all at a time when German forces were desperate for resupplies and when German industry was desperate for manpower.[155]

This irrationality extended down to the "normal" concentration camps, initially created to neutralize the internal enemies of the Reich and later used to provide forced labor. At these camps, the SS was unable to preserve the physical fitness of the prisoners marked for such labor because of inadequate food, barbarous working conditions and totally deficient housing and hygiene. Nor could threats of punishment in such cases increase productivity. The average work done by concentration camp inmates, Speer estimated in this regard, amounted to only one-sixth of that achieved by free civilian labor. "In short," one historian summed up the fundamental irrationality, "it turned out that a system based on repression, dehumanization and physical extermination could hardly be converted into a high-productivity economic enterprise."[156]

Weapons of Mass Destruction

Hitler's involvement in the invention of new weapons had mixed results. On the one hand, when his interest was aroused in any piece of technology like the V-1 flying bomb or the V-2 rocket, his energy and enthusiasm, focused through his dominant decision-making position, could quickly impel a developmental project to the top of any priority list. On the other hand, the Nazi leader's lack of discipline coupled with his initial enthusiasm for new weapons resulted in a total lack of prioritization which meant, in turn, that Germany suffered from an excess of "secret weapons" throughout the war. For example, his commitment to the huge V-2 rocket was, as David Irving has noted, "an extravagant irrelevance" that siphoned off a large amount of industrial capacity which could have been used in manufacturing more mundane and useful weapons such as the surface-to-air rocket, first developed in 1942. That weapon, with its ability to hit enemy bombers at 50,000 feet, as Speer pointed out, might have beaten back the 1944 Allied spring air-offensive, particularly if it had been combined with the new jet fighter, its development also a victim of Hitler's interference.[157]

As a consequence of all this, German scientists were deeply involved in the political process, with the life of projects often dependent on the most outrageous whims of the Führer or his inner circle. Under these circumstances, it was almost impossible for the scientists, whose expertise was supposed to be in either rational calculation or the analysis of data designed to aid rational calculation, to be distinguished from the party politicians who gained their official status based on entirely different criteria. Many scientists simply retreated into their own world, conducting investigations and experiments with no possible application to the war effort, despite the Nazi slogan: "German science in the cause of war." The head of the uranium project, for instance, did not hesitate to use funds and special privileges accorded that project in the expectation of some military benefit, for the general furthering of German theoretical science. "Academically a very satisfying pursuit," David Irving has pointed out, "it was not the way to win wars."[158]

Nevertheless, German scientists in the Nazi era made many scientific breakthroughs with military applications, one of the most significant of which was Tabun, a new nerve gas that penetrated the filters of all known gas masks and produced fatalities after even limited contact.[159] Why Hitler did not use this and other gases can only be a matter of conjecture. Certainly there was his own experience in a gas attack before Ypres on 13 October 1918, from which he "stumbled and tottered back with burning eyes."[160] And there was always the abysmal German intelligence trumpeting the chemical warfare edge of the Western Allies that blinded the Nazi leader to Germany's unique advantage with nerve gas. Or perhaps he realized that at the tactical and operational level, where he increasingly retreated as the war went on, chemical warfare was of as dubious value as at the strategic level.[161]

What is certain is that Hitler generally opposed the offensive use of gas, despite some consideration of its use during the war, and that he was concerned about Allied retaliation. How concerned he was on this matter was demonstrated by the fact that he prohibited the transfer of any gas munitions outside the pre-war Reich borders for fear of accidental use. And as German forces retreated on both fronts in the closing months of the war, there was the personal order over his signature that no toxic chemical dumps were to be blown up in case such actions might be construed as initiation of gas warfare. But if the costs were so great that even Hitler could see the irrational ends-means disconnect, it should not be forgotten that where poison gas could be used without this cost, it was, as millions of genocidal victims attest.[162]

Nuclear weapons, however, were another matter. In June 1942, Speer discussed the German uranium project with Hitler as one item on a long agenda. No other documentary evidence exists to show the Nazi leader's involvement in that project, although there are indications that he certainly would have approved the construction of an atomic bomb if he had been convinced of its feasibility. He was interested, for example, in the most advanced conventional explosives, once boasting that the V-1 flying bomb used an explosive "2.8 times as powerful as normal bomb explosives."[163] And there were vague allusions to atomic weapons in a talk with the Romanian leader in August 1944, when Hitler described the most recent work on "new explosives whose development has been advanced to the experimental stage," adding that in his view the leap from modern explosives to the new development was the biggest since the invention of gunpowder. Moreover, when the subject turned to such secret weapons as the V-2 rocket, the Nazi leader pointed out that "another of these weapons ... has such colossal force that all human life is destroyed within three or four kilometers of its point of impact."[164]

By that time, the matter had long since been decided by a number of factors. To begin with, Hitler's anti-Semitic policies had depleted the ranks of nuclear scientists in his regime early on; and the entire atomic bomb effort still had Jewish associations in the mind of the Nazi leader who occasionally referred to nuclear physics as "Jewish physics."[165] Then there was the drafting of scientists into the *Wehrmacht* which increased so markedly by 1943 that one professor associated with the Reich Research Council warned that "while 3,000 fewer soldiers would not weaken the armed forces, 3,000 more physicists might well decide the war."[166] Most importantly, the German uranium project was directed by scientists throughout its existence, not by the military as was the case in the United States with the Manhattan Project. Acting alone, the nuclear scientists consistently failed to impress the Nazi leaders with anything but what Field Marshal Milch called their "artlessness and naivete."[167] As a result, the rapport between the scientists and the elites of government and industry, so necessary for the full and expeditious implementation of the nuclear program, was never established.[168]

All this notwithstanding, the German uranium project would have been a very near thing if it had not been for Allied commando attacks in Norway on the German production of heavy water, the key ingredient for the type of reactors being researched in Berlin. The February 1943 raid on the Vemork facility, east of Oslo, was a conspicuous success, resulting in the loss of a ton of heavy water and the temporary closing of the plant. Such attacks, culminating a year later in the sinking of a Norwegian ferry with large supplies of heavy water, played the dominant role in ending German hopes of building an atomic reactor, much less an atomic bomb. How effectively the heavy water could have been used in Berlin was indicated by the Chief of the US Scientific Intelligence Mission who inspected the underground uranium reactor bunker constructed to hold the first pilot reactor. "It looked as if it had once been excellently equipped," he wrote in July 1945. "I remembered the primitive setup with which Enrico Fermi had started in a basement room at Columbia University. By contrast, this Berlin laboratory, even empty, gave an impression of high-grade achievement."[169]

The Way Ahead

"Probably none of us is entirely 'normal,'" Hitler noted rather plaintively in 1941.[170] This, of course, is true. It is also true that all crazy states are not equally crazy. But by using the factors identified in the "black box" of Nazi Germany as a referent, the extent of a nation's craziness can be roughly identified.

To begin with, authoritarian, non-democratic rule in a nation state is more likely to produce craziness from a Western democratic perception. Add charismatic, one-man rule, as was the case in Nazi Germany, or charismatic, collective rule, as is the case of the Mullah-led theocracy in Iran, and the stage is set for varying degrees of craziness. All this, in turn, depends on the impact such rule has on decision-making variables involved with bureaucratic organizations, governmental politics and the personal characteristics of the leader. In this construct, the operational code of any leader, but particularly one "touched by God," is extremely important. How such a leader views the world must be the logical startpoint. In this regard, as we have seen in the detritus of Hitler's mind, early speeches or writings by a leader as well as psychoanalytical studies can provide some insight into his cognitive trail, touching on such factors as ethnocentrism, threat perception and long-term goals.

In particular, a leader's goals and his commitment to them are important dimensions of craziness, especially if the leader is, like Hitler, one of those "terrible simplifiers" of ideas.[171] At one end of the scale might be goals that do not go much beyond the status quo; or commitment that requires no more than minor parts of the budget and limited manpower being devoted to external goals, which in turn are subordinated to internal ones. The other end would involve extreme external goals up to the destruction of another nation state and the absorption or liquidation of its population. Or, in terms of commitment, an all-out devotion of resources to external goals, with internal objectives only perceived as means for external ends; and with the acceptance of these external goals perceived as a national mission up to the ultimate readiness to sacrifice national existence to achieve them.

This craziness intensifies, as we have seen in Hitler's Germany, if such a commitment is focused through charismatic authority derived from neo-feudal personal loyalty. When that authority is superimposed upon the bureaucratic structures of a modern nation state, the corrosive effect on those structures leaves the power of the state intact for the pursuit of crazy goals while destroying institutional restraints. The result is progressive and cumulative radicalization of a regime in which more and more power resides with the leaders and the loyal paladins of his inner circle.

This type of organization can only feed the ego and hubris of the leaders perched at the very top. A Kim Il Sung or a Saddam Hussein, in this regard, is no more likely to suffer adversarial inputs into his decision-making process than Hitler. In the case of Hussein, for instance, there is the story of a Cabinet meeting held in the dark summer of 1982 when Iraq was bracing for an Iranian invasion. At that meeting, the Minister of Health suggested that the Iraqi ruler resign temporarily from office in order to facilitate a negotiated cease-fire. Hussein demonstrated no sign of irritation, offering instead to discuss the matter with the minister in another room. Shortly after they left, a shot was heard, and the Iraqi President returned alone to the Cabinet showing no sign that anything had happened. The issue of his resignation was not raised again.[172]

All this can affect a leader's propensity to take risks, an important dimension of craziness. It is a dimension that brings the analysts full circle back to the cognitive trail of the crazy-state leader, since, as was the case at times with Hitler, what is deemed risky by other nations and even by internal military and diplomatic elites, may not be so perceived by the leader. To be aware of cultural and/or perceptual differences on the part of a crazy-state leader, however, does not lessen the importance of this subjective factor in evaluating craziness. At one extreme are "normal" leaders who tend to avoid risky policies. At the other extreme, there could be the preference for particularly risky policies that might vary from ideological commitment to adventurism and risk-taking as preferable life styles. That these are key considerations in examining a leader like Saddam Hussein is self-evident. What makes the Iraqi president even more dangerous, some analysts have pointed out in this regard, is that rather than compromise or retreat when he encounters problems as a result of running risks, he tends to double his bets.[173]

Risk perception, as we have seen, has a great deal to do with instrumental rationality. Total irrationality in the sense of a complete rejection of the need to relate ends to means is rarely encountered even in the craziest of national leaders. In the early stages of his career before succumbing to the nemesis of hubristic success, Hitler adjusted to mistakes like the 1923 *putsch* which resulted from an imbalance in the means-ends relationship. And in 1939 with the Nazi–Soviet Non-aggression Pact, the Nazi leader was willing to jettison fundamental ideological animosities to further means-ends commensurability. More recently, in the fall of 1990, Saddam Hussein demonstrated a similar rational tendency with the Iraqi–Iranian Pact, by which the Iraqi leader returned most of the territory acquired in the savage eight-year war with Iran in order to ensure the security of Iraq's border with that country. Finally, there is the picture of Hitler during the later stages of the Second World War retreating into cognitive dissonance or down to a lower level of warfare when faced by too large an irrational gap at the strategic level between means and ends.

In the end, all these factors for evaluating craziness overlap with a leader's style. At the personal level, as Hitler's non-bureaucratic habits demonstrated, decision-making style may affect the entire functioning of government. In terms of the international arena, style may reflect the no-holds-barred, *coup de main*, revolutionary-type approach that is part of the normal path to power for leaders within the type of parties that seek to gain national dominance by other than rational-legal means.

At its most extreme form of craziness, this factor can encompass a complete rejection of diplomatic norms and the adoption of methods that could include eco-terrorism, mass assassination of opposing leaders, systematic sabotage, counter-value terror – all the way up to genocide as a style of operation. Strategies, in turn, will incorporate these stylistic methods, depending on external action capabilities. For example, the socio-psychological instrument of external action, as the Nazis effectively demonstrated, lends itself to deception and fifth column-type infiltration and erosion from within – the latter to the extent that the name of the Norwegian fascist, Vidkum Quisling, has passed into the English language as a synonym for such activity. On the other hand, a nuclear capability may cause a crazy state to adopt strategies ranging from blackmail to occupation or destruction of neighboring states.

The Munich Metaphor

Containment is not the strategic answer to crazy states, since it is based on an assumption that if foreign policy successes are denied aggressive nations, then domestic constituencies will insist on change and thus moderate expansion. Although this "cure" worked for the Soviet Union after 45 years, an increasingly multipolar world in an era of advanced weapons technology is simply too dangerous to wait for crazy-state constituencies to rise up against the pervasive terror that normally forms a part of the authoritarian rule in such states.

The inner logic of such terror was revealed in 1956 at the 20th Congress of the Soviet Communist Party, when Nikita Khrushchev revealed for the first time the full scope of Stalin's atrocities. As the premier spoke, one account goes, he received a note passed up from the audience. Khrushchev glanced at the note and then read it aloud: "If Stalin was such a monster, why didn't you and the Soviet leadership stand up to him?" The Soviet leader added: "This is an excellent question. I would be grateful if the comrade who asked it would rise so that I can answer him face to face." There was no movement in the auditorium. "Well," Khrushchev finished, "there's the answer to your question."[174]

Unlike containment, deterrence assumes no cure for an aggressive nation – a much more realistic method of dealing with crazy states. Concessions to an aggressive state, according to this model, sometimes named after the 1938 Munich Conference, simply lead that state to expect further concessions. In opposition, the so-called "spiral" model looks to the outbreak of the First World War and focuses on the security perceptions of the other state, arguing that correctly executed concessions to another state lead to reciprocation from that state. Perceptions of friendship can become, in other words, self-fulfilling prophecies given enough time and the right environment. Thus, there is the British Ambassador to Germany cabling London in February 1939:

> My instinctive feeling is that this year will be the decisive one, as to whether Hitler comes down on the side of peaceful develop- ment and closer cooperation with the West or decides in favour of further adventures eastward .... If we handle him right, my belief is that he will become gradually more pacific. But if we treat him as a pariah or mad dog we shall turn him finally and irrevocably into one.[175]

This example demonstrates, of course, that the most obvious problem with the concession approach is when an aggressive state does not respond in kind to conciliation. In such instances, minor concessions – the willingness to treat individual problems separately, and even an offer for negotiations – can convince an aggressor that the status quo power is operating from a position of weakness. Moreover, as Hitler's foreign policy repeatedly demonstrated in the 1930s, once an aggressive state comes to believe in the weakness of the defenders of the status quo, it may be impossible to alter this image short of war.

Another particularly serious aspect of this problem is when nation A makes concessions to aggressive nation B in the incorrect belief that B is a status quo power. Such concessions are especially likely to be misinterpreted if B does not understand that A's policy is based on a false image of B. In this case, aggressive nation B will often think that its intentions are obvious to A and, as a consequence, will conclude that any concessions made by A must be the product of fear and weakness.[176]

All this was evident with Hitler as the Munich Conference began to run its full course. At that time, the Nazi leader appeared to believe that the British were aware that his ambitions did not stop at the borders of areas inhabited by Germans. As a result, he concluded that Chamberlain's conciliatory efforts reflected not so much a feeling that Germany would be satiated by these concessions, but the fact that Britain lacked the resolve to go to war in order to oppose German hegemony on the Continent. Since Hitler did not perceive that Chamberlain's policy was based on an analysis of German intentions that was later altered by that country's seizure of non-German Czechoslovakia, the Nazi leader in September 1939 could not under- stand why British policy would be any different from what it had been in September 1938.

The problem in all this was summed up by Nikolai Bukharin, the

Communist ideologue before the great purges in the mid-1930s. "Imperialism is a policy of conquest," he wrote. "But not every policy of conquest is imperialism."[177] In other words, there is a distinction between a policy of conquest operating within the existing status quo and one seeking to overthrow it. The distinction is not academic. "Appeasement is a foreign policy," Hans Morgenthau has pointed out in this regard, "that attempts to meet the threat of imperialism with methods appropriate to a policy of the status quo."[178]

To make such a distinction in a given situation is extremely difficult, as we have seen in the case of Hitler. How could anyone gauge with any degree of certainty what his ultimate objectives were? From 1935 on, the Nazi leader piled demand upon demand, each of which taken by itself could be fully reconciled with a policy of status quo within adjustments of the Versailles territorial provisions; yet each of which might be a step on the path to empire. The individual moves in themselves were ambiguous and, as a result, did not reveal the actual nature of the policy of which they were components and which were best summarized in a confidential conversation by the Nazi leader during this period. "The struggle against Versailles," he said, "is the means, but not the end of my policy. I am not in the least interested in the former frontiers of the Reich. The recreation of pre-war Germany is not a task worthy of our revolution."[179]

The distinction is too fine for the modern world. As soon as a state begins to exhibit some of the characteristics of craziness, deterrence should be used to ensure not only that the state does not commit aggression, but that it does not significantly improve its external action capabilities. In order for this concept to work, there has to be at least a modicum of instrumental rationality on the part of the nation to be deterred. Absent this, deterrence factors such as fear and uncertainty would be irrelevant, since they suggest consequences – a link between means and ends. But such extremes, as demonstrated for most of the peacetime history of the Third Reich, hardly ever obtain even in the craziest of states.

At the same time, there should be some appreciation of the fundamental values of a crazy state, no matter how bizarre in Western perceptions, if there is to be effective deterrence. Without this appreciation, deterrence may be attempted by threatening punishment directed at values which, while important to the deterring nation, may be quite irrelevant to the crazy state. A boycott threat, for example, might be viewed as welcome aid by a crazy-state leader in isolating his country from outside influence while concurrently increasing the threat perception of the general population. Even limited military threats may help such a state mobilize its population by confirming the image of the external world that its leader wants the people to have.

In all this, credibility remains a key ingredient of deterrence that will have to be achieved through obvious actions which cannot be ignored even by highly-biased, ideologically-shaped perceptions of what constitutes reality. The consequences of not being deterred, in other words, must not be perceived by a crazy state as just possible or even highly probable. Those consequences must be assured, particularly if there is a high propensity for risk on the part of the nation to be deterred. In such cases, Thomas Schelling's concepts of clear thresholds and automated reactions in a conventional sense might apply, with visible, irreversible commitments to undertake clearly defined actions in clearly defined circumstances. In such circumstances, at some point in the build-up of deterrence forces, no matter where a state lies on the spectrum of craziness and no matter how great the differences in culture and mindset, attention must finally be paid, as Willy Loman's wife reminds the world in *Death of a Salesman*. "It is sometimes stated," Herman Kahn has pointed out in this regard,

> that even an adequate ... deterrent would not deter an irrational enemy. This might be true if irrationality were an all-or-nothing proposition. Actually, irrationality is a matter of degree and if the irrationality is sufficiently bizarre, the irrational decision-maker's subordinates are likely to step in. As a result, we should want a safety factor in ... deterrence systems so large as to impress even the irrational and irresponsible with the degree of their irrationality and therefore the need for caution.[180]

The Vietnam Metaphpor

The Vietnam experience has left two major lessons that are important in dealing with crazy states if deterrence fails. The first is that if action is decided upon, there should be clear purposes and definable military objectives. "No one starts a war," Clausewitz pointed out, "... without first being clear in his mind what he intends to achieve by that war and how he intends to conduct it."[181] The second lesson is that if ends are clear, means must serve them without succumbing to gradualism, or war by degree, as occurred in Vietnam. That conflict confirmed Clemenceau's observation that war is too serious to be left to generals in the Clausewitzian sense that military means must be governed by the political ends to which they are applied. But Vietnam also added the codicil that war is too serious to be left to the politicians in the sense that when ends are established, the military means must be used in a professional and decisive manner.[182]

Added to all this in terms of crazy states is the need to recognize their nature early on and to take direct action even before they become a visible threat. This imperative, in turn, may require that traditional approaches in international relations be jettisoned or, at the very least, altered. This applies particularly to the concept of national sovereignty, since most strategies to limit the growth of undeterrable crazy states may require, at one extreme, intervention in the domestic affairs of these countries. For example, external support for the German military conspirators in the late 1930s, before Hitler's foreign policy successes undermined their resolve and efforts, might have altered the course of events in Nazi Germany. At the other extreme, strategies against crazy states may require the outright destruction of particularly dangerous potential external action capabilities in peacetime, as was accomplished in the 1981 Israeli bombing of the Osiraq reactor, the wartime analogue of which were the Vemork heavy water raids in 1943–44.[183]

For the United States, however, it is more likely in the near future that military action against a crazy state will be conducted after that state has committed an act of aggression or, as in the limited case of Libya in 1986, an act of terrorism. Moreover, as the war with Hitler's Germany and the Gulf conflict both illustrate, such full-scale fighting against a crazy state is no different in most respects from war with a "normal" foe, given the usual problems ranging from coalition maintenance and strategy at the highest level to "beans and bullets" necessities at the lowest. Irrational fanaticism and martyr tendencies, of course, constitute important differences; but where the fighting actually occurs, a manifestation of these tendencies, such as human wave attacks, cannot stand up to modern technology, as the Japanese learned in the Second World War and the Chinese in Korea. At the highest level of political–military decision-making, however, crazy tendencies can have a vital impact in terms of ending the war.

The process of conflict termination can be superimposed on top of the pure rationality model, with all belligerents as unitary actors acting on all necessary information to make the proper cost–benefit calculations concerning the decision to end a war. "As war is no act of blind passion, but is dominated by the political objective," Clausewitz summed up the instrumental rationality of this approach,

> therefore the value of that objective determines the measure of sacrifices by which it is to be purchased ... not only as regards

extent, but also as regards duration. As soon, therefore, as the
required outlay becomes so great as that the political object is no
longer equal in value, the object must be given up, and peace will
be the result.[184]

The reality, of course, is much different, particularly in the case of
crazy states where a leader's *Weltanschauung* may completely govern
events. In victory, for instance, behavior may vary from Hitler's
establishment of Vichy France to his brutal occupation of Poland or
Hussein's similar treatment of occupied Kuwait. In defeat, there can
be an all-encompassing success through failure, a last stand approach.
"Remember," one Arab analyst has pointed out in this regard, "all
our heros were defeated but brave."[185]

In a similar manner, there is the familiar romantic, *Götterdäm-
merung*-like vision of conflict termination associated with Hitler. Key
to that vision was national resistance until the final *Untergang* of an
undeserving nation, a belief to which Hitler held with remarkable
consistency. "Unfortunately," he wrote in 1925, "the military defeat
of the German people is not an undeserved catastrophe, but the
deserved chastisement of eternal retribution. We more than deserved
this defeat."[186] And 20 years later in the closing days of the the Second
World War, the Nazi leader again returned to this theme. "If the war
is to be lost, the nation will also perish," he stated, adding that ". . . it is
better to destroy it and to destroy it ourselves."[187]

Nevertheless, there is a danger in allowing Hitler's vision to dominate an analysis of anticipated behavior by the leader of a crazy state
in the face of military defeat. Apparently irrational resistance, for
example, may be nothing more than holding out for better terms or
until the situation changes. How else to explain the British resistance
in 1940 against overwhelming odds before the United States and the
Soviet Union had entered the war; or the Russian defense in late 1941
with German troops on the outskirts of Moscow.[188]

Moreover, there is evidence that within his own cognitive construct,
no matter how far that construct was removed from reality, Hitler had
rational reasons for delaying the end of the war. To begin with, there
was the hope that his "miracle" surprise weapons could turn the tide –
a hope, as we have seen, unfulfilled because of the instrumentally
irrational impact of his leadership on the generation of external action
capabilities. Added to this was the expectation that the allied Grand
Alliance would fall apart, that sooner or later the English-speaking
Allies would join with Germany to keep the Soviet "invaders" out of
Europe. "We'll fight," he stated in August 1944 in this regard, "until

we get a peace which secures the life of the German nation." And later that same day, the Nazi leader rationally weighed the political–military variables that might allow such an outcome. "The time hasn't come for a political decision," he concluded, adding that "... it is childish and naive to expect that at a moment of grave military defeats the moment for favorable political dealing has come. Such moments come when you are having success."[189]

The Summing Up

The picture of Hitler at the end is not that of a raging maniac as was sometimes portrayed by his generals, but instead of a man roughly in control of his senses, making reasoned calculations.[190] There was, for example, his prescient view in his last testament in April 1945 of the post-war world. "With the defeat of the Reich and pending the emergence of the Asiatic, the African, and perhaps the South American nationalisms," he stated, "there will remain in the world only two Great Powers capable of confronting each other – the United States and Soviet Russia. The laws of both history and geography will compel these two powers to a trial of strength, either military or in the fields of economies and ideology."[191] To point all this out is in no way a vindication of the Nazi leader; rather it is a stronger indictment, for as Martin van Creveld has pointed out, "a maniac is not responsible for his actions. Hitler was."[192]

This picture of the Nazi leader also demonstrates how mixed the signals can be from a crazy state. Here again, the Munich analogy is a reminder of how difficult it is for nations, much less collective security organizations like the League of Nations, to come to agreement on the nature or even the existence of a threat from a crazy state. The lesson from the Gulf conflict, in this regard, is that obvious aggression must already have occurred by a crazy state and involve key strategic resources before global action is taken – and even then only with determined US leadership. That this pattern is not the optimal way to deal with crazy states in the future is self-evident, particularly when a score or more of potentially crazy nations are already or are becoming major military powers just as both superpowers are declining in conventional strength.[193]

All this suggests preventive action while the crazy state is in *status nascendi*. But in terms of collective security, there is no reason to expect that the United Nations will be any more successful in coming to agreement on the ambiguous signals of early craziness than its predecessor. If anything, with the proliferation of states and the decline of international standards of behavior in the post-Second

World War era, the United Nations is likely to be more tolerant of early signs of craziness in a particular nation.

In the end, the United States will have to continue to take the lead in terms of setting the pattern in the post-cold war era for dealing with crazy states. Part of that effort will involve building up global or regional coalitions specifically targeted on crazy or rogue nations. These coalitions might focus on preventing the conditions which spawn such states, or on ensuring the cessation of imports that might improve their external action capabilities. The most important ingredient in these organizations will be deterrence – the obvious and credible willingness to use military power to prevent the growth and expansion of crazy states. The lesson from Vietnam, as we have seen in this regard, is not that this element of power has little utility in the modern world, but that it has to be used with other power elements as an instrumentally rational link to clearly defined political objectives.

Future cutbacks in US military capabilities will make it harder to maintain that link in dealing with crazy states. Nevertheless, the effort will have to be made. The problem of crazy states will not disappear. There will always be leaders like Hitler, able to bring a kind of rational order to their nations based on force, action and brutality, while ignoring the basic fabric of their own societies. For the majority of their populations, as it was for the citizens of Nazi Germany, this ordered framework will be enough even absent the societal fabric. The relationship between the strands that make up that fabric have grown too complicated in modern societies; and the great simplifiers in the tradition of Adolf Hitler stand ready in potential crazy states to abandon the weaving of these strands into a meaningful whole, slicing instead through the Gordian knot to offer apparent answers to what seem otherwise insoluble problems.

NOTES

This essay was first published in 1991 by the Strategic Studies Institute, US Army War College.

1. Henry Kissinger, *A World Restored* (Gloucester, MA: Peter Smith, 1964), pp. 2–3. Robert Jervis, *Perception and Misperception in International Politics* (Princeton: Princeton University Press, 1976), p. 271.
2. Joseph Frankel, *The Making of Foreign Policy. An Analysis of Decision Making* (London: Oxford University Press, 1968), p. 171. "In the nineteen-thirties Hitler established his position by a series of carefully planned movements," Eden wrote to Eisenhower during the Suez crisis. "... Similarly the seizure of the Suez Canal is, we

are convinced, the opening gambit in a planned campaign." Fred C. Ikle, *How Nations Negotiate* (New York: Harper & Row, 1964), p. 158. Looking back on the Korean War, Truman wrote:

> In my generation, this was not the first occasion when the strong had attacked the weak. I recalled some earlier instances: Manchuria, Ethiopia, Austria. I remembered how each time that the democracies failed to act it had encouraged the aggressors to keep going ahead. Communism was acting in Korea just as Hitler, Mussolini, and the Japanese had acted ten, fifteen, and twenty years earlier. I felt certain that if South Korea was allowed to fall Communist leaders would be emboldened to override nations closer to our own shores. If the Communists were permitted to force their way into the Republic of Korea without opposition from the free world, no small nation would have the courage to resist threats and aggression by stronger Communist neighbors. If this was allowed to go unchallenged it would mean a third world war, just as similar incidents had brought on the second world war.

Harry S. Truman, *Memoirs*, Vol. 2: *Years of Trial and Hope* (Garden City, NY: Doubleday, 1956), pp. 332–3.

3. David Halberstram, *The Best and the Brightest* (Greenwich, CT: Fawcett, 1973), p. 729. For British analogies to Hitler during the Falklands War, see Richard Ned Lebow, "Miscalculation in the South Atlantic: The Origins of the Falklands War," *Psychology and Deterrence*, eds. Robert Jervis, Richard Ned Lebow, Janice Gross Stein (Baltimore: Johns Hopkins University Press, 1985), pp. 115–16.

4. Ann Devroy, "Bush Denies Preparing the U.S. for War," *The Washington Post*, 2 November 1990, p. A1. See also Charles Krauthammer, "Nightmare From the '30s," ibid., 27 July 1990, p. A27. George F. Will, "Tribes with Flags," ibid., 8 August 1990, p. A21. For a counterview, see Rowland Evans and Robert Novak, "Overkill on Saddam," ibid., 8 August 1990, p. A21.

5. Donald Kaul, "We're still trying to stop Hitler," *The Carlisle Sentinel*, 30 August 1990, p. C3.

6. Yehezkel Dror, *Crazy States. A Counterconventional Strategic Problem* (Lexington, MA: Heath, 1971).

7. For a similar model, see Yehezkel Dror, *Public Policymaking Reexamined* (Scranton, PA: Chander Publishing, 1968), p. 134. It should be noted that the pure rationality model implies a perfect strategy (in the game-theory sense) which, prior to any action being implemented, also includes a means for readjusting the policy to the extent that the predicted results require. Thus no feedback mechanism is required in pure rational policy-making. In the real world, however, where actual results often are significantly different from expected results, feedback is also necessary during and after policy-making ends and may even be continuous and iterative in order to revise policies and improve the process. Ibid., p. 161. See also James G. March and Herbert A. Simon, *Organizations* (New York: Wiley and Sons, 1958), pp. 137ff. and James E. Dougherty and Robert L. Pfaltzgraff, Jr., *Contending Theories of International Relations* (New York: Lippincott, 1971), p. 331. The model incorporates Max Weber's three criteria for rational decision-making. See Max Weber, *The Theory of Social and Economic Organizations*, trans. A.M. Henderson and Talcott Parsons (New York: Norton, 1947), pp. 115–18 and *Methodology of the Social Sciences*, trans. E.A. Shils and H. Finch (Glencoe, IL: Free Press, 1949), pp. 52–3.

8. Weber, *Theory*, p. 92. Exceptions are some problems that are susceptible to quantification such as those concerning inventory or replacement policies. Dror, *Public Policymaking*, p. 133. See also G.H. Snyder, *Deterrence and Defense* (Princeton: Princeton University Press, 1961), p. 25, and Sidney Verba, "Assumptions of Rationality and Non-Rationality in Models of the International System," *The International System. Theoretical Essays*, eds. Klaus Knorr and Sidney Verba, 2nd printing (Princeton: Princeton University Press, 1969), p. 107.

9. On ideology, see Seweryn Braler, *Stalin's Successes: Leadership, Stability and Change*

in the Soviet Union (New York: Cambridge University Press, 1980); on culture and national character, see Gabriel Almond, *The American People and Foreign Policy* (New York: Praeger, 1960).

10. Harold and Margaret Sprout, *Foundations of International Politics* (Princeton, NJ: Van Nostrand, 1966), and Nicholas J. Spykman, America's *Strategy in World Politics: The United States and the Balance of Power* (New York: Harcourt Brace, 1942).

11. Graham T. Allison, *Essence of Decision. Explaining the Cuban Missiles Crisis* (Boston: Little, Brown, 1971), and Roger Hilsman, *The Politics of Policy Making in Defense and Foreign Affairs* (Englewood Cliffs, NJ: Prentice-Hall, 1987).

12. Carl von Clausewitz, *On War*, trans. Michael Howard and Peter Paret (Princeton: Princeton University Press, 1976), p. 605. Charles Reynolds, *The Politics of War. A Study of the Rationality of Violence in Inter-State Relations* (New York: St. Martin's Press, 1989), p. 69. Colonel (Ret.) Arthur Lykke has schooled an entire generation of U.S. Army War College students on his ends-ways-means paradigm for analyzing strategy at any level. For the strategists, this conceptual approach offers a much more comprehensive way to examine strategic problems and risk than the normal ends-means terminology used in strategic literature in which "means" normally incorporates Lykke's use of "ways" and "means." Thus, the definition of strategy as the calculated relationship of ends and means is also in instrumental terms the definition of rationality. Strategy, then, if executed properly, is by definition rational. Arthur F. Lykke, "Defining Military Strategy," *Military Review*, Vol. LXIX, No. 5 (May 1989), pp. 2–8.

13. Dror, *Crazy States*, p. 19. But see Reynolds, p. 49, who argues that as long as those entrails contributed to Caesar's reasoning, helping to connect means and ends in a commensurate pattern, the instrumental sense of rationality was present.

14. Ken Booth, *Strategy and Ethnocentrism* (New York: Holmes & Meier, 1979), p. 64.

15. "Muslim perceptions would be guided more by passions than by Western expectations of logic and rationality." Mowahid H. Shah, "Perils of War: A Muslim View," *Christian Science Monitor*, 5 November 1990, p. 18. Michael I. Handel, *The Diplomacy of Surprise. Hitler, Nixon, Sadat* (Cambridge, MA: Center for International Affairs, Harvard University, 1981), pp. 92–3. Richard K. Betts, *Surprise Attack. Lessons for Defense Planning* (Washington, DC: Brookings Institution, 1982), p. 130. Warfare is "an extension of culture, as well as of politics." Booth, p. 74. Reynolds, pp. 54, 56 and 90. It should be noted that a cultural failure to perceive what objectively is instrumental rationality on the part of one nation does not lessen that rationality. It is incumbent on other nations to move beyond ethnocentric limits if the rationality is to be appreciated.

16. Roberta Wohlstetter, *Pearl Harbor: Warning and Decision* (Stanford: Stanford University Press, 1962), pp. 349 and 354.

17. Booth, p. 64.

18. Reynolds, p. 57; Frankel, p. 174.

19. Abraham Kaplan, "Some Limitations on Rationality," *Rational Decision. Nomos VII. Yearbook for the American Society for Political and Legal Philosophy*, ed. Carl J. Friedrich (New York: Atherton Press, 1964), p. 57.

20. Ibid.

21. Felix E. Oppenheim, "Rational Decisions and Intrinsic Valuations," ibid., p. 220. Dror, *Public Policymaking*, p. 166. Carl J. Friedrich, "On Rereading Machiavelli and Althusius: Reason, Rationality, and Religion," ibid., p. 179. See also Philip Green, *Deadly Logic. The Theory of Nuclear Deterrence* (Columbus, OH: Ohio State University Press, 1966), p. 191, who points out the difficulty in even finding a "national value set."

22. Lord Vansittart, *The Mist Procession* (London: Gollancz, 1958), pp. 468–9 and 536. Frankel, p. 164.

23. Isaiah Berlin, *Historical Inevitability* (London: Oxford University Press, 1955), pp. 76–7. Green, p. 224. Frankel, p. 163. Berlin concludes that there have to be some values, however general and however few, "that enter into the normal definition of

what constitutes a sane human being" and that absent these values or ends, beings "can scarcely be described as human; still less as rational." Thus, in extreme cases, "pursuit, or failure to pursue, certain ends can be regarded as evidence of – and in extreme cases part of the definition of – irrationality." Isaiah Berlin, "Rationality of Value Judgments," *Nomos VII*, p. 223.

24. Dror, *Crazy States*, p. 23; Booth, p. 69.
25. O.R. Holsti cited in Margaret G. Hermann, "Effects of Personal Characteristics of Political Leaders on Foreign Policy," *Why Nations Act. Theoretical Perspectives for Comparative Foreign Policy Studies*, eds. Maurice A. East, Stephen A. Salmore, Charles F. Hermann (London: Sage, 1978), p. 50.
26. G.D. Paige, *Political Leadership* (New York: Free Press, 1972), p. 69.
27. Norman Rich, *Hitler's War Aims*, Vol. I (New York: Norton, 1973), p. 11. Klaus Hildebrand, *The Foreign Policy of the Third Reich*, trans. Anthony Fathergill (Berkeley: University of California Press, 1973), and Andreas Hillgruber, *Hitlers Strategie. Politik und Kriegsführung 1940–1* (Frankfurt: Bernard and Graefe, 1965). See also Ian Kershaw, *The Nazi Dictatorship. Problems and Perspectives of Interpretation* (London: Arnold, 1985), pp. 61–2.
28. Joachim C. Fest, *The Face of the Third Reich. Portraits of the Nazi Leadership* (New York: Pantheon Books, 1970), p. 302. The earlier works were: Ernst Fraenkel, *The Dual State* (New York: Oxford University Press, 1942), and Franz Neumann, *Behemoth. The Structure and Practice of National Socialism* (London: Oxford University Press, 1942). See also Kershaw, p. 65.
29. Hans Mommsen, *Beamtentum im Dritten Reich* (Stuttgart: DVA, 1966); Peter Diehl-Thiele, *Partie und Staat im Dritten Reich* (Munich: Beck, 1969); Peter Hüttenberger, *Die Gauleiter. Studie zum Wandel des Machtgefüges in der NSDAP* (Stuttgart: DVA, 1969), and Edward N. Petersen, *The Limits of Hitler's Power* (Princeton: Princeton University Press, 1969). See also Kershaw, p. 66.
30. O.R. Holsti, "The Belief System and National Images: A Case Study," *Journal of Conflict Resolution*, No. 6, 1962, p. 244.
31. Alexander L. George, "The 'Operational Code': A Neglected Approach to the Study of Political Leaders and Decision-Making," *International Studies Quarterly*, Vol. 13, No. 2, June 1969, p. 197.
32. Ibid., p. 221. See also ibid., pp. 201–217.
33. Ibid., p. 200.
34. Ibid., pp. 195 and 197. See also Alfred L. Rowse, *Appeasement* (New York: Norton, 1961), pp. 116–17, who pointed out in respect to British statesmen:

> Not one of these men in high place [sic] in those years ever so much as read *Mein Kampf* or would listen to anybody who had. They really did not know what they were dealing with, or the nature and degree of the evil thing they were up against. To be so uninstructed – a condition that arose in part from a certain superciliousness, a lofty smugness, as well as superficiality of mind – was in itself a kind of dereliction of duty.

35. Walter C. Langer, *The Mind of Adolf Hitler. The Secret Wartime Report* (New York: Basic Books, 1972), p. 74.
36. Frankel, p. 115. One example of motivation could be emotional insecurity or low self-esteem on the part of the leader who then compensates with a drive for power. Harold Lasswell, *Power and Personality* (New York: Norton, 1948). In this construct, the greater this need for power by the national policy-maker, the more forceful and aggressive his nation's foreign policy will be. D.G. Winter, *The Power Motive* (New York: Free Press, 1973).
37. Langer, p. 126. "Probably none of us is entirely 'normal'," Hitler noted in 1941. Adolf Hitler, *Hitler's Secret Conversations, 1941–1944*, trans. Norman Cameron and R.H. Stevens (New York: Farrar, Straus, and Young, 1953), 28–29 December 1941, p. 127. William Carr, "The Hitler Image in the Last Half-Century," H.W. Koch, ed., *Aspects of the Third Reich* (London: Macmillan, 1985). Examples of Hitler psycho-histories:

Rudolf Binion, *Hitler Among the Germans* (New York: Elsevier, 1976), and Robert Waite, *Adolf Hitler. The Psychopathic God* (New York: Basic Books, 1977). In terms of Hitler, one historian has noted that even

> if the finds were less dependent on conjecture and speculation, it is difficult to see how this approach could help greatly in explaining how such a person could become ruler of Germany and how his ideological paranoia came to be implemented as government policy by nonparanoids and nonpsychopaths in a sophisticated bureaucratic system.

Kershaw, p. 63. Another critic has inquired sarcastically, in this regard:

> Does our understanding of National Socialist politics really depend on whether Hitler had only one testicle?...Perhaps the Fuehrer had three, which made things difficult for him – who knows?...Even if Hitler could be regarded irrefutably as a sadomasochist, which scientific interest does that further? ... Does the "final solution of the Jewish question" thus become more easily understandable or the "twisted road to Auschwitz" become the one-way street of a psychopath in power?

Hans-Ulrich Wehler, "Psychoanalysis and History," *Social Research*, No. 47, 1980, p. 531.

38. Booth, p. 13.
39. Ibid., p. 20. The problem of projecting assumptions about rational behavior is compounded when organizational models and governmental politics are added to the unitary actor model. Bureaucracies, for instance, have their own cultural idiosyncratic characteristics, and there is always the danger of projecting one norm of bureaucratic behavior onto those of another nation. Ibid., p. 65.
40. Ibid., p. 94. On "unrealistic, exaggerated diabolism," see Ralph K. White, *Nobody Wanted War: Misperceptions in Vietnam and Other Wars* (Garden City, NY: Doubleday, 1968), p. 318. See also Alix Strachey, *The Unconscious Motives of War* (New York: International Universities Press, 1957), pp. 202–3.
41. Booth, p. 95. "In general the art of all truly great national leaders at all times consists among other things primarily in not dividing the attention of a people, but in concentrating it upon a single foe." Adolf Hitler, *Mein Kampf*, trans. Ralph Manheim (Boston: Houghton Mifflin, 1943), p. 118.
42. Herman Rauschning, *Hitler Speaks. A Series of Political Conversations with Adolf Hitler on his Real Aims* (London: Thornton Butterworth, 1939), p. 137.
43. Chester Wilmot, *The Struggle for Europe* (London: Collins, 1952), p. 21. Booth, pp. 85–6. Ikle, pp. 82 and 104.
44. Booth, p. 60. Robert Jervis, *Perceptions and Misperceptions in International Politics* (Princeton: Princeton University Press, 1976), p. 193, and *The Logic of Images in International Relations* (Princeton: Princeton University Press, 1970), p. 205. Hans J. Morgenthau, *Politics Among Nations. The Struggle for Power and Peace* (New York: Knopf, 1967), p. 63.
45. Alan Bullock, *Hitler. A Study in Tyranny* (New York: Harper & Row, 1962), p. 528.
46. For a similar model, see David Braybrooke and Charles E. Lindblom, *A Strategy of Decision. Policy Evaluation as a Social Process* (London: Free Press of Glencoe, 1963), p. 67. "Most of the actions taken by bureaucracies involve doing again or continuing to do what was done in the past. In the absence of some reason to change their behavior, organizations keep doing what they have been doing." M.H. Halperin, *Bureaucratic Politics and Foreign Policy* (Washington, DC: Brookings Institution, 1974), p. 99.
47. Gordon A. Craig, "The German Foreign Office from Neurath to Ribbentrop," *The Diplomats*, Vol. 2, *The Thirties*, eds. Gordon A. Craig and Felix Gilbert (New York: Antheneum, 1963), p. 410.
48. Ibid., p. 413.
49. Braybrooke and Lindblom, p. 68.

50. Albert Speer, *Inside the Third Reich*, trans. Richard and Clara Winston (New York: Macmillan, 1970), p. 169.

51. Ole R. Holsti, *Crisis Escalation War* (Montreal: McGill-Queen's University Press, 1972), p. 208. Booth, p. 116. See also Leon Festinger, *Theory of Cognitive Dissonance* (Stanford: Stanford University Press), 1957. "As long as Prussia limited herself to purely European foreign policy aspirations," Hitler wrote as early as 1928, "she had no serious danger to fear from England." Adolf Hitler, *Hitler's Secret Book*, trans. Salvator Attanasio (New York: Bramhill House, 1986), p. 151.

52. Bullock, p. 550. See also D.C. Watt, "The Debate Over Hitler's Foreign Policy – Problems of Reality or Faux Problèmes?" *Deutsche Frage und europaisches Gleichgewicht*, eds. Klaus Hildebrand and Reiner Pommerin (Cologne: Boehlau, 1985), p. 165.

53. *Secret Conversations*, 4 August 1942, p. 496. See also ibid., 9 February 1942, p. 249; Bullock, pp. 771–2, and Reynolds, p. 126. See also Martin van Creveld, "War Lord Hitler: Some Points Reconsidered," *European Studies Review*, January 1974, p. 75, who points out that "one should keep in mind that Hitler never intended to fight a World War in the first place." As late as 31 August 1944, Hitler still could not get over British actions in 1939 as well as the British refusal to compromise after the German victories in the west in 1940. "I have proved that I did everything to come to some understanding with the English," he concluded. Felix Gilbert, ed., *Hitler Directs His War* (New York: Octagon Books, 1982), p. 105.

54. *Secret Conversations*, 25 September 1941, p. 35.

55. Heinz Guderian, *Panzer Leader*, trans. Constantine Fitzgibbon (New York: Dutton, 1952), p. 407; Bullock, p. 775.

56. There is a natural tendency to see decision-makers who lose as unreasonably tied to their views, disregarding correct information. But in most cases, those judged by history to have been right have demonstrated no more openness to new information nor willingness to modify their images than those like Hitler who were wrong. For example, the anti-appeasement policy of Robert Vansittart, the Permanent Undersecretary in the British Foreign Office, has generally been viewed as an example of foresight and courage. In fact, like Chamberlain, he fitted each bit of information about the Nazi regime, no matter how ambiguous, into his own hypothesis without an openminded analysis to see if the explanations proffered by the appeasers accounted better for the information than did his own preconceptions. It is often forgotten, in this regard, that Hitler in the 1930s was the exception that made the anti-appeaser's case. Under conservative leaders, Germany would not have run a high risk in order to regain a powerful European position and thus would have been appeased. "Had Hitler not come to power," Robert Jervis has pointed out, "many of the Englishmen who now seem wise would have been dangerous warmongers." Jervis, *Perception*, pp. 25 and 180. See also ibid., pp. 119–20, 128–30, 138, 151–3, 172 and 176; Verba, pp. 94–5; Ian Colvin, *Vansittart in Office* (London: Gollancz, 1965), p. 23; and Martin Gilbert and Richard Gott, *The Appeasers* (London: Weidenfeld and Nicolson), 1963, p. 34.

57. Walter Schellenberg, *The Schellenberg Memoirs*, ed. and trans. Louis Hagen (London: Deutsch, 1956), p. 209. Betts, p. 133.

58. Barton Whaley, *Codeword BARBAROSSA* (Cambridge: MIT Press, 1973), pp. 14–15, 50 and 251. See also Eberhard Jäckel, *Hitler's Weltanschauung. A Blueprint for Power* (Middletown, CT: Wesleyan University Press, 1972), p. 45, and Wolfgang Michalka, "From the Anti-Comintern Pact to the Euro-Asiatic Bloc: Ribbentrop's Alternative Concept of Hitler's Foreign Policy Programme," in Koch, p. 283, who points out that pivotal role of Hitler's perception of Britain's irrational stubbornness: Hitler was only able to explain the decision of the British government to continue the war despite its hopeless position and not to act upon his suggestions of an alliance by thinking that London was still counting on two allies, the United States and the Soviet Union, so that in time an enormous anti-Hitler coalition could be formed, aimed at gradually starving out and finally conquering the German Reich, in much the same way as in the last phase of the First World War.

59. Alan Clark, *Barbarossa: The Russo-German Conflict, 1941–45* (New York: Morrow, 1965), p.23. Betts, pp.133ff. Schellenberg, p.208. Wilhelm Keitel, *The Memoirs of Field Marshal Keitel*, ed. Walter Gorlitz, trans. David Irving (New York: Stein and Day, 1966), p.122.

60. Martin van Creveld, *Supplying War: Logistics from Wallenstein to Patton* (Cambridge: Cambridge University Press, 1977), pp.150–1 and 154.

61. Dror, *Crazy States*, p.60. R.C. Snyder and J.A. Robinson, *National and International Decision-Making* (New York: Institute for International Order, 1961), p.164.

62. Rauschning, p.280.

63. Ibid., p.81.

64. Ibid., pp.277, 280–1.

65. *Mein Kampf*, pp.516, 660; Rauschning, p.134.

66. Kurt Ludecke, *I Knew Hitler* (London: Unwin, 1938), pp.217ff. Joachim Fest, "On Remembering Adolf Hitler," *Encounter*, Vol.XLI, No.4 (October 1973), p.30. "I shall attain my purpose without a struggle, by legal means." Rauschning, p.107. "When I was younger," Hitler declared in 1941, "I thought it was necessary to set about matters with dynamite. I've since realized that there's room for a little subtlety." *Secret Conversations*, 13 December 1941, p.117.

67. Fest, "On Remembering," p.31, and *Face*, p.41.

68. Handel, pp.31–2.

69. *Secret Conversations*, 2 November 1941, p.88. "The basic notions that served us in the struggle for power have proved that they are correct, and are the same notions as we apply today in the struggle we are waging on a world scale." Ibid., 19 November 1941, p.110.

70. John Thomas Emmerson, *The Rhineland Crisis* (Ames, IA: Iowa University Press, 1977), p.101. Handel, p.52.

71. Rauschning, p.11. See also ibid., pp.109–10:

> I am willing to sign anything. I will do anything to facilitate the success of my policy. I am prepared to guarantee all frontiers and to make non-aggression pacts and friendly alliances with anybody. It would be sheer stupidity to refuse to make use of such measures merely because one might possibly be driven into a position where a solemn promise would have to be broken. There has never been a sworn treaty which has not sooner or later been broken or become untenable. There is no such thing as an everlasting treaty. Anyone whose conscience is so tender that he will not sign a treaty unless he can feel sure he can keep it in all and any circumstances is a fool. Why should one not please others and facilitate matters for oneself by signing pacts if the others believe that something is thereby accomplished or regulated? Why should I not make an agreement in good faith today and unhesitatingly break it tomorrow if the future of the German people demands it?

72. Emmerson, p.291. Handel, p.58.

73. Handel, pp.58, 60–1.

74. Fest, *Face*, p.183.

75. Herbert Feis, *The Road to Pearl Harbor* (Princeton: Princeton University Press, 1950), p.202. Jervis, *Logic*, p.106.

76. Frankel, p.172. Jervis, *Perception*, p.278. "When I first met him," the British ambassador remarked of Hitler, "his logic and sense of realities had impressed me, but as time went on he appeared to me to become more and more... convinced of his own infallibility and greatness." Sir Nevile Henderson, *Failure of a Mission* (New York: Putnam's, 1940), p.177.

77. Frankel, p.172. Fest, *Face*, pp.39 and 50, and "On Remembering," p.31. "I couldn't say whether my feeling that I am indispensable has been strengthened during this war. One thing is certain, that without me the decisions to which we today owe our existence would not have been taken." *Secret Conversations*, 13–14 October 1941, p.48.

78. Karl Dietrich Bracher, *Adolf Hitler* (Berne/Munich/Vienna: Droste, 1964), p. 12. *Fest*, "On Remembering," p. 26.
79. Fest, *Face*, p. 44.
80. *Secret Conversations*, 26–27 February 1942, p. 276.
81. Fest, *Face*, p. 83.
82. Jeremy Noakes and Geoffrey Pridham, *Documents on Nazism* (London: Unwin, 1974), p. 256. Kershaw, pp. 74–5.
83. Noakes and Pridham, p. 261. Kershaw, p. 75. Fest, *Face*, pp. 127–8 and 131. "What would happen to me if I didn't have around me men whom I completely trust, to do the work for which I can't find time?" *Secret Conversations*, 13–14 October 1941, p. 48.
84. Koch, pp. 184–5.
85. Noakes and Pridham, p. 245. Joseph Nyomarkay, *Charisma and Factionalism Within the Nazi Party* (Minneapolis: University of Minnesota Press, 1967), pp. 70–1. "The best organization is not that which inserts the greatest, but that which inserts the smallest, intermediary apparatus between the leadership of a movement and its individual adherents." *Mein Kampf*, p. 346.
86. "I am often urged to say something in praise of bureaucracy – I can't do it." *Secret Conversations*, 1–2 August 1941, p. 15. See also ibid., 26 August 1942, p. 336. "In olden times it was the strolling player who was buried in the public refuse-heap; today it is the lawyer who should be buried there. No one stands closer in mentality to the criminal than the lawyer." Ibid., 22 July 1942, p. 475.
87. Hugh R. Trevor-Roper, *The Last Days of Hitler* (London: Macmillan, 1947), p. 202.
88. Kershaw, p. 73. On the feudal aspects, see Robert Koehl, "Future Aspects of National Socialism," *Nazism and the Third Reich*, ed. Henry A. Turner, (New York: Quadrangle Books, 1972), pp. 151–74. See also Hans Mommsen, "National Socialism: Continuity and Change," *Fascism. A Reader's Guide*, ed. Walter Laquer, (London: Penguin, 1979), pp. 176–8, and Jane Coplan, "Bureaucracy, Politics and the National Socialist State," *The Shaping of the Nazi State*, ed. Peter D. Stachura, (London: Croom Helm, 1978), pp. 234–56.
89. John M. Collins, *Grand Strategy* (Annapolis, MD: Naval Institute Press, 1973), p. 277.
90. Dror, *Crazy States*, p. 17.
91. Jervis, *Perception*, p. 52.
92. Speer, p. 72. General von Blomberg pointed out after the war that if the French had resisted, the Germans would "have to have beat a hasty retreat." And Keitel confided that "he wouldn't have been a bit surprised" if three battalions of French troops had flicked the German forces right off the map. G.M. Gilbert, *The Psychology of Dictatorship* (New York: Ronald Press, 1950), p. 211. But see also Emmerson, p. 105, who concludes that the "French army of 1936 had no strike force capable of marching as far as Mainz, to say nothing of occupying the whole of the demilitarized zone. Nor did it possess a single unit which could be made instantly combat ready."
93. "In the past the French produced a Talleyrand and a Fouché; today they have become humdrum and circumspect, a nation of dried-up clerks. They will venture to play for halfpence, but no longer for a great stake." Rauschning, p. 277. For additional measures Hitler took to lower the risk by keeping the operation as unprovocative as possible, see Emmerson, p. 97, and Handel, p. 61.
94. This did not mean, of course, that Hitler was not nervous. "The forty-eight hours after the march," he stated, "... were the most nerve-wracking in my life." Bullock, p. 345. But see *Secret Conversations*, 27 January 1942 and 21 May 1942, pp. 211–12 and 406–7.
95. Jervis, *Logic*, p. 205, and *Perception*, p. 52.
96. Original emphasis. Franz Halder, *The Halder Diaries. The Private War Journals of Colonel General Franz Halder*, ed. Arnold Lissance (Boulder, CO, and Dunn Loring, VA: Westview Press and T.N. Dupuy Associates, 1976), Vol. I, p. 8.
97. A.J.P. Taylor, "War Origins Again," *The Origins of the Second World War*, ed. Esmonde Robertson (London: Macmillan, 1971), p. 139.
98. Jervis, *Perception*, p. 53.

99. *Secret Conversations*, 5–6 January 1942, p. 150.
100. Ibid., 17–18 October 1941, p. 58.
101. Betts, p. 130.
102. Ibid., p. 145. Telford Taylor, *The March of Conquest. The German Victories in Western Europe, 1940* (New York: Simon & Schuster, 1958), p. 158.
103. *Secret Conversations*, 5–6 January 1942, p. 150.
104. Ibid., 26 August 1942, p. 537. "The more we see of conditions in Russia, the more thankful we must be that we struck in time." Ibid., 22 July 1942, p. 476. "If to-day you do harm to the Russians, it is so as to avoid giving them the opportunity of doing harm to us." Ibid., 23 September 1941, p. 33.
105. David Irving, *Hitler's War* (New York: Viking Press, 1977), p. 180. See also H.W. Koch, "Hitler's Programme and the Genesis of Operation Barbarossa," in Koch, pp. 319–20, Betts, pp. 143–4, and Watt, p. 163.
106. Betts, pp. 137–8 and 144.
107. Ibid., p. 144. See also Samuel p. Huntington, *The Common Defense* (New York: Columbia University Press, 1961), p. 206, who noted that "risks cannot be calculated, they can only be felt."
108. *Fuehrer Conferences on Naval Affairs 1939–1945* (London: Greenhill Books, 1990), pp. 119 and 136.
109. *Secret Conversations*, 13 October 1941, p. 46.
110. Bullock, p. 754.
111. For a similar diagram and description, see Dror, *Public Policymaking*, pp. 149, 151–2. See also Annatol Rapoport and A.M. Gharmmah, *Prisoner's Dilemma* (Ann Arbor: University of Michigan, 1965).
112. Eugene E. Jennings, *An Anatomy of Leadership* (New York: Harper & Brothers, 1960), p. 98. Dror, *Public Policymaking*, p. 158.
113. Rauschning, p. 6.
114. Langer, pp. 33 and 159. *Mein Kampf*, p. 29.
115. Original emphasis. Rauschning, p. 181.
116. Ibid., p. 181.
117. Ibid.
118. Langer, p. 29. Bullock, p. 378.
119. *Secret Conversations*, 17 September 1941, p. 26. Jennings, pp. 73–4. "Only the man who acts becomes conscious of the real world." Rauschning, p. 224.
120. Fest, *Face*, p. 187.
121. Fest, "On Remembering," p. 24.
122. Ibid., p. 25. *Mein Kampf*, pp. 234 and 459. Fest, *Face*, p. 300. See H.A. Trevor-Roper, "The Mind of Hitler," *Secret Conversations*, pp. xxv-xxvi, who points out that "Hitler was more practical than his own doctrinaire followers," believing "the general truth of the doctrine," but lacking any "patience with the theological niceties" of such priests as Himmler and Rosenberg. Rosenberg never realized that the ideology he believed in so fervently carried no weight in the centers of power, as shown by Hitler's musings during the war:

> I must insist that Rosenberg's "The Myth of the Twentieth Century" is not to be regarded as an expression of the official doctrine of the Party. The moment the book appeared, I deliberately refrained from recognising [sic] it as any such thing. ... It is interesting to note that comparatively few of the older members of the Party are to be found among the readers of Rosenberg's book, and that the publishers had, in fact, great difficulty in disposing of the first edition. ... It gives me considerable pleasure to realise [sic] that the book has been closely studied only by our opponents. Like most of the Gauleiters, I have myself merely glanced cursorily at it. It is in any case written in much too abstruse a style, in my opinion.

Ibid., 11 April 1942, p. 342.
123. Weber, *Essays in Sociology*, pp. 116–17. Hilsman, pp. 66–7.

124. One manifestation of this trend was grotesquely illustrated long before the *Machtüber-nahme* by death notices in which Hitler replaced God. And after 1933, there was the prayer used in Nazi day-care centers that ended: "Fuehrer, my Fuehrer, my faith, my light." Fest, *Face*, pp. 41, 188, 367. Nevertheless, there was more than just a simplistic deistic connection between the *Führer* and the *Geführten*. For it was chiefly in an "inner" sense, as Karl Dietrich Bracher has pointed out,

> that Hitler was presented as the revealer of a new meaning to life, one which absorbed his followers' needs to surrender themselves, to serve him and to submit to him, to shed their weariness of responsibility, and as one who alone was capable of translating this need into the release of political action. He was the incarnation of the 'people's community'; thanks to his intuition and leadership talent he was always right, he was the unchallengeable interpreter of their interests. For this reason he was not subject, even vis-a-vis his own followers, to any rules of law.

Karl Dietrich Bracher, *The Age of Ideologies, A History of Political Thought in the Twentieth Century* (New York: St Martin's, 1984), p. 122.

125. Fest, *Face*, p. 46.
126. Ibid., p. 49. See also Dorothy Thompson's description quoted in Langer, p. 57.

> At Garmisch I met an American from Chicago. He had been at Oberammer-gau, at the Passion Play. "These people are all crazy," he said. "This is not a revolution, it's a revival. They think Hitler is God. Believe it or not, a German woman sat next to me at the Passion Play and when they hoisted Jesus on the Cross, she said, 'There he is. That is our Fuehrer, our Hitler.' And when they paid out the thirty pieces of silver to Judas, she said: 'That is Roehm, who betrayed the Leader.'

127. *Secret Conversations*, 20 August 1942, p. 524.
128. Kershaw, p. 76. Petersen, p. 7. See also Lothar Kettenacker, "Social and Psychological Aspects of the Führer's Rule," in Koch, pp. 101–2, who emphasizes the link between charismatic rule and the internal chaos of the Third Reich. It was Hitler's role of "overlord" of the underbureaucratic, almost neo-feudalistic system of government that "appealed to some atavistic instincts of a still strongly dynastic oriented society."
129. Herman Kahn, *On Escalation* (New York: Praeger, 1965), p. 57. Reynolds, p. 106.
130. Thomas Schelling, *Arms and Influence* (New Haven: Yale University Press, 1966), p. 37 and *The Strategy of Conflict* (Cambridge: Harvard University Press, 1960), p. 17. See, however, Green, p. 164, who was obviously not considering the concept of crazy states when he noted of Schelling's examples that "one can only say that anything is possible but that ordinarily we hope for a little more assurance about the way the arts of statesmanship are being practiced than we would find in an institution for the mentally ill."
131. *Mein Kampf*, p. 18.
132. Bullock, p. 528. By this time, Sir Nevile Henderson, the British Ambassador, believed Hitler was a psychopath, given to extreme and sudden fits of rage and often influenced by sudden bursts of intuition. Felix Gilbert, "Two British Ambassadors: Perth and Henderon," *The Diplomats*, p. 543. On Hitler's reputation with his subordinates as a *Teppichfresser*, see William L. Shirer, *Berlin Diary* (New York: Knopf, 1941), p. 137. All of Hitler's anger, of course, was not just pure calculation. In 1938, for instance, Hitler built up an extraordinary rage against Czechoslovakia, marked by the violence not only of his language but even of his internal communications. Watt, p. 164.
133. Dror, *Crazy States*, pp. 23–24.
134. Hillgruber, Jäckel, and Hildebrand. As an example, Jäckel states that by the completion of his 1928 unpublished second book, Hitler's conception of foreign policy, although "still riddled with contradictions and absurdities," was complete. It was a concept already demonstrating "a high degree of purposeful orientation, consistency, and coherence. Here we have clearly defined political goals and an indication

of the means which might be used to strive for and possibly attain these goals." Jäckel, pp. 40 and 42. Kershaw, pp. 63–4 and 108.

135. Karl Dietrich Bracher, "The Stages of Totalitarian Integration (*Gleichschaltung*)," in Hajo Holborn, ed., *Republic to Reich. The Making of the Nazi Revolution* (New York: Praeger, 1973), p. 128, and *The German Dictatorship*, trans. Jean Steinberg (New York: Praeger, 1970), pp. 247–59.

136. Martin Broszat, *The Hitler State. The Foundation and Development of the Internal Structure of the Third Reich*, trans. John Hiden (London: Longman, 1981), p. 9, and "Soziale Motivation und Führer-Bindung des Nationalsozialismus," *Vierteljahrshefte für Zeitgeschichte*, 18 (1970), pp. 407–9. Kershaw, pp. 67 and 110–11.

137. Ibid., p. 85. Lucy Dawidowicz, *The War Against the Jews, 1933–1945* (New York: Harmondsworth, 1975), pp. 198–208.

138. Martin Broszat, "The Genesis of the 'Final Solution'," in Koch, p. 405. Hans Mommsen, "National Socialism: Continuity and Change," p. 179. Kershaw, p. 87.

139. *Secret Conversations*, 12 May 1942, p. 380. Kettenacker, pp. 129–30.

140. Kershaw, p. 100. For the speech, see Jäckel, p. 61. But see also Koch, p. 459.

141. Kershaw, p. 105.

142. Ibid., p. 104.

143. For the basic concept, see Dror, *Crazy States*, p. 43.

144. Ibid., pp. 39–45.

145. On the primacy of political decades over economics, see Tim Mason, "The Primacy of Politics – Politics and Economics in National Socialist Germany," in Turner, p. 175, who concludes "that both the domestic and foreign policy of the National Socialist government became, from 1936 onward, increasingly independent of the influence of the economic ruling classes, and even in some essential aspects ran contrary to their collective interests." For similar opinions, see Kershaw, p. 51, and Richard J. Overy, *The Nazi Economic Recovery 1932–1938* (London: Oxford University Press, 1982), p. 58. But see William Carr, *Arms, Autarky and Aggression* (London: Oxford University Press, 1979), 3rd ed., p. 65, who points out that "Ideological, strategic and economic factors are too closely intermeshed in a country's foreign policy to permit a clinical separation."

146. Burton H. Klein, *Germany's Economic Preparation for War* (Cambridge, MA: Harvard University Press, 1959), pp. 76–82.

147. Rauschning, p. 20. "Even to Schacht, I had to begin by explaining this elementary truth: that the essential cause of the stability of our currency was to be sought for in our concentration camps. The currency remains stable when the speculators are put under lock and key." *Secret Conversations*, 15 October 1941, p. 54. Klein, pp. 76–82.

148. Ibid. p. 46, and Koch, "Introduction to Part III," in Koch, pp. 328–329.

149. Ibid. Tim Mason, "The Legacy of 1918 for National Socialism," in Anthony Nicholls and Erich Matthias, eds., *German Democracy and the Triumph of Hitler. Essays in Recent German History* (London: Allen & Unwin, 1971), p. 231.

150. Rauschning, pp. 212–13. Fest, *Face*, p. 43. Mason, "Legacy," p. 230.

151. *Secret Conversations*, 28 July 1942, p. 487. Karl Hardach, *The Political Economy of Germany in the Twentieth Century* (Berkeley, 1980), pp. 76–9. Kershaw, p. 78. "The Fuehrer understands the Navy's difficulties.... He explains in detail how he must first of all prevent a collapse of any front where the enemy could substantially injure home territory." *Fuehrer Conferences*, 22 December 1942, p. 304.

152. Speer, p. 214. Mason, "Legacy," pp. 226–7 and 239.

153. Jäckel, p. 67.

154. Karl Manheim, *Man and Society in an Age of Reconstruction* (New York: Paterson, 1940), pp. 49–75. Green, pp. 216–17 and 315.

155. Mason, "Primacy," p. 195. Broszat, "Genesis," pp. 409–10. Kershaw, pp. 57–8.

156. Bernd Wegner, "The 'Aristocracy of National Socialism': The Role of the SS in National Socialist Germany," in Koch, p. 444.

157. David Irving, *The Mare's Nest* (Boston: Little, Brown, 1964), p. 304. See also Peter G. Cooksley, *Flying Bomb: The Story of Hitler's V Weapons in World War II* (New York:

Charles Scribner's Sons, 1979), and Basil Collier, *The Battle for the V-Weapons 1944–1945* (New York: Morrow, 1965). The code name for the SAM was "Waterfall": Speer, pp. 364–65. The Me-262 fighter plane, with two jet engines and a speed of over 500 miles per hour, was the most valuable of the German "secret weapons." In September 1943, however, without explanation, Hitler ordered preparations for large-scale production of the jet to stop. In January 1944, he ordered resumption of production, but directed that the plane, which was built to be a fighter, was to be used as a fast bomber. Despite objections, Hitler, not surprisingly, prevailed. But the element of useful surprise had been lost through this change in role and the concomitant production delays. Ibid., pp. 362–3.

158. David Irving, *The German Atomic Bomb. The History of Nuclear Research in Nazi Germany* (New York: Simon and Shuster, 1967), pp. 231 and 299. Green, p. 187. For similar problems in Britain concerning intragovernmental disputes over radar and strategic bombing, see C.P. Snow, *Science and Government* (Cambridge, MA: Harvard University Press, 1961).

159. Until the chemical industry was bombed in 1944, German production amounted to 3,100 tons of mustard gas and 1,000 tons of Tabun per month. Speer, p. 413. See also Stephen L. McFarland, "Preparing for What Never Came: Chemical and Biological Warfare in World War II," *Defense Analysis*, Vol. 2, June 1986, p. 113, and John Ellis van Courtland Moon, "Chemical Weapons and Deterrence: The World War II Experience," *International Security*, Vol. 8, No. 4, Spring 1984, p. 25.

160. *Mein Kampf*, pp. 201–2. See also ibid., p. 279, for the seeds of later policy in this experience.

> If at the beginning of the War and during the War twelve or fifteen thousand of these Hebrew corrupters of the people had been held under poison gas, as happened to hundreds of thousands of our very best German workers in the field, the sacrifice of millions at the front would not have been in vain. On the contrary: twelve thousand scoundrels eliminated in time might have saved the lives of a million real Germans, valuable for the future.

161. But in conversations during the early 1930s, Hitler pointed out that a "nation denied its rights may use any weapon, even bacterial warfare. I have no scruples, and I will use whatever weapon I require. The new poison gases are horrible. But there is no difference between a slow death in barbed-wire entanglements and the agonized death of a gassed man or one poisoned by bacteria." Rauschning, pp. 3–4.

162. Speer, pp. 413–14, and McFarland, p. 114. "Anyone," the German Army Chief of Gas Operations testified after the war, "who suggested in Germany that chemical warfare should be initiated would have been called 'an idiot' and 'crazy.'" Ibid., p. 113. Green, pp. 208–9.

163. Irving, *German Atomic Bomb*, p. 241.

164. Ibid., pp. 241–2. "I am sure that Hitler would not have hesitated for a moment to employ atom bombs against England." Speer, p. 227.

165. Ibid., p. 228.

166. Irving, *German Atomic Bomb*, p. 255.

167. Ibid., p. 295.

168. Ibid. One result was that Speer consistently underestimated Germany's nuclear effort. "At best," he wrote, "with extreme concentration of all our resources, we could have had a German atom bomb by 1947 ...." Speer, p. 229.

169. Irving, *German Atomic Bomb*, p. 233. "When one considers that right up to the end of the war in 1945," the deputy director of the German project recalled, "there was virtually no increase in our heavy water stocks in Germany, and that for the last experiments in 1945 there was in fact only two and a half tons of heavy water available, it will be seen that it was the elimination of German heavy water production in Norway that was the main factor in our failure to achieve a self-sustaining atomic reactor before the war ended." Ibid., p. 211.

Churchill and Hitler

170. *Secret Conversations*, 28–29 December 1941, p. 127.
171. Trevor-Roper, "Mind of Hitler," p. xxiii.
172. Efraim Karsh, "In Baghdad, Politics is a Lethal Game," *New York Times Magazine*, 30 September 1990, p. 42. George Lardner, Jr., "Saddam's Inner Circle Seen as Unquestioning," *Washington Post*, 3 December 1990, p. A1.
173. Patrick E. Tyler, "Saddam, in Grasping, Tends to Overreach," *Washington Post*, 7 August 1990, p. A9.
174. Karsh, p. 39.
175. Roger Parkinson, *Peace for Our Time, Munich to Dunkirk – The Inside Story* (New York: McKay, 1971), p. 103. Kenneth Boulding, "National Images and International Systems," *Journal of Conflict Resolution*, 3 June 1959, p. 127. Jervis, *Perception*, pp. 82–3. Betts, p. 142. A self-fulfilling prophecy is "a false definition of the situation which makes the originally false conception come true." Robert Merton, *Social Theory and Social Structure* (Glencoe, IL: Free Press, 1957), p. 423.
176. Jervis, *Perceptions*, p. 85.
177. N.I. Bukharin, *Imperialism and World Economy* (New York: H. Fertig, 1929), p. 114.
178. Morganthau, p. 59. The problem in assessing motivation is summed up in Glen Snyder and Paul Diesing, *Conflict Among Nations. Bargaining, Decision Making, and System Structure in International Crises* (Princeton: Princeton University Press, 1977), p. 254:

> Whether to be firm and tough toward an adversary, in order to deter him, but at the risk of provoking his anger or fear and heightened conflict, or to conciliate him in the hope of reducing sources of conflict, but at the risk of strengthening him and causing him to miscalculate one's own resolve, is a perennial and central dilemma of international relations. A rational resolution of this dilemma depends most of all on an accurate assessment of the long run interests and intentions of the opponent. If his aims are limited, conciliation of his specific grievances may be cheaper than engaging in a power struggle with him. If they are possibly unlimited, the rational choice is to deter him with countervailing power and a resolve to use it.

179. Rauschning, p. 118. Morganthau, pp. 65–6.
180. Herman Kahn, *Thinking about the Unthinkable* (New York: Horizon Press, 1962), pp. 111–12. Dror, *Crazy States*, pp. 80–3. Schelling, *Strategy of Conflict*, pp. 11–13 and 197ff. Marion J. Levy, Jr., *The Structure of Society* (Princeton: Princeton University Press, 1952), pp. 242–6. Hitler, of course, understood the importance of credible deterrence. As early as 1928 he commented in his unpublished book on the capability of the "inner value" of the "iron fist" to "make its appearance so visibly . . . that merely the actuality of its existence must compel a regard for and an appraisal of this fact." *Secret Book*, p. 124.
181. *On War*, p. 579.
182. Alexander M. Haig, Jr., "Gulf Analogy: Munich or Vietnam?", *New York Times*, 10 December 1990, p. A-19.
183. Schelling, *Arms and Influence*, p. 86, and *Strategy of Conflict*, p. 20, the latter of which suggests the ancient institution of hostage taking may have to be reevaluated along with other devices, adding:

> We tend to identify peace, stability and the quiescence of conflict with notions like trust, good faith, and mutual respect. To the extent that this point of view actually encourages trust and respect it is good. But where trust and good faith do not exist and cannot be made to by our acting as though they did, we may wish to solicit advice from the underworld, or from ancient despotisms, on how to make agreements work when trust and good faith are lacking and there is no legal recourse for breach of contract. The ancients exchanged hostages, drank wine from the same glass to demonstrate the absence of poison, met in public places to inhibit the massacre of one by the other, and even deliberately exchanged spies to facilitate transmittal of authentic information. It seems

likely that a well-developed theory of strategy could throw light on the efficacy of some of those old devices, suggest the circumstances to which they apply, and discover modern equivalents that, though offensive to our taste, may be desperately needed in the regulation of conflict.

184. *On War*, p. 125.
185. Georgie Anne Geyer, "Saddam has history of defeated heroes to build on," *Harrisburg Patriot*, 5 December 1990, p. A-13.
186. *Mein Kampf*, p. 229.
187. Trevor-Roper, *Last Days of Hitler*, p. 92. For Langer's prediction of Hitler's suicide, see Langer, p. 211.
188. Michael I. Handel, *War, Strategy and Intelligence* (London: Cass, 1989), p. 464.
189. Both 31 August 1944 quotes from Gilbert, *Hitler Directs His War*, pp. 105–6. See also ibid., pp. xxiv-xxv. The air of unreality, of course, grew stronger as the end drew near. Thus, there is the picture of Hitler's drawing encouragement from the news of President Roosevelt's death or through the rereading of Carlyle's work on Frederick the Great. Trevor-Roper, *Last Days*, pp. 112–13 and 214. How divorced from reality Hitler's minions could be was shown in Himmler's greeting to the representative of the World Jewish Congress who came to see him in April 1945: "Welcome to Germany, Herr Masur," the SS chief stated. "It is time you Jews and we National Socialists buried the hatchet.' Fest, *Face*, p. 123.
190. Felix Gilbert noted that "in reading Hitler's case as it emerges from the records, it would appear that it is an oversimplification to view the contrasts between him and the Generals as one between professionals who act on the basis of rational considerations and a madman acting on the basis of intuition." Gilbert, *Hitler Directs His War*, p. xxiii. See also van Creveld, "War Lord Hitler," pp. 74–5. For a typical example of the military description of Hitler, see Guderian, pp. 414–15:

> His fists raised, his cheeks flushed with rage, his whole body trembling, the man stood there in front of me, beside himself with fury and having lost all self-control. After each outburst of rage Hitler would stride up and down the carpet-edge, then suddenly stop immediately before me and hurl his next accusation in my face. He was almost screaming, his eyes seemed about to pop out of his head and the veins stood out on his temples.

191. Francois Genoud, ed., *The Testament of Adolf Hitler: The Hitler–Bormann Documents. February–April 1945*, trans. R.H. Stevens (London: Cassell, 1961), p. 107.
192. Van Creveld, "War Lord Hitler," p. 79.
193. "By the year 2000," Secretary of Defense Cheney has pointed out in this regard, "more than two dozen developing nations will have ballistic missiles, 15 of those countries will have the scientific skills to make their own, and half of them either have or are near to getting nuclear capability, as well. Thirty countries will have chemical weapons and ten will be able to deploy biological weapons." Secretary of Defense Dick Cheney, *Address to the Conservative Leadership Conference*, Washington, DC, 9 November 1990.

6

Landmarks in Defense Literature: *The World Crisis, The Unknown War, The Aftermath, Marlborough: His Life and Times*

Winston Churchill earned his living for much of his life by writing. Beginning with his account of the 1897 Malakand Field Force campaign in India and including his only novel, *Savrola*, the British statesman produced an enormous output over the years, culminating in the last decades of his long life with his memoirs of the Second World War and his *History of the English-Speaking Peoples*. At no time was he more productive, however, than the inter-war years, in which he wrote an endless stream of articles, an autobiography and a collection of perceptive sketches of famous contemporaries. But his greatest achievements were the two massive studies he undertook during that period. Between 1922 and 1925, Churchill wrote *The World Crisis*, which together with *The Unknown War* (UK – *The Eastern Front* – 1931) and *The Aftermath* (1929) comprised his six-volume study of the First World War. And from 1932 to 1937, he produced a multi-volume biography of his famous ancestor, the first Duke of Marlborough.

Both of these works reflect Churchill, the writer, at his best – the massive mastery of detail, the depiction of myriad personalities, the descriptive linkage of conflicts in distant theaters, and the skillful interlocking of complex themes. Above all, there is the marvelous incisive prose rolling across the written page. "The terrible 'Ifs' accumulate," he wrote of the opening events of the Great War.[1] And there is the British fleet dispatched northward to its wartime position in 1914, "squadron by squadron, scores of gigantic castles of steel wending their way across the misty, shining sea, like giants bowed in anxious thought."[2] In Churchill's hand, Admiral Jellicoe at Jutland is "the only man on either side who could lose the war in one afternoon"; and Admiral von Spee, who is unable to repair or refuel his ships while conducting German raids in the Pacific, becomes "a cut flower in a vase; fair to see, yet bound to die, and to die very soon if the water was not constantly renewed."[3] Even the future British leader's vaunted magnanimity failed him in the case of General Morro who, at

Gallipoli, was "an officer of swift decision. He came, he saw, he capitulated."[4] And finally, there is Churchill's description of his ancestor, reflecting the struggles of the 1930s against appeasement. "One man, still carrying with him the British Island in its most remarkable efflorescence of genius and energy," he wrote, "stood against this kind of accommodation. Marlborough, harassed and hampered upon every side, remained unexhausted and all-compelling."[5]

The significance of *The World Crisis* and *Marlborough*, however, goes far beyond their literary and historical qualities. Both works reflect Churchill's experience in the crucible of the First World War, his first encounter with total war, a far cry from the limited wars in the "palmy days of Queen Victoria" in which he had spent his formative years.[6] From October 1911 to May 1915, the British statesman had a front-row strategic seat in the preparation and conduct of the war as First Lord of the Admiralty in the Asquith government; and from July 1917 to November 1918 as Minister of Munitions under Lloyd-George. Between these assignments he honed his appreciation of the tactical and operational levels of modern war as a battalion commander on the Western Front. As a consequence of all that, there is a degree of truth in one colleague's conclusion that "Winston has written an enormous book about himself and called it *The World Crisis*."[7] But it is this very application of the lessons from these personal experiences in his analyses and commentaries in both *The World Crisis* and *Marlborough* volumes that makes them so unique in defense literature – for rarely has a future national leader in wartime revealed so much about his approach to war and politics as Churchill did in those works.

Most important, there emerged from those writings an appreciation by Churchill for a larger strategic approach in dealing with the new totality of warfare – the emergence of certain sombre facts, "solid, inexorable, like the shapes of mountains from drifting mist," which would directly affect his approach to grand strategy as a leader in the next conflict. "There are many kinds of manoeuvers in war, ... in time, in diplomacy, in mechanics, in psychology," he concluded; "all of which are removed from the battlefield, but react often decisively upon it."[8] This broader picture of wartime strategy, requiring the use of all elements of national power, would have been understood by Marlborough, Churchill believed, who "was not only Commander-in-Chief of the English and Dutch armies, but very largely a Prime Minister as well."[9]

A major part of the political connection to grand strategy would stand Churchill in good stead in preparation for the parliamentary votes of confidence in his government during the early dark days of the Second World War. That the political foundations of his nation were inextricably linked with the fortunes of war at all levels was by the time he wrote his major inter-war works a truism for Churchill who had been thrust from office as First Lord by the operational defeats on the Gallipoli Peninsula in 1915. "At home," he could thus note sympathetically of Marlborough's situation two centuries before, "... the wolves were always growling" – a reference to the violent attacks by the High Tory Party in 1703 on the Duke's conduct of the war. The answer to these attacks, as it would also be in the Second World War for the British statesman, was military victory – in this case the Blenheim campaign, in which Marlborough's operational art produced political–military results on a grand strategic scale. "The wide plain, bathed in the morning sunlight," Churchill wrote of the Blenheim battlefield, "was covered with hostile squadrons and battalions. ... But behind this magnificent array ... were the shapes of great causes and the destinies of many powerful nations .... All these had brought their cases before the dread tribunal now set up in this Danube plain."[10]

The study of Marlborough's Blenheim campaign also reinforced other political lessons from the First World War that Churchill would put to good use as warlord in the formation of the Grand Alliance in the Second World War – in particular, the role of government in recruiting allies. "The manoeuver which brings an ally into the field," he wrote in *The World Crisis*, "is as serviceable as that which wins a great battle."[11] But that study also revealed Churchill's concern with the problems of coalition warfare, once allies were recruited. Those impressions were reinforced in the 1930s as he examined Marlborough's efforts to coalesce the disparate Dutch, British and German forces that formed the core of the Grand Alliance against Louis XIV's France in the War of the Spanish Succession. "It was never in his power to give orders which covered the whole field of war," Churchill observed in this regard, "and in many quarters ... his command was disputed."[12] Added to this were the centrifugal tendencies of sovereign states with their own national interests, a phenomenon with which he would have to deal in the closing years of the next world conflict. "The history of all coalitions,'" he concluded at the end of the study of his ancestor's Grand Alliance, "is a tale of the reciprocal complaints of allies."[13]

In addition to the political aspects of grand strategy, both works are also keys to understanding Churchill's mastery of other elements of national power during the Second World War. In particular, *The*

World Crisis demonstrated his awareness of how the Great War had required a total mobilization with its rigors not just confined to frontline combatants, but to entire populations on what came to be called the "Home Front," no matter how distant from the battle area. "It is established that henceforth whole populations will take part in war," he wrote, "all doing their utmost, all subjected to the fury of the enemy."[14] The key to this new "Front," Churchill came to realize as he wrote his analysis of the First World War, was national will molded by strong leadership and propaganda. In the 1930s, he returned in his study of Marlborough to the dominant influence this psychological aspect of grand strategy could have, even entitling one chapter of that biography, "The Home Front." And in his analysis of a possible British conflict with Spain in the closing years of the War of the Spanish Succession, he admitted that even his ancestor could be wrong concerning that influence. "Marlborough himself considered that a single campaign would suffice," he wrote. "It may well be that he greatly underrated the resisting power of a nation, and thought of it in terms of professional armies."[15]

The understanding of this concept would be a mainstay for the British leader in the darkest days of the Second World War; for this type of will meant even the most crushing military defeats in the field would not necessarily be decisive. On the other hand, as Churchill also demonstrated in his two major inter-war works, if the national will was weakened or lacking, the most trifling defeat at the tactical or operational level could be decisive. The continuum of war could no longer be divorced from its parent society. Nowhere was this phenomenon better illustrated than in Churchill's analysis of the 1918 Ludendorff offensive in which the psychological and economic instruments of grand strategy worked in adverse synergism against Germany. "Here then was the wearing down," he wrote, "which coming at the moment when the German national spirit was enfeebled by its exertions during four years and by the cumulative effects of the blockade, led to the ... sudden final collapse of German resistance in November 1918."[16]

All these aspects of total war led in Churchill's inter-war analyses to the issue of overall centralized government direction of grand strategy. Those analyses are absolutely essential to an understanding of why Churchill in 1940 combined the office of Prime Minister with the newly created one of Minister of Defence in order to become, as he had written of Marlborough, "the central link on which everything was fastened."[17] All this, he came to believe in the inter-war years, was particularly important in ensuring that policy dominated strategy in dealing with the military at the highest levels of leadership. "The distinction between politics and strategy," he wrote in words he would repeat in his Second World War memoirs, "diminishes as the point of view is raised. At the summit true politics and strategy are one."[18] But the civilian apparatus for control at that level in Britain during the First World War was lacking. After the Dardanelles, Churchill recorded in *The World Crisis*, "no one had the power to give clear brutal orders which would command unquestioning respect. Power was widely disseminated among the many important personages who in this period formed the governing instrument."[19]

Even the more highly controlled and centralized government under Lloyd-George, who took office in 1916, was not adequate to the task as far as Churchill was concerned. For by that time the dispute between the military, or "brass hats," and the politicians, or "frocks," was public knowledge. In fact, the people were becoming more and more influential in the dispute since the general public impression was that the military leaders must be right on matters of war. That impression, Churchill pointed out, "was not entirely in accordance with the facts, and facts, especially in war, are stubborn things."[20] The basic fact, as he consistently demonstrated in *The World Crisis*, was that broad issues of grand strategy were so complex and far reaching that the coordination of ends, ways and means for the war effort could only be accomplished at the highest policy level. "The General no doubt was an expert on how to move his troops," he observed, "and the Admiral upon how to fight his ships . . . . But outside this technical aspect they were helpless and misleading arbiters in problems in whose solution the aid of the Statesman, the financier, the manufacturer, the inventor, the psychologist, was equally required."[21]

Equally far reaching in terms of Churchill's role in the Allied grand strategy of the Second World War was the effect on him of the horrific indecisive blood baths of the Great War, which he admitted in his memoirs of the next conflict "were not to be blotted out by time or reflection."[22] That sense of indecisiveness, of being unable to mold tactical battles and engagements at the operational level of war into campaigns that could meet strategic objectives, permeated both *The World Crisis* and *Marlborough*. Churchill depicted the Tories demanding of Marlborough after the storming of the Schelenberg: "What was the sense of capturing a hill in the heart of Germany at such a loss? Were there not many such hills?"[23] In a similar manner, there was the destruction in 1914 of Kitchener's Army, the "flower of that generous manhood" struggling "forward through the mire and filth of the trenches, across the corpse-strewn crater fields, amid the flaring, crashing, blasting barrages and murderous machine-gun fire, ... unconquerable except by death, which they had conquered."[24]

One solution to the problem that Churchill often returned to in his inter-war studies was force multipliers. *The World Crisis*, for instance, clearly outlined his extensive involvement in the codebreaking successes of the First World War that would begin the British statesman's lifelong enthusiasm for signal intelligence, culminating in his extensive and masterful use of ULTRA throughout the Second World War. In a similar manner, Churchill's fascination with machines and technological surprise (most notably the role of the tank) is clearly in evidence throughout that work, demonstrating the genesis of his abiding interest during the Second World War in technology for use along the continuum of war ranging from the strategic "Mulberry" harbors, to the "Window" clutter at the operational level, down to the tactical "sticky bomb." Finally, the inter-war studies held the key to Churchill's primary role during the Second World War in encouraging and centralizing deception activities at all levels of war, the most important outcome of which was the strategic "bodyguard of lies" in the form of the Fortitude operation that protected the cross-Channel invasion. During the 1930s, as he immersed himself in Marlborough's eighteenth century campaigns, for example, Churchill lingered in great detail over every deception operation conducted by his ancestor. At Elixem in 1705, Marlborough's deception plan, which included the construction of eleven bridges that were never to be used, was not only designed to fool the French but his cautious Dutch allies as well. And in the 1711 campaign, marked by "the artifices and stratagems which he used," Churchill detailed a series of operational-level deceptions implemented by Marlborough's forces as they "traversed those broad undulations between the Vimy Ridge and Arras which two centuries later were to be dyed with British and Canadian blood."[25]

For Churchill in the inter-war years, however, force multipliers were only one part of the solution for returning decisiveness to warfare. The primary necessity, he repeatedly demonstrated in *The World Crisis*, had been the need for new leaders who "realized the blunt truth – quite obvious to common soldiers – that bullets kill men," and who as a consequence could enlarge their perspectives beyond the operational stalemate in one theater.[26] "He never ceased to think of the war as a whole," he later wrote of Marlborough. "To him the wide scene of strife and struggle ... was but one."[27] And so it was with Churchill, who would apply these lessons to his Mediterranean strategy in the next war. "The essence of the war problem was not changed by its enormous scale," he wrote in his justification of the Gallipoli campaign. "But once the view was extended to the whole scene of the war, and that vast war conceived as if it were a single battle, ... turning movements of a most far-reaching character were open to the Allies."[28]

Strategic maneuver notwithstanding, however, both *The World Crisis* and *Marlborough* repeatedly demonstrated Churchill's realization that modern warfare would never return to the decisive, "sporting" conflicts of the Victorian era. It is how he came to grips with that realization which ultimately makes both works so compelling and vital to an understanding of the future British leader. For despite the inroads of impersonal technology on modern conflict, Churchill never lost his belief in the need for the man of action at the vortex of war. Thus, there was his visit to General Tudor and his Ninth Division just before the Ludendorff offensive. "The impression I had of Tudor," he wrote, "was of an iron peg hammered into the frozen ground, immovable. And so indeed it proved." Before he left the battlefield that day, Churchill turned and once again looked back on the men of the Ninth Division. "I see them now, serene as the Spartans of Leonidas on the eve of Thermopylae."[29]

In the 1930s as he researched and wrote his history of Marlborough, Churchill's studies renewed his faith in the man of action, whose every word "was decisive. Victory often depended upon whether he rode half a mile this way or that."[30] And in the fourth volume of his biography, Churchill lingered over the aftermath of the battle of Elixem in which Marlborough had pierced the Lines of Brabant. Even as the battle neared its end, his grateful troops responded with spontaneous mass affection. When Marlborough rode up sword in hand to take his place in the final cavalry charge, the soldiers and their officers broke into cheering, extremely unusual considering the formal military etiquette of the times. Afterwards, when he moved along the front of his army, the veterans of Blenheim, as Churchill describes it, "cast discipline to the winds and hailed him everywhere with proud delight" – almost identical to a description by Churchill to his wife a few years later of his reception in North Africa by men of the Eighth Army.[31]

Here, then, is the ultimate essence in the Second World War of Churchill who, unlike the other Allied and enemy leaders in that conflict, travelled freely within the war zones. This allowed him to solve major military issues by face-to-face contact with his operational commanders. Most important, however, the visits to the front lines meant that he could fulfill his conceptions, so vividly etched in his inter-war works, of a war leader at the scene of action. Wherever he went, whether in the fighter control rooms of 1940, at the triumphal victory parade in Tripoli, on the beaches of Normandy or at the Rhine crossings, Churchill's visible, inspirational presence in the most outrageous of *ad hoc* uniforms was a key factor that contributed not only to the prosecution of the war, but to the genuine affection in which he was held by the officers and men throughout the services.

In this, as in so many cases during the Second World War, the key to the great British leader's actions can be found in his two monumental studies from the inter-war period. *The World Crisis* and *Marlborough* are truly landmarks in defense literature that are absolutely indispensable to an understanding of Britain's future pivotal role in history's most ferocious cataclysm.

NOTES

This essay was first published in *Defense Analysis*, Volume 9, Number 2 (August 1993), pp. 235–42.

1. Winston S. Churchill, *The World Crisis 1911–1914* (New York: Charles Scribner's Sons, 1928), p. 274.
2. Ibid., p. 225.
3. Winston S. Churchill, *The World Crisis 1915* (New York: Charles Scribner's Sons, 1929), p. 318.
4. Ibid., pp. 515–16. Churchill later spoke of General Morro to Violet Asquith. "I should like him to starve," he told her, "to starve without a pension in a suburban hovel facing a red-brick wall." Violet Bonham Carter, *Winston Churchill, An Intimate Portrait* (New York: Harcourt, Brace & World, 1965), p. 348.
5. Winston S. Churchill, *Marlborough, His Life and Times*. Vol. V. *1705–1708* (New York: Charles Scribner's Sons, 1937), p. 242.
6. Churchill, *The World Crisis 1911–1914*, p. 1.
7. Virginia Cowles, *Winston Churchill, the Era and the Man* (New York: Harper, 1953), p. 246.
8. Churchill, *The World Crisis 1915*, p. 6.
9. Winston S. Churchill, *Marlborough, His Life and Times*. Vol. III. *1702–1704* (New York: Charles Scribner's Sons, 1935), p. 353.
10. Winston S. Churchill, *Marlborough, His Life and Times*. Vol. IV. *1704–1705* (New York: Charles Scribner's Sons, 1935), p. 84. "Surveying the general war," Churchill wrote of the situation just prior to Blenheim, "we can see that matters had now come to such a pitch that without a great victory in two or three months, the Grand Alliance was doomed." Churchill, *Marlborough*, Vol. III, p. 355. See also Churchill, ibid., p. 290: "The Grand Alliance quivered at this moment in every part of its fragile organization. Marlborough saw that without some enormous new upholding force it must come clattering down."
11. Churchill, *The World Crisis 1915*, p. 6.
12. Winston S. Churchill, *Marlborough, His Life and Times*. Vol. VI. *1708–1722* (New York: Charles Scribner's Sons, 1938), p. 124. "Neither in his headquarters at the front nor behind him at home did he have that sense of plenary authority which gave to Frederick the great and to Napoleon their marvelous freedom of action." Churchill, *Marlborough*, Vol. V, p. 6.
13. Ibid.

14. Winston S. Churchill, *The Aftermath* (New York: Charles Scribner's Sons, 1929), p. 483.
15. Churchill, *Marlborough*, Vol. V, p. 246.
16. Winston S. Churchill, *The World Crisis 1916–1918*, Part 1 (London: Thornton Butterworth, 1927), p. 58.
17. "It is not until we reach Napoleon, the Emperor-statesman-captain, that we see the threefold combination of functions – military, political and diplomatic – which was Marlborough's sphere, applied again upon a Continental scale." Churchill, *Marlborough*, Vol. III, pp. 13–14.
18. Churchill, *The World Crisis 1915*, p. 6. "It is not possible in a major war to divide military from political affairs. At the summit they are one." Winston S. Churchill, *The Second World War. Vol. III. The Grand Alliance* (Boston: Houghton Mifflin, 1950), p. 28.
19. Churchill, *The World Crisis 1915*, p. 526. "There was no supreme authority."
20. Churchill, *The World Crisis 1916–1918*, Part 1, p. 244.
21. Ibid., p. 243.
22. Winston S. Churchill, *The Second World War. Vol. V. Closing the Ring* (Boston: Houghton Mifflin, 1951), p. 38.
23. Churchill, *Marlborough*, Vol. IV, p. 40.
24. Churchill, *The World Crisis 1916–1918*, Part 1, pp. 195–196.
25. Churchill, *Marlborough*, Vol. VI, pp. 421–8. For the Elixem operation, see Winston S. Churchill, *Marlborough, His Life and Times*, Vol. II. *1688–1702* (New York: Charles Scribner's Sons, 1933), p. 209. See also Churchill's description of Marlborough's feint before Tournai in 1709 where the "surprise was complete and the fortress was caught with barely five thousand men." Churchill, *Marlborough*, Vol. VI, p. 111.
26. Winston S. Churchill, *The Unknown War* (New York: Charles Scribner's Sons, 1931), p. 287.
27. Churchill, *Marlborough*, Vol. III, p. 258.
28. Churchill, *The World Crisis 1915*, p. 14.
29. Churchill, *The World Crisis 1916–1918*, Part 2, pp. 410, 412 and 435.
30. Churchill, *Marlborough*, Vol. III, p. 116.
31. Churchill, *Marlborough*, Vol. IV, p. 215. The Duke of Marlborough wrote to his wife: "*The kindness of the troops to me has transported me.*" Original emphasis. Churchill, *Marlborough*, Vol. IV, p. 220. See also Martin Gilbert, *Winston Churchill. Vol. VII. Road to Victory, 1941–1945* (Boston: Houghton Mifflin, 1986), p. 167.

7

Röhm and Hitler: The Continuity of Political–Military Discord

The time: Sunday, 1 July 1934; the place: a cell in Munich's Stadelheim prison. Two men leave the cell after handing a Browning to a rotund, almost squat figure, whose pudgy face is laced with bullet scars. The two men return in ten minutes. The occupant of the cell, Ernst Röhm, has not used the pistol as they directed. Instead, he has removed his shirt and is standing with his bare chest thrust defiantly at the men. They oblige him by executing him on the spot.[1]

The death of Ernst Röhm at the indirect hand of Adolf Hitler during the so-called "Night of the Long Knives" was the culmination of many events and forces, not the least of which was the dichotomy in their views concerning the relationship of the political and military arms of the Nazi Party. Simply put, Hitler demanded that the military must be subordinated to the political aims and objectives of the party. Röhm, on the other hand, believed in the "primacy of the soldier over the politician"[2] and conceived of a national military organization into which all military and paramilitary groupings would be assimilated as the prerequisite for any Nazi political endeavours.

These differing concepts sprang from the two roots of the National Socialist movement: the party and the *Wehrverbände* – the latter being the nationalistic, paramilitary defence leagues, consisting primarily of veterans of the organized nihilism of the First World War who had emerged from that experience with a vision of the future grounded in the heroic myth of the front-line soldier, the *Frontkämpfer*.[3] Chief among them was Röhm, who was primarily responsible for the idea, so crucial to the new movement, of combining the military organizations with the nascent political grouping, then known as the German Workers' Party (DAP). But it was Hitler who implemented the idea and by mid-1921 he dominated the DAP. Shortly thereafter, he merged the political organization, renamed the National Socialist German Workers' Party (NSDAP), with the defence leagues, which that same year changed from 'Defence Sections' to 'Sport Sections' and finally to 'Storm Sections' (*Sturmabteilung* – SA).[4]

With the establishment of both the party and the SA, the twin conductors of future political–military tensions were in place. Ironically, however, it would not be until both organizations were hounded out of existence for sixteen months beginning in the autumn of 1923 that the conflicting political–military concepts of Hitler and Röhm would crystallize. In order to understand that process, it is necessary to examine in detail how these two men interacted during that period of proscription. Only then is it possible to demonstrate how these concepts, formed during the lowest ebb of the NSDAP in the *Kampfzeit*, emerged with remarkable continuity a decade later on a collision course that could only end inside the grey walls of Stadelheim prison.

In 1923, the NSDAP occupied a leftist position in the right-wing *völkisch* movement, the tenets of which included pan-German nationalism, an extreme form of anti-semitism based on racial beliefs, and an uncompromising hostility to liberalism, Marxist socialism, and parliamentary democracy. At the head of the party was Adolf Hitler, who was again elected First Chairman in late January at a national conference in Munich. The result of that conference was a consolidation of both Hitler's personal rule of the party and his conception of the NSDAP as a *Kampforganisation* designed to overthrow the political system of the Weimar Republic.

The conception was not unique to Hitler. In Bavaria it provided the basis for the creation of umbrella organizations, under which the NSDAP and other radical groups operated, unified only in their demands for a violent revolution. In early February, the party joined with some of these groups at Röhm's suggestion to form the *Arbeitsgemeinschaft der vaterländischen Kampfverbände*.[5] But the unity of the new organization did not survive its involvement in the Munich May Day demonstrations of 1923. Thanks to Röhm's acquisition of *Reichswehr* weapons, over 1300 armed *Arbeitsgemeinschaft* troops, primarily the NSDAP's SA, were poised on the outskirts of Munich on the morning of 1 May to march against socialist celebrants in the city. Before they could move, however, the troops were confronted by large contingents of police, and after standing around for hours, they turned over their weapons and dispersed. Later that evening, Hitler gave a defiant speech to a National Socialist rally. But there was no disguising the débâcle. "Hitler fastened on his helmet!" one writer commented. "He took his helmet off …. He put it on again. … He took it off again …. He marched home."[6]

After the May Day incident, the NSDAP remained relatively quiet for the rest of the spring and summer. On 1 and 2 September, however, the National Socialists joined with other *völkisch* groups at an enormous "German Day" rally in Nuremberg to celebrate the victories of Sedan and Tannenberg. Among the guests of honour from both battles was General Erich von Ludendorff, who had settled near Munich after the war, fulminating against what he perceived as a world conspiracy of Jews, Catholics and Freemasons to take over Germany. Ludendorff was extremely pleased to be the focus of right-wing adulation and consequently was active in the meetings, social events and massive parades in the old city that, in Röhm's description, "rocked in a sea of flags."[7]

On the last day of the rally, Ludendorff also witnessed the founding of the *Deutscher Kampfbund*, a relatively smaller umbrella organization that replaced the *Arbeitsgemeinschaft*. There was nothing basically new in the *Kampfbund* goals, which included the crushing of Jews, Marxism, international capital and the Weimar constitution. What was new, however, was that Hitler subordinated the SA directly to the new umbrella organization, thus treating the Brown Shirts for the first time as a completely separate entity from the political organization of the NSDAP. The situation was altered somewhat later the same month when Hitler was appointed political leader of the *Kampfbund*.[8] Nevertheless, the fact remained that the National Socialist leader had allowed the military organization of his movement to be absorbed into a rightist organization over which neither he, personally, nor the party, institutionally, had much control.

As the autumn progressed, an almost uncontrollable momentum within the *Kampfbund* developed for a military move against the state and national governments. For years, Hitler had ruled out participation in parliamentary elections and preached the need for violent action. By November, amid mounting tensions and confusing rumours, he could not very well waver when there was, as Röhm described it, "the expectation of a putsch almost every day."[9] The result was the so-called "Hitler *Putsch*" that began in the evening of 8 November in Munich and almost immediately spread from the Bürgerbräukeller throughout the city. The next morning, when it appeared that the scattered *putschists* were in danger of being defeated piecemeal by the *Reichswehr* and State Police, Hitler concurred with Ludendorff's suggestion to stage a propaganda march into the city. Shortly before noon, a long column of armed *Kampfbund* members headed by Hitler and Ludendorff crossed over the Isar River and moved toward the centre of Munich. Half an hour later, the *putsch* collapsed before the gunfire from a police cordon. Within two days, following a state-wide ban on the NSDAP and the *Kampfbund*, Hitler, Röhm and most of the other *putsch* leaders were in jail.

The major trial of Hitler and the other *putsch* leaders began on 26 February 1924 under unusually heavy security. Unlike other failed *putschists* of the era, the National Socialist leader did not plead his innocence during the twenty-four days of the trial. Instead, he took full responsibility, frankly outlined his political programme, and indicted the Weimar government's "November criminals." At the same time, he courted the military by exonerating the *Reichswehr's* lack of support for the *putsch*. "The army has not become tarnished," Hitler concluded. "The day will come when the Reichswehr will be on our side."[10]

The first day of April was the last day of the trial. Röhm and three other defendants were sentenced to fifteen months imprisonment for abetting high treason and were then immediately paroled. Ludendorff, to his consternation, was acquitted. Only Hitler and three other *Kampfbund* leaders were given five-year sentences and even these held out the possibility of parole in six months. Hitler had also requested that he not be deported to his native Austria, although the law was specific in this regard for any foreigner who caused trouble. Nevertheless, the presiding judge was happy to accede to the request from a man "who thinks and feels like a German."[11]

The entire proceedings were a triumph for the National Socialist leader who had recouped much of the position lost by the *putsch* through his skilful use of words. "How that man Hitler spoke!" a party member recalled 15 years later. "Those days of his trial became the first days of my faith in Hitler."[12] Despite this success, however, Hitler was well aware that the ban on the NSDAP had rendered the party increasingly ineffective, particularly in terms of the underground SA, whose wrangling factions in Germany were only loosely controlled by Hermann Göring and other party leaders in Austrian exile. Consequently, the Nazi leader moved to re-establish authority through the *Kampfbund*, the umbrella organization in which he still retained legal political leadership. On the day of his sentencing, Hitler gave Röhm a handwritten note.

> Captain Röhm is *military leader* of the Kampfbund. I command, therefore, that his ordinances be obeyed by members and particularly leaders of the NSDAP's SA. Those who cannot unconditionally follow the commands of Captain Röhm are to be considered as no longer belonging to the SA.[13]

The new military leader's initial efforts were directed at obtaining legal status for the SA, but he was not able to persuade the Bavarian authorities to revoke the ban on that organization. In addition, there was new competition. In early January, the Bavarian government had surreptitiously supported the creation of the *Notbann*, an apolitical paramilitary organization specifically designed to attract members of the various right-wing defence leagues. There was, in fact, much in the *Notbann* that appealed to Röhm, particularly – as his frustration increased over the SA's illegal status – the idea of a *Wehrverbände* umbrella organization completely divorced from the political leadership. Moreover, there was the example at the national level of the *Reichsbanner*, a new and successful socialist-inspired paramilitary umbrella structure.[14]

These impressions formed the basis of discussions that Röhm conducted throughout April with several *völkisch* paramilitary groups near Munich. On 6 May he was elected to the Reichstag and used the first-class rail pass, authorized to him as a national deputy, to travel extensively to visit other defence leagues throughout Germany. One of these trips took him to Halle, where on 10 and 11 May there was a huge "German Day" rally attended by 200,000 people. It was a heady experience for Röhm. Almost 4,000 flags were assembled on the parade field the first day as Ludendorff trooped the line, and that evening Röhm addressed enthusiastic crowds packed into four different beer halls. At the conclusion of all this activity, he decided to form a supra-regional defence league that he believed would build on the unity still existing in areas where the *Kampfbund* organizations were banned.[15]

But first there was the matter of the SA. On 17 and 18 May, Hermann Göring hosted a summit conference in Salzburg, attended by SA leaders from every district in Germany and Austria. Göring was aware that Hitler had authorized Röhm to rebuild the SA and consequently appointed him as his deputy with complete power in Germany. Röhm then took charge, and the conference proceeded smoothly. The new SA was to continue as a paramilitary organization under the political direction of the NSDAP and Röhm, always a superb organizer, issued regulations and directives on matters ranging from uniform and ranks to enlistments and formations.[16] Despite this ostensible success, however, Röhm's focus remained firmly fixed on a new umbrella organization for *völkisch* defence leagues; and immediately after the Salzburg conference, he approached Ludendorff with his concept. The result was a draft plan for a strictly centralized national paramilitary structure called the *Frontbann*, which the general agreed to take under his protection after Röhm assured him that it was "unpolitical" and not a continuation of one of the banned organizations.[17]

On 31 May, Röhm journeyed to Landsberg to discuss his plan with the imprisoned *Kampfbund* leaders. The negative potential in a centralized military structure organized by the dynamic Röhm was not lost on Hitler, particularly since it could involve defence leagues that had never been closely associated with the NSDAP. Moreover, although Röhm's plan acknowledged that each of the *Frontbann* military leader's "relationship to the political leader of his organization is an internal matter of the organization concerned," it also emphasized that the "pre-condition for admission is ... the unconditional subordination to the military leadership of Ludendorff."[18] And if this were not enough for the isolated National Socialist leader, there was always the possibility that involvement with the *Frontbann* might harm his chances for probation when he became eligible in October, or even cause his deportation.[19] As a consequence, Hitler opened his meeting with Röhm by voicing his fundamental objection to the *Frontbann*. Röhm, who had already received a negative response from the other *Kampfbund* leaders, replied that Hitler was in no position to judge the matter while in prison. After another somewhat acrimonious exchange in which Hitler refused to allow Röhm to go into details concerning the military control of the new organization, the Nazi leader ended the interview.[20]

Although he was aware of the opposition from the prisoners at Landsberg, Ludendorff continued to support the *Frontbann* concept.[21] Thus encouraged, Röhm pressed on with his plan, at times denying that he lacked the support of Hitler and the other imprisoned leaders. On 12 July, he attempted once again to obtain approval. But Hitler could not be convinced about the plan, which he advised Röhm "was completely wrong."[22] When the two met again for a short period on 17 June, Röhm sought once more to discuss his new organization. This time Hitler terminated the strained conversation by remarking that he was withdrawing as an active political leader during his imprisonment and therefore no longer wished to know anything else about the *Frontbann*.[23]

Hitler's decision to withdraw from politics was made public in early July. This decision, coupled with the Nazi leader's almost simultaneous rejection of the *Frontbann*, ended, in effect, all legitimation from that source for Röhm, to include the 1 April authorization for rebuilding the SA. As a consequence, Röhm increasingly emphasized Ludendorff's role as protector of the *Frontbann* and began to involve the general in more of the day-to-day *Frontbann* activities. The result within a month was an expansion of membership to over 30,000 centred on an organizational framework similar to the SA restructuring plan established at Salzburg in May. By the end of July, it was possible for the *Frontbann* leader to communicate throughout Germany from four geographical group commands down through a hierarchical layer of provincial, divisional and district leaders to his local commanders.[24]

This structure was meaningless, however, if the national and state governments moved against the *Frontbann*. On 20 July, Röhm dispatched his draft plan for that organization to the Bavarian Interior Minister, along with a note soliciting an official opinion. Once he had the Minister's views on the draft, he concluded in his note, he intended to publicize the *Frontbann* plan widely.[25] The Interior Minister returned the draft to Röhm in early August with a note expressing his personal satisfaction with the plan. Whether it would be possible for Röhm to implement such a plan, he concluded vaguely, would depend on further developments.[26] This governmental rebuff, however gently applied, only impelled Röhm to seek a closer and more dependent relationship with Ludendorff by increasing the *Frontbann*'s focus on and allegiance to the general at the expense of whatever tenuous links were still perceived as existing to the NSDAP leader in Landsberg.[27]

Ludendorff was less successful in the political sphere. At a three-day national *völkisch* unity conference held at Weimar in mid-August, he was able to establish only the most general leadership over the disparate and bickering rightist factions, and even then only after invoking Hitler's name as part of the political hierarchy. On the last day of the conference, Ludendorff submerged his growing political frustration by addressing a vast array of *völkisch* defence leagues assembled at the municipal airport on the outskirts of Weimar. After his opening speech, the general was visibly moved as he inspected the line of units. To Röhm's satisfaction, Ludendorff paid particular attention to *Frontbann* units, speaking with members as he decorated their flags and praising the vitality of their organization before he closed the ceremony.[28]

The ending of the Weimar conference was symptomatic of Röhm's success during the summer of 1924. Despite his break with Hitler, the *Frontbann* continued to expand under the "protection" of Ludendorff. Moreover, the general's claims for himself and Hitler concerning the leadership of the *völkisch* movement provided means for Röhm to reconcile his allegiance to his protector with the bond that still existed between him and the Nazi leader. The result of these developments was renewed activity at the end of the month by the *Frontbann* commander. On 23 August, he issued directions for the

High Command in which *Frontbann* policy was dealt with in excruciating detail.[29] This document was followed by a training manual that clearly demonstrated Röhm's enthusiasm for organizational planning, and at the same time his lack of realism in a year that had already shown the diminished utility of military force. *Frontbann* members, for instance, were ordered to keep their weapons and equipment ready for combat at short notice; and there was an elaborate total force concept using active, reserve, and replacement troops that never had a chance for implementation by the unsophisticated and, in many cases, half-organized local *Frontbann* commands.[30] Nevertheless, Röhm was encouraged enough to step up his proselytizing campaign with other organizations and once again to solicit advice and help from the Bavarian Minister of the Interior.[31] More importantly, he was ready to go public in his efforts to unite rightwing paramilitary groups. On 29 August, in a ringing proclamation that was headlined in most *völkisch* newspapers, Röhm called on Germans of all classes to join the *Frontbann*. Only by means of this organization could dissension among the various *völkisch* groups be avoided. It was therefore necessary, he concluded, for the *Frontbann* to remain "unpolitical" in the sense of not being oriented toward any political party.[32]

Röhm's efforts to disassociate his organization from political parties took a peculiar and frustrating turn in the autumn. On 16 September, the Munich police raided the homes of two key *Frontbann* deputies. The yield included not only complete files of the organization's membership and correspondence, but heavy machine-guns and artillery aiming circles as well.[33] Based on this material, the police began to arrest personnel associated with the *Frontbann*, and by the end of the day most of the High Command was in prison. Röhm, who possessed immunity as a Reichstag deputy, appealed immediately to the Bavarian Minister of the Interior. The correspondence referring to "unconditional obedience," he explained, "had only a military-technical meaning and not a political one." Soldiers, not parties, were the key, and the *Frontbann* would provide a finished product when its young men were called to the active army. "It is taken for granted," he concluded, "that when I build an organization, it will be a military organization and not a singing group."[34] The Interior Minister was unmoved. "We will tolerate no private army of General Ludendorff's," he declared to the Bavarian parliament shortly after his interview with Röhm.[35]

There was another, more ominous twist to these events as rumours began to circulate concerning Hitler's linkage to the *Frontbann*. Röhm fully realized the negative aspect of such an association for the imprisoned NSDAP leader who would be eligible for probation in two weeks. On 21 September, he issued his strongest declaration in this regard. "I am the creator and founder of the *Frontbann*," he proclaimed, "and ... as the confiscated files will show, the only one who should be arrested – not Hitler to whom I report neither orally nor in writing."[36] This was followed by an equally strongly worded letter from Ludendorff to the Munich state court on behalf of Hitler. That same day, however, the Bavarian state police recommended that Hitler be deported if he were set free. And the following day the Munich police, citing Hitler's connection with the *Frontbann*, recommended against his probation at the beginning of the next month. Nevertheless, Hitler's lawyer maintained the legal pressure, and on 25 September the state court decided in favour of the 1 October probation for the Nazi leader.[37]

With this problem apparently resolved, Röhm turned to the *Reichswehr* for support, but was rebuffed as soundly by his former organization as he had been by the Bavarian government.[38] This was not his only setback in the closing days of the month. On 29 September, Ludendorff was the guest of honour at a parade staged by the Lower Saxony *Frontbann* command. During the ceremony, Ludendorff refused to allow the 125-man Bremen contingent to pass in review. The general's decision was made at the instigation of his deputy, who had a personal feud with the Bremen commander. But that commander had been appointed by Röhm; and the subsequent discussions engendered considerable ill-feeling on the part of Ludendorff toward Röhm, who resisted the decision "up to the surrender of my position."[39] As if this incident were not bad enough for the beleagured *Frontbann* leader, the high state court decided that same day to honour the state prosecutor's request to keep Hitler and the other *Kampfbund* leaders in prison, until the *Frontbann* matter could be more fully investigated.[40] This reversal of the 25 September decision was a serious blow to Röhm, since it highlighted the *Frontbann* as the cause. Now, in addition to lack of support from the Bavarian government and the *Reichswehr* and his growing estrangement from Ludendorff, Röhm's organization and its activities were firmly associated with the continued imprisonment of the *Kampfbund* leaders in the minds of most *völkisch* groups as well as the prisoners themselves.[41]

For the remainder of autumn, Röhm's energies were focused on disassociating Hitler from the *Frontbann*. In October, he provided a supporting memorandum to statements by Ludendorff and various prisoners attesting to Hitler's antagonism at Landsberg to the idea of a paramilitary umbrella organization.[42] In mid-November, he gave formal testimony on Hitler's behalf, during which he pointed to the innocuousness of his ordinances and regulations for what, he emphasized, was a purely military organization that had never been intended to subvert the Republic.[43] Ludendorff's testimony at the same time also stressed this point: the *Frontbann* was not a vehicle for continuing the organizations banned by the Bavarian government. Nevertheless, the general concluded disingenuously, "It is understandable that like-minded persons will always try to get together again."[44] Both testimonies were reinforced by that of the Landsberg prison director, who stated that Hitler had not carried on any political activity while in prison and that visitors such as Röhm and Ludendorff arrived without any solicitation on the part of the NSDAP leader.[45] Finally, there were the statements by the *Kampfbund* leaders that complemented each other and the testimonies of the other witnesses who had emphasized Hitler's rejection of the *Frontbann*.[46]

The result was that by December there was no other course for the state prosecutor to take. There existed, he reported at the time, "neither a cause nor a possibility" for the prosecution of Hitler.[47] On 19 December, the Munich state court decreed that Hitler was to begin his probation immediately. The next day at noon Heinrich Hoffmann, the party photographer, and a companion drove to Landsberg to pick up Hitler. The taking of photographs was forbidden in the prison; but before leaving Landsberg, Hitler posed before one of the town's ancient gates with his hat in his hand and a raincoat covering his *Lederhosen*. Two hours later he was in his Munich apartment, where he was greeted by many of his followers as well as his dog who, it was faithfully reported, "immediately recognized his master after thirteen months separation."[48]

Bavarian authorities lifted the ban on the NSDAP on 16 February 1925. Ten days later, Hitler gave the first indication of how the new party would be structured in three articles for the newly-refounded party newspaper.[49] The next evening he elaborated on these themes in his first public address in over a year. For several days, the familiar red posters announced his speech throughout Munich. Although the meeting was not scheduled to begin at the Bürgerbräukeller until 8.00 p.m., crowds were already demanding entrance into the hall early in the afternoon. By 5.00 p.m., the hall was jammed with over 3,000 people, and the crowd that continued to grow outside caused the police to block off the entire area around the beer hall and to turn away over 2,000 people.[50]

Ludendorff was notably absent from the meeting. He soon reconciled himself to the new organization, however, when Hitler moved quickly to support him in a fruitless bid for the Reich presidency after Friedrich Ebert's death in mid-February. Röhm also missed the meeting, but he was not so easily reconciled. In his articles and his 27 February speech, Hitler had emphasized that the SA was to become more closely attuned to the political goals of the party. The principles that had applied until the founding of the *Arbeitsgemeinschaft* in February 1923 would apply again in the restructuring of the para-military organization. In particular, the political laws of the party must be strictly followed by SA members who were not to provide the authorities with excuses to attack the movement.

Once again, the fundamental political–military dichotomy between the two men was sharply etched. Röhm was disturbed by the new party restrictions, and he could not fail to notice that Hitler had not mentioned the *Frontbann*. On 28 February, he convened a meeting of that organization's leadership at a nearby castle that belonged to one of the group. For several days, while the swastika flag fluttered next to the Prussian colours on top of the castle, the members of the High Command discussed their future and indulged their penchant for military ceremonies with a 250-man *Frontbann* contingent quartered nearby. On 2 March, the meeting ended with an announcement by the leader of their support for Hitler as the leader of the National Socialist movement and of Ludendorff as the protector of the *Frontbann*.[51]

On 16 April, Röhm presented Hitler with a memorandum which they then discussed in detail. Röhm's major point was that his umbrella organization was the answer to the defence needs of the NSDAP, but that in order for it to be effective, Hitler must make a decision that placed the *Frontbann* leader in a position where he could "give clear and definite orders." The *Frontbann*, Röhm acknow-ledged, would be independent of the political leadership; but the entire question of political–military co-operation was, in any case, "whether you trust me to lead the organization according to your conception."[52] Hitler's response was to demand the complete subor-dination of the *Frontbann* to the party leadership and the incorpora-tion of that organization into the NSDAP framework. At the same time, however, he offered Röhm the leadership of the SA. This presented, of course, no real choice from Röhm's perspective. In either case, the political dominance of the party prevailed. On the one hand, there was the leadership of what appeared to the ex-*Frontkämpfer* as an emasculated SA; on the other, there was what would in effect amount to the dissolution of the *Frontbann*. And yet there was still the residual bond to Hitler that made it unthinkable for

Röhm to take his organization into opposition.[53] The next day he rejected the leadership of the SA and resigned as commander of the *Frontbann*. There was no response from Hitler. On 1 May, Röhm publicly announced his decision and at the same time wrote to his subordinate commands, explaining that he could not lead "an SA like the one Hitler felt was necessary." Nevertheless, he concluded, "I have the conviction that Adolf Hitler will be successful ... in leading the National Socialists ... to victory."[54]

Almost six years passed before Hitler and Röhm met again. During that period, Hitler's attitude toward the SA was consistent with the position he had maintained against Röhm concerning political–military relations within the NSDAP. As early as autumn 1926, in the new statutes for the SA, the National Socialist leader stressed the role of that organization as an unarmed instrument for propaganda and mass intimidation directly responsive to his political aims. "The formation of the SA," he wrote to his newly appointed SA supreme commander, "does not follow a military standpoint, but what is expedient to the party."[55] With this approach, Hitler was not only able to circumvent a possible threat to his absolute control over the NSDAP, but to use a dynamic force in the service of the party without straying from the path of legality, which his experiences since the aborted *putsch* had convinced him was necessary if he were ever to achieve power.

There was resistance to this approach within the SA. But until the upswing of party fortunes as the depression deepened in 1930, the discord inherent in the relationship between the SA and the party leadership was veiled by the common effort to survive. In the summer of that year, however, that discord bubbled to the surface in a chain of events that was eventually to reunite Hitler and Röhm. In August, the supreme commander of the SA resigned. Like Röhm, he resented the continued growth of the party at the expense of the SA, and was increasingly unable to work within Hitler's paradoxical framework that required the storm-troopers to respect the law and forgo the use of weapons, while simultaneously encouraging the romanticism of insurgency and armed conflict. That same month, a revolt by the Berlin SA against that city's party leadership threatened to spread to other areas. Only Hitler's personal intervention with a small contingent of *Schutzstaffeln* (SS) troops, an organization ostensibly subordinate to the SA, ended the uprising. On 2 September 1930, in the aftermath of the revolt, Hitler assumed the post of supreme SA commander.[56]

By this move, Hitler made himself the complete master of his movement, particularly when a few days later he required all SA members to take a personal oath to him. Bolstered by this success, as well as by his personal role in ending the Berlin revolt, Hitler apparently felt confident and self-assured enough to recall Röhm from Bolivia where the former *Frontbann* commander was employed as a military instructor. Other factors undoubtedly played a role in this decision. There was, for instance, the deliberate policy of ambiguity that governed all of Hitler's actions. There was also at least residual gratitude on his part toward a man who had greatly contributed to his rise.[57] Certainly, the stunning Nazi advances in the national elections on 14 September were also important. On the one hand, the SA had played a key role in pre-election activities; on the other, there was the frustrating knowledge in the aftermath of those elections that renewed efforts would still be necessary if the NSDAP were to achieve power.

In terms of those efforts, Hitler was prepared, at least for the moment, to look upon the SA and not the party as the critical centre of gravity for the movement. But the SA had become increasingly decentralized and difficult to control since its refounding. Moreover, Hitler did not have time to run its day-to-day activities. If this unruly instrument were to expand as the struggle for power intensified, it would require masterful organizational skills and firm discipline. In this context alone, the risk of recalling Röhm must have appeared worthwhile to the NSDAP leader. The irony of the fact that the discipline that Hitler had in mind ultimately would include enforcement of the Nazi Party's domination of the SA was not lost on Röhm, who was still dedicated to the idea of an above-party national organization of paramilitary *Wehrverbände*. Nevertheless, he was being given considerable organizational latitude, and he realized that some compromise was necessary. On 4 January 1931, Röhm assumed the post of SA Chief of Staff.[58]

Röhm did not disappoint Hitler. Within two years, he improved the SA's organization, standardized its training on a national scale, and, aided by the deepening depression, transformed the paramilitary membership into an army of 300,000. In the final drive to power that resulted in the *Machtübernahme* of January 1933, Hitler owed a great debt to Röhm and the SA. Röhm was well aware of this debt and consequently became increasingly dissatisfied as the pseudo-legal domestic *Gleichschaltung* of Hitler's "cold" revolution proceeded through 1933 and into the new year. The world was thus divided into "men who make revolutions and men who do not."[59] The political leaders of the NSDAP were clearly in the second category; and Röhm, resentful that the seizure of power had led to their dominance, organized huge demonstrations and parades that emphasized the military power of his organization and began to talk of a "second revolution."

What disturbed Hitler most, however, was not talk of revolution, but Röhm's associated demand in March 1934 that the SA, now grown to a force of almost four million members, should absorb the *Reichswehr* and become a national people's army for the new state. Once again the dichotomy of views on the role of the military that had taken form a decade earlier dominated their relationship. This time, however, the issue was intensified. In the days of the *Frontbann*, Röhm had conceived of a ruling duumvirate consisting of himself, if not Ludendorff, as the generalissimo of a great armed force operating as a co-equal with Hitler, the political leader and agitator. This conception remained unacceptable to Hitler. Now Röhm was raising the stakes. For as the NSDAP leader well realized, to allow the SA to absorb the regular army in a *Volksheer* would have given Röhm and his storm-troopers a dominant position in the Third Reich at the expense of the political leadership.[60]

Equally important, Röhm's plan was also unacceptable to the *Reichswehr* leaders, who understandably were not enthusiastic about Röhm's assertion that the "grey rock must be submerged by the brown flood."[61] Hitler had no illusions about this reactionary element, but since his seizure of power, the *Heeresleitung* had proved more tractable than he had expected. Moreover, the great lesson for the Nazi leader from the 1923 *putsch* was that it must never again come to the point where he was in open conflict with the *Reichswehr*, whose tolerance had also assured his victory in 1933. In any case, he was aware that meaningful rearmament of Germany would largely depend on the highly specialized technological expertise in the regular army corps. The support of the army would also be instrumental if he were to succeed President Hindenburg, who was not expected to outlive the year. For these and a host of associated reasons, Hitler decided in favour of the *Reichswehr*.[62] The result was the Blood Purge that began on 30 June 1934.

Although Hitler emerged significantly strengthened from the events of 30 June, the true victor of the Blood Purge appeared to be the army. In a speech to the Reichstag on 13 July, Hitler reconfirmed that the *Reichswehr* was the sole bearer of arms for the German state and that it would remain a non-political instrument in the Third Reich. But as

Hitler's dealings with Röhm on political–military relationships had demonstrated consistently over the years, the army had little cause for rejoicing. Seven days after the Reichstag speech, in a decree of far-reaching consequence, Hitler thanked the SS for its key role in the purge by making it independent of the now-neutered SA. Unlike Röhm, however, the leaders of the SS made no claims of independence from political direction. Instead, the SS was a totally loyal instrument of the Führer's will. In the aftermath of the Blood Purge, it began its inexorable expansion that ultimately would penetrate all existing institutions, weaken their political power and then replace them.[63]

Equally significant, army complicity or at least passive acquiescence to a purge in which two of its generals had been murdered, sowed the seeds of dissension in an officer corps whose united social structure was based above all on a common ethical attitude. It only remained for Hitler to merge the offices of President and Chancellor upon Hindenburg's death on 1 August 1934 to end the traditional *Überparteilichkeit* of the army. That the military did not appreciate the significance of this act was demonstrated the next day when the leaders and men of the armed services swore allegiance to Hitler as Supreme Commander of the *Wehrmacht*. In the wake of this oath in the years to come, the army leadership would be no more successful than Ernst Röhm in its political–military dealings with Adolf Hitler who, as Gordon Craig has pointed out, "was to impose upon the army a control more rigid than any in its long existence and to compel the obedience of its officers even to commands which violated their historical traditions, their political and military judgement and their code of honour."[64]

NOTES

This essay was first published in *Journal of Contemporary History*, Volume 23, 1988, pp. 367–86.

1. Post-Second World War description by the Governor of Stadelheim prison. Jeremy Noakes and Geoffrey Pridham, eds., *Documents on Nazism, 1919–1945* (New York, 1975), p. 215.
2. Ernst Röhm, *Die Geschichte eines Hochverräters* (München, 1934), p. 349.
3. Joachim C. Fest, *The Face of the Third Reich* (New York, 1970), p. 136. For Röhm, the *Frontkämpfer* offered the best hope for Germany's regeneration. "I am no longer a member of this nation," he declared, "I can only remember that at one time I served in the German Army." Konrad Heiden, *A History of National Socialism* (New York, 1935), p. 8.

4. Hermann Mau, "The 'Second Revolution' – June 30, 1934," in Hajo Holborn, ed., *Republic to Reich. The Making of the Nazi Revolution* (New York, 1972), pp. 225–6.
5. This was not the first umbrella organization for the NSDAP. On 9 November 1922, the party had joined the *Vereinigte Vaterländische Verbände Bayerns* (VVVB) in a ceremony held in the Munich Hofbräuhaus. But its organization was too weak and its membership too diverse to hold the NSDAP. On 14 January 1923, the party broke with the VVVB over the umbrella organization's decision to support the government's policy of passive resistance to the French occupation of the Ruhr. Hanz Volz, *Daten der Geschichte der NSDAP* (Berlin, 1939), 10th edn., p. 10.
6. Quote from Friedrich Wilhelm Heinz's book, *Sprengstoff*, in Albert Krebs, *The Infancy of Nazism. The Memoirs of Ex-Gauleiter Albert Krebs 1923–1933*, William Sheridan Allen, ed. (New York, 1976), p. 84. For the May Day demonstrations, see Munich Policy Directory Reports Nr. 1840 and 974VId, both dated 3 May 1923 and both contained in *NSDAP Hauptarchiv*, Reel 35A, Folder 1817. (All *Hauptarchiv* material hereafter designated as *HA*, followed by the reel number or letter, then the folder number and finally the page number if applicable; thus *HA* 35A/1817.)
7. Röhm, p. 209. The official programme from the *Deutscher Tag* contained a signed greeting from Ludendorff, *HA* 4/105. The final march-by of *völkisch* troops reviewed by Ludendorff on 2 September lasted over two hours. Munich Police Report Nr. 427, undated, *HA* 65/1481, p. 1.
8. Volz, p. 12 and Röhm, p. 213. For the 1–2 September *Kampfbund* proclamation, see *HA* 82/1639, *HA* 81/1636, and the *Völkischer Beobachter*, Nr. 188, 14 September 1923.
9. Röhm, p. 230. On 27 February 1924, Hitler testified that "it was impossible any longer to keep in suspense the people who, night-by-night and morning-by-morning, had been imbued in the barracks with nothing but ideas of war." *The Hitler Trial Before the People's Court in Munich*, trans. H. Francis Freniere, Lucie Karcie and Philip Fandek (Arlington, VA, 1976), Vol. 1, p. 137.
10. Ibid., Vol. III, p. 365. For security precautions see Munich Police Directory Instructions, 15 February 1924, *HA* 69/1499, and Bernd Steger, "Der Hitlerprozess und Bayerns Verhältnis zum Reich 1923/1924," *Vierteljahrshefte für Zeitgeschichte*, 4 (1979), p. 441.
11. *Hitler Trial*, Vol. III, p. 395. The judge's decision may have been influenced by the fact that no replies had been received by the Bavarian police to their January and March inquiries about Austria's acceptance of Hitler if he were deported. D.C. Watt, "Die bayerischen Bemühungen um Ausweisung Hitlers," *Vierteljahrshefte für Zeitgeschichte*, 6 (1958), p. 272.
12. Theodore Abel, *The Nazi Movement. Why Hitler Came to Power* (New York, 1965), pp. 69–70.
13. Hitler's emphasis. *HA* 53/1258.
14. The full title was *Das Reichsbanner Schwartz-Rot-Gold, Bund der Republikanischen Kriegsteilnehmer*. The *Notbann* was headed by General von Epp, the rightwing hero credited with the destruction of the 1919 Bavarian Soviet Republic. Epp accepted the command on 3 January 1924 with great reluctance. Walter Frank, *Franz Ritter von Epp; Der Weg eines deutschen Soldaten* (Hamburg, 1934), pp. 118 and 121.
15. Röhm, pp. 324–5. For Röhm's rationalization of parliamentary participation, see ibid., pp. 311 and 314.
16. For accounts of the meeting see ibid., p. 324, and Röhm's "Command," 20 May 1924, *HA* 68/1497A. For examples of Röhm's efforts, see his "Directions for the New Organization of the SA of the NSDAP," 24 May 1924, *HA* 73/1549 and the SA High Command Directives of 17 and 20 May 1924, *HA* 68/1497A.
17. Ludendorff Testimony, 10 November 1924, pp. 2 and 4, and 12 November 1924, p. 2 – both in *HA* 16A/1633.
18. Röhm's one-page "Plan for a Defence Organization of the Völkisch Movement," undated, *HA* 82/1639. Röhm spoke for 15 minutes alone with Hitler and for 45 minutes with the other *Kampfbund* leaders. Landsberg Prison Director Leybold's Telegram, 1 December 1924, *HA* 16A/1634.

19. Austrian authorities had already agreed on 20 April to recognize Hitler's citizenship and accept his deportation to Austria. Upper Austrian State Government – Linz Letter Nr. a/2Z1.2335/2, 20 April 1924, *HA* 25A/1760. Hitler was undoubtedly aware of the negotiations, which were highly publicized. See, for example, the headline article entitled "Hitler's Deportation," *Völkisches Echo*, 22 April 1924. And on 8 May the Munich Police had submitted a detailed recommendation to the Bavarian Ministry of the Interior, concluding that "Hitler represents a permanent danger to the internal and external security of the state .... With regard to the public welfare of the state, his extradition from Bavaria ... appears necessary." Munich Police Directory Report Nr. 20323.43, 8 May 1924, *HA* 25A/1760, p. 6.
20. Adolf Hitler Testimony, 1 December 1924, *HA* 16A/1634, pp. 1–2. For the resistance by the other *Kampfbund* leaders to Röhm's *Frontbann* concept, see Hermann Kriebel Testimony, 1 December 1924, p. 1 and Friedrich Weber Testimony, 2 December 1924, p. 1 – both in *HA* 16A/1634. Röhm later stated that he received no resistance to his plan from the three prisoners. Röhm, p. 325. But the testimony of all participants, including Ludendorff's, contradicts this.
21. Hitler Testimony, 1 December 1924, *HA* 16A/1634, p. 3. An alternative plan, hurriedly written by Kriebel and accepted by Hitler and Weber at lunch on 31 May, emphasized that the political leader of each organization was self-reliant and not subject to unified military direction. The alternative plan was given to Ludendorff during his visit that same afternoon for delivery to Röhm. It made no impression on either man. Ibid., pp. 2–3, and Kriebel Testimony, 1 December 1924, ibid., pp. 2–3.
22. Hitler Testimony, 1 December 1924, *HA* 16A/1634, p. 3.
23. "I delivered as an answer ... that I hereby laid down the political leadership." Ibid.
24. Röhm, p. 326. There were, of course, constant organizational revisions. On 3 August, for instance, the *Frontbann* High Command established thirteen divisional commands under the Bavarian Provincial Command. Frontbann High Command Order I.a Nr.138, 3 August 1924, *HA* 15A/1631. In September a fourteenth command, Western Upper Franconia, was added. Group Command South Order Nr. 3, 14 September 1924, ibid.
25. Röhm's letter was a typical combination of obsequiousness, bravado and lies. At one point he assured Interior Minister Stützel that even his closest friends had no knowledge of the draft plan, one of the most widely circulated of all *Frontbann* documents since that organization's creation in the spring. Röhm Letter, 20 July 1924, *HA* 53/1258. Röhm's purpose was to obtain at least a neutral stance from the Bavarian government. Röhm, p. 326.
26. Interior Minister Stützel Letter, 8 August 1924, *HA* 53/1258.
27. "I vow," the oath of the *Frontbann* read, "to be true and obedient to my leader Ludendorf [sic] ... and to my flag until death." Appendix 1, "Rules for Frontbann," undated, *HA* 81/1636, p. 7.
28. Erich von Ludendorff, *Vom Feldherrn zum Weltrevolutionär und Wegbereiter Deutscher Volksschöpfung. Meine Lebenserinnerungen von 1919 bis 1925* (München, 1940), pp. 351–2; Röhm, p. 327; *Berliner Tageblatt*, Nr. 391, 18 August 1924; *Völkischer Kurier*, Nr. 165, 19 August 1924.
29. An adviser to Röhm, for instance, could sign High Command orders "For the High Command"; whereas the *Frontbann* adjutant could only sign "On Command." "Service Directions for the High Command," 23 August 1924, *HA* 81/1636. These directions were expanded even further in "Rules for the High Command," 20 August 1924, ibid.
30. Training Service Manual, undated, *HA* 15A/1631, pp. 4–9.
31. Röhm Letter, 27 August 1924, *HA* 53/1258.
32. *Völkischer Kurier*, Nr. 174, 29 August 1924, and *Deutscher Presse*, Nr. 179, 29 August 1924.
33. Munich Police Directory VIa Report, 16 December 1924, *HA* 68/1497A, and Munich Police Directory list of confiscated correspondence, undated, *HA* 81/1636. Earlier that same day Röhm had cautioned his followers in a directive from Berlin not to give the authorities an excuse to ban the *Frontbann*. Röhm, p. 330.

34. Röhm Testimony, 14 November 1924, *HA* 16A/1633, pp. 4–5, and 18 November 1924, *HA* 16A/1634, p. 10.
35. Frank, p. 122.
36. *Völkischer Kurier*, Nr. 192, 21/22 September 1924.
37. For the series of letters and reports leading up to the 1 October decision, see Ludendorff Letter, 22 September 1924, *HA* 69/1501; Robert M.W. Kemper, "Blueprint of the Nazi Underground – Past and Future Subversive Activies," *Research Studies of the State College of Washington*, No. 2 (1945), Document A, pp. 54–55; Munich Police Directory Letter Nr. VIa/2427, 23 September 1924, *HA* 69/1501, p. 2; Lorenz Roder Letter to the State Court, 24 September 1924, ibid., and Third Chamber of the State Court Munich I Decision Nr. XIX, 421/1923, ibid.
38. Reichswehr Ministry Letter, Nr. 343, 13 November 1924, *HA* 16A/1633, p. 1.
39. Röhm, p. 335. For an account of the incident in the town of Walsrode, see Lieutenant Lindenberg's Report, 1 October 1924, *HA* 44/895, pp. 3–4.
40. High State Court Decision, 29 September 1924, *HA* 69/1501. See also the State Prosecutor's Letter of 29 September to Hitler's lawyer, Lorenz Roder. Ibid.
41. Röhm found the condemnation of his organization by *völkisch* groups oriented on Hitler to be particularly galling. "The Movement," he recalled years later, "should have stood up as one man for the ... Frontbann." Röhm, p. 332. For Hitler's reaction, see Werner Jochmann, ed., *Nationalsozialismus und Revolution. Ursprung und Geschichte der NSDAP in Hamburg 1922–1923* (Frankfurt am Main, 1963). Document No. 52, p. 167.
42. Röhm and Ludendorff Memoranda, both 15 October 1924, *HA* 69/1501. For an example of statements provided by Landsberg prisoners, see Rudolf Hess's Letter, 9 October 1924, ibid. These and other supportive documents were collected by Hitler's lawyer and sent with a formal letter of appeal to the State Supreme Court. Roder Letter, 17 October 1924, *HA* 15A/1632.
43. Röhm Testimony, 14 November 1924, *HA* 16A/1633, p. 5. A few days later, Röhm also provided a 15-page memorandum in which he not only described in detail the political creation and purely military evolvement of his organization, but emphasized the continued opposition to it by Hitler and the other *Kampfbund* leaders. Röhm Memorandum, 18 November 1924, *HA* 16A/1634.
44. Ludendorff Testimony, 12 November 1924, ibid., p. 7.
45. Leybold Letter, 13 November 1924, *HA* 69/1501.
46. Testimonies of Hitler and Kriebel, both 1 December 1924, and Weber, 2 December 1924 – all in *HA* 16A/1634.
47. State Prosecutor Memorandum Nr. 734/24, 1 December 1924, p. 8 – attached as an enclosure to the Chief State Prosecutor's Report, 5 December 1924, *HA* 69/1501.
48. Fritz Maier-Hartmann, ed., *Dokumente der Zeitgeschichte*, Vol. 1 (München, 1942), p. 211 and Heinrich Hoffmann, *Hitler was my Friend* (London, 1955), p. 61. For the 19 December decision to free Hitler, see Munich State Court Telegram to Landsberg Prison Director Leybold, 20 December 1924, *HA* 3/67.
49. *Völkischer Beobachter*, No. 1, 26 February 1925. The *VB* re-appeared as a weekly; it became a daily again on 4 April 1925.
50. Munich Police Directory Report VIa to Bavarian Interior Ministry, undated, *HA* 87/1835.
51. Röhm, pp. 338–40.
52. Ibid., pp. 340–1.
53. Röhm demonstrated this continued allegiance to Hitler in a letter of 30 April to the NSDAP leader. "I use this opportunity," he wrote, "in remembrance of the good and hard times we went through together, to thank you heartily for your comradeship and to ask you not to give up your personal friendship with me." Röhm, p. 341.
54. Röhm Letter, 1 May 1925, *HA* 69/1505, pp. 2–3.
55. Geoffrey Pridham, *Hitler's Rise to Power: The Nazi Movement in Bavaria, 1923–1933* (New York, 1973), p. 54.
56. Joseph Nyomarkay, *Charisma and Factionalism in the Nazi Party* (Minneapolis, 1967),

pp. 114 and 116–18.

57. Years later, Hitler told a group of high-ranking party leaders that if ever the history of the movement were written, Röhm would have to rank as the second man behind himself in that history. Mau, p. 232.

58. Nyomarkay, p. 11; Fest, p. 144; Dietrich Orlow, *The History of the Nazi Party: 1919–1933* (Pittsburgh, 1969), pp. 212–14.

59. Quoted in Mau, p. 233. See also Hermann Rauschning, *Hitler Speaks* (London, 1939), pp. 154–5.

60. Mau, p. 234. Nyomarkay, pp. 126–7. Noakes and Pridham, p. 207. A people's army would also remain a militia-type force, essentially an instrument of defence without the offensive power Hitler required. Joachim C. Fest, *Hitler* (New York, 1975), p. 452.

61. Quoted in Fest, *The Face of the Third Reich*, p. 147.

62. Hitler was in the midst of his domestic *Gleichschaltung* process in the spring of 1934 and was concerned that the unruly and arbitrary behaviour of the SA might endanger Germany's stability, so necessary for economic recovery and the concomitant attainment of popular support. Without these twin pillars, he would have had to delay his main objective of mobilizing Germany's economy and national will in order to achieve his diplomatic and strategic ends. In addition, there was always the prospect that the SA might alienate other key groups besides the army such as the civil service and German industry, or inspire intervention from France, which regarded the growing SA ranks as a violation of the Versailles military limitations. Noakes and Pridham, pp. 203–4; Gordon A. Graig, *The Politics of the Prussian Army 1640–1945* (London, 1970), pp. 475–7; Mau, pp. 237–8; Nyomarkay, pp. 126–7.

63. Mau, pp. 245–6 and Fest, *Hitler*, pp. 472–3.

64. Craig, p. 469.

Lightning Source UK Ltd.
Milton Keynes UK
17 April 2010

152944UK00001B/4/P